Probability and Statistics: Theories and Applied Principles

Probability and Statistics: Theories and Applied Principles

Editor: Derek Beaven

NY RESEARCH PRESS
New York

Published by NY Research Press
118-35 Queens Blvd., Suite 400,
Forest Hills, NY 11375, USA
www.nyresearchpress.com

Probability and Statistics: Theories and Applied Principles
Edited by Derek Beaven

International Standard Book Number: 978-1-63238-580-2 (Hardback)

Cataloging-in-Publication Data

Probability and statistics : theories and applied principles / edited by Derek Beaven.
p. cm.
Includes bibliographical references and index.
ISBN 978-1-63238-580-2
1. Probabilities. 2. Statistics. I. Beaven, Derek.
QA273 .P76 2018
519.2--dc23

Contents

Preface

In my initial years as a student, I used to run to the library at every possible instance to grab a book and learn something new. Books were my primary source of knowledge and I would not have come such a long way without all that I learnt from them. Thus, when I was approached to edit this book; I became understandably nostalgic. It was an absolute honor to be considered worthy of guiding the current generation as well as those to come. I put all my knowledge and hard work into making this book most beneficial for its readers.

The subjects of statistics and probability are interrelated and are generally studied together. They can be applied in a number of fields such as finance, machine learning, game theory, etc. The book presents researches and studies performed by experts across the globe. From theories to research to practical applications, case studies related to all contemporary topics of relevance to this field have been included in this book. Through it, we attempt to further enlighten the readers about the new concepts in this field.

I wish to thank my publisher for supporting me at every step. I would also like to thank all the authors who have contributed their researches in this book. I hope this book will be a valuable contribution to the progress of the field.

Editor

1

Kumaraswamy Transmuted Exponentiated Additive Weibull Distribution

Zohdy M. Nofal[1], Ahmed Z. Afify[1], Haitham M. Yousof[1], Daniele C. T. Granzotto[2] & Francisco Louzada[3]

[1] Department of Statistics, Mathematics and Insurance, Benha University, Egypt

[2] Department of Statistics, State University of Maringá, Maringá, Brazil

[3] Institute of Mathematical and Computing Sciences, University of São Paulo, São Carlos, Brazil

Correspondence: Department of Statistics, State University of Maringá, Maringá, Brazil, Av. Colombo, 5790, Jd. Universitário, Maringá, CEP 87020-900, Brazil. E-mail: dctgranzotto@uem.br

Abstract

This paper introduces a new lifetime model which is a generalization of the transmuted exponentiated additive Weibull distribution by using the Kumaraswamy generalized (Kw-G) distribution. With the particular case no less than **seventy nine** sub models as special cases, the so-called Kumaraswamy transmuted exponentiated additive Weibull distribution, introduced by Cordeiro and de Castro (2011) is one of this particular cases. Further, expressions for several probabilistic measures are provided, such as probability density function, hazard function, moments, quantile function, mean, variance and median, moment generation function, Rényi and q entropies, order estatistics, etc. Inference is maximum likelihood based and the usefulness of the model is showed by using a real dataset.

Keywords: Additive Weibull distribution, order statistics, maximum likelihood estimation, Rényi and q entropies, goodness of fit, moment generating function.

1. Introduction

Aiming improve the modeling of survival data, there has been a growing interest among statisticians and applied researchers in constructing flexible lifetime models. As a result, significant progress has been made towards the generalization of some well-known lifetime models, which have been successfully applied to problems arising in several areas of research.

There are many distributions for modeling such data among the known parametric models, the most popular are the gamma, lognormal and the Weibull distributions, which is the most popular ones. However, with a limited hazard shapes, monotonic increase, decrease and constant, the Weibull distribution is not able to fit data sets with different hazard shapes as bathtub or upside down bathtub shaped (unimodal) failure rates, often encountered in reliability, engineering and biological studies. For many years, researchers have been developing various extensions and modified forms of the Weibull distribution, with number of parameters ranging from 2 to 7, see for example: Pham and Lai (2007) that present a review of some of the generalizations or modifications of Weibull distribution; the two-parameter flexible Weibull extension of Bebbington *et al.* (2007) that the hazard function can be increasing, decreasing or bathtub shaped; a three parameter model, called exponentiated Weibull distribution, introduced by Mudholkar and Srivastava (1993); another three-parameter one, introduced by Marshall and Olkin (1997) and called extended Weibull distribution; proposed by Xie *et al.* (2002); Xie and Lai (1995), a three parameter modified Weibull extension and a four parameter additive Weibull (AW) distribution, both with a bathtub shaped hazard function; the transmuted additive Weibull introduced by Elbatal and Aryal (2013) and Al-Babtain *et al.* (2015) introduced a new seven parameter model called the Kumaraswamy transmuted exponentiated modified Weibull distribution.

In this paper we introduce a new eight parameters model as a competitive extension for the Weibull distribution using the Kumaraswamy-generalized (Kw-G) distribution. The new model is very flexible in accommodating all forms of the hazard rate function by changing its parameter values, so it seems to be an important distribution that can be used. Another importance of the proposed model that it is very flexible model that approaches to different distributions when their parameters are changed. The new distribution is reffered to as the *Kumaraswamy transmuted exponentiated additive Weibull* (Kw-TEAW) distribution which extends all recent developments on the additive Weibull such as the transmuted exponentiated additive Weibull, Kumaraswamy transmuted exponentiated modified Weibull, transmuted modified Weibull introduced by Khan and King (2013), modified Weibull introduced by M. and Zaindin (2013) and additive Weibull introduced by Xie and Lai (1995) among others.

This paper is outlined as follows. In Section 2 we demonstrate the subject distribution and the mixture representation of its probability density function (pdf), cumulative distribution function (cdf), reliability function, hazard rate and cumulative hazard rate. The graphical presentation and sub-models of the Kw-TEAW are also provided in this section. The statistical properties include quantile functions, moments, moment generating functions, incomplete moments, mean deviations, moments of the residual life and moments of the reversed residual life are derived in Section 2.1. The order statistics and their moments are investigated in Section 4. In Section 5, We discuss maximum likelihood estimation of the model parameters. In Section 6, the Kw-TEAW distribution is applied to a real data sets to illustrate the potentiality of the new distribution for lifetime data modeling. Finally, we provide some concluding remarks in Section 7.

2. The Kw-TEAW Distribution

In this section, we present the Kw-TEAW distribution and its sub-models as follows:

Proposition 2.1. *Let X a positive random variable with Kw-TEAW distribution with vector parameters* $\upsilon = (\alpha, \beta, \gamma, \theta, \delta, \lambda, a, b)$*. The cumulative distribution function is defined as*

$$F(x) = 1 - \left\{1 - \left(1 - e^{-(\alpha x^{\theta} + \gamma x^{\beta})}\right)^{a\delta}\left[1 + \lambda - \lambda\left(1 - e^{-(\alpha x^{\theta} + \gamma x^{\beta})}\right)^{\delta}\right]^{a}\right\}^{b}, \tag{1}$$

where α and γ are scale parameter representing the characteristic life, θ, β, δ, a and b are the shape parameters representing the different patterns of the Kw-TEAW and λ is the transmuted parameter.

Proof: *The proof is imediately as follows: A new six parameter additive Weibull was introduced recently. Let X be a positive random variable with additive Weibull, it cumulative distribution function (cdf) is given by*

$$F(x) = \left(1 - e^{-(\alpha x^{\theta} + \gamma x^{\beta})}\right)^{\delta}\left[1 + \lambda - \lambda\left(1 - e^{-(\alpha x^{\theta} + \gamma x^{\beta})}\right)^{\delta}\right], \tag{2}$$

where $\alpha, \beta, \gamma, \theta \geq 0$ with $\theta < 1 < \beta$ (or $\beta < 1 < \theta$), θ and β are the shape parameters and α and γ are scale parameters.

The corresponding pdf of (2) is

$$\begin{aligned} f(x) &= \delta\left(\alpha\theta x^{\theta-1} + \gamma\beta x^{\beta-1}\right)e^{-(\alpha x^{\theta} + \gamma x^{\beta})}\left[1 - e^{-(\alpha x^{\theta} + \gamma x^{\beta})}\right]^{\delta-1} \\ &\quad \times\left\{1 + \lambda - 2\lambda\left(1 - e^{-(\alpha x^{\theta} + \gamma x^{\beta})}\right)^{\delta}\right\}. \end{aligned} \tag{3}$$

Cordeiro and de Castro (2011) defined the Kw-G distribution by following general construction. For an arbitrary baseline cdf, $G(x)$, of a positive random variable X, the generalized class of distributions can be defined by

$$F(x) = 1 - \{1 - G(x)^{a}\}^{b}, \tag{4}$$

where $g(x)=dG(x)/dx$ and a and b are two extra positive shape parameters which govern skewness and tail weights. The Kw-G distribution can be used quite effectively even if the data are censored. Correspondingly, its density function is distributions has a very simple form

$$f(x) = abg(x)G(x)^{a-1}\{1 - G(x)^{a}\}^{b}. \tag{5}$$

Hence, each new Kw-G distribution can be generated from a specified G distribution.

Thus, as a proof, the equations (2) and (3) are inserted in equations (4) and (5), respectively, and we obtain the Kw-TEAW distribution.

□

The corresponding pdf of the Kw-TEAW is given by

$$\begin{aligned} f(x) &= ab\delta e^{-(\alpha x^{\theta} + \gamma x^{\beta})}\left(\alpha\theta x^{\theta-1} + \gamma\beta x^{\beta-1}\right)\left[1 - e^{-(\alpha x^{\theta} + \gamma x^{\beta})}\right]^{a\delta-1}\left\{1 - \left[1 - e^{-(\alpha x^{\theta} + \gamma x^{\beta})}\right]^{a\delta}\left[1 + \lambda - \lambda\left(1 - e^{-(\alpha x^{\theta} + \gamma x^{\beta})}\right)^{\delta}\right]^{a}\right\}^{b-1} \\ &\quad \times\left\{1 + \lambda - 2\lambda\left[1 - e^{-(\alpha x^{\theta} + \gamma x^{\beta})}\right]^{\delta}\right\}\left\{1 + \lambda - \lambda\left[1 - e^{-(\alpha x^{\theta} + \gamma x^{\beta})}\right]^{\delta}\right\}^{a-1}. \end{aligned} \tag{6}$$

Furthermore, the reliability function $R(x)$, hazard rate function $h(x)$ and and cumulative hazard rate function $H(x)$ of the random variable X are given, respectively, by

$$R(x) = \left[\left\{ 1 - \left(1 - e^{-(\alpha x^{\theta} + \gamma x^{\beta})}\right)^{a\delta} \left[1 + \lambda - \lambda \left(1 - e^{-(\alpha x^{\theta} + \gamma x^{\beta})}\right)^{\delta} \right]^{a} \right\}^{b} \right],$$

$$\begin{aligned} h(x) &= ab\delta \left[\alpha\theta x^{\theta-1} e^{-(\alpha x^{\theta} + \gamma x^{\beta})} + \gamma\beta x^{\beta-1} e^{-(\alpha x^{\theta} + \gamma x^{\beta})}\right] \left\{ 1 - \left(1 - e^{-(\alpha x^{\theta} + \gamma x^{\beta})}\right)^{a\delta} \left[1 + \lambda - \lambda \left(1 - e^{-(\alpha x^{\theta} + \gamma x^{\beta})}\right)^{\delta} \right]^{a} \right\}^{-1} \\ & \left[1 - e^{-(\alpha x^{\theta} + \gamma x^{\beta})}\right]^{a\delta-1} \left\{ 1 + \lambda - 2\lambda \left[1 - e^{-(\alpha x^{\theta} + \gamma x^{\beta})}\right]^{\delta} \right\} \left\{ 1 + \lambda - \lambda \left[1 - e^{-(\alpha x^{\theta} + \gamma x^{\beta})}\right]^{\delta} \right\}^{a-1}, \end{aligned}$$

$$H(x) = -\ln \left[\left\{ 1 - \left(1 - e^{-(\alpha x^{\theta} + \gamma x^{\beta})}\right)^{a\delta} \left[1 + \lambda - \lambda \left(1 - e^{-(\alpha x^{\theta} + \gamma x^{\beta})}\right)^{\delta} \right]^{a} \right\}^{b} \right]. \quad (7)$$

2.1 Mixture Representation for cdf and pdf

Expansions for equations (1) and (6) can be derived using the series expansion

$$(1-z)^{k} = \sum_{j=0}^{\infty} \frac{(-1)^{j} \Gamma(k+1)}{j! \Gamma(k-j+1)} z^{j}, \quad |z| < 1, \ k > 0.$$

Then, the cdf of the Kw-TEAW in (1) can be expressed in the mixture form

$$F(x) = \sum_{j,i,l,w=0}^{\infty} s_{j,i,l,w} e^{-w(\alpha x^{\theta} + \gamma x^{\beta})}, \quad (8)$$

where

$$s_{j,i,l,w} = \frac{(-1)^{j+i+l+w} \Gamma(bj+1) \Gamma(ai+1) \Gamma(\delta ai + l + 1) \lambda^{l} (1+\lambda)^{ai-l}}{j! i! l! w! \Gamma(2-j) \Gamma(bj+1-i) \Gamma(ai+1-l) \Gamma(\delta ai + l + 1 - w)}.$$

The pdf of the Kw-TEAW can be expressed in the mixture form

$$f(x) = \sum_{j,i,lw=0}^{\infty} \zeta_{j,i,l,w} \left[\alpha\theta x^{\theta-1} + \gamma\beta x^{\beta-1}\right] e^{-(w+1)(\alpha x^{\theta} + \gamma x^{\beta})}, \quad (9)$$

where

$$\zeta_{j,i,l,w} = ab\delta \frac{(-1)^{j+i+l+w} \Gamma(b) \Gamma(aj+a) \Gamma[\delta(aj+a+i+l)] 2^{l} \lambda^{l+i} (1+\lambda)^{aj+a-i-l}}{j! i! l! w! \Gamma(b-j) \Gamma(aj+a-i) \Gamma(2+l) \Gamma[\delta(aj+a+i+l) - w]}. \quad (10)$$

Further, the Kw-TEAW density function can be expressed as a mixture of additive Weibull densities. Thus, some of its mathematical properties can be obtained directly from the properties of the additive Weibull distribution. Therefore equation (6) can be also expressed as

$$f(x) = \sum_{j,i,lw=0}^{\infty} \frac{\zeta_{j,i,l,w}}{w+1} g(x; \theta, \beta, \alpha(w+1), \gamma(w+1)), \quad (11)$$

where $g(x; \theta, \beta, \alpha(w+1), \gamma(w+1))$ denotes to the AW pdf i.e., $X \sim \mathrm{AW}(\theta, \theta, \beta, \alpha(w+1), \gamma(w+1))$.

The Kw-TEAW model is very flexible model that approaches to different distributions. It includes as special cases seventy nine sub-models when its parameters vary as presented in Table 1. Simply by replacing the values of the parameters of Kw-TEAW as indicated in Table 1, is possible to write the particlar cases of this model, wheter these are new or known.

Figure B provides some plots of the Kw-TEAW density and hazard curves for different values of the parameters $a, b, \alpha, \beta, \gamma, \theta, \delta$ and λ.

3. Some Statistical Properties

The statistical properties of the Kw-TEAW distribution including quantile and random number generation, moments, moment generating function, incomplete moments, mean deviations and Rényi and q entropies are discussed in this section.

3.1 Quantile Function

The quantile function (qf) of X, where $X \sim$ Kw-TEAW$(\alpha, \beta, \gamma, \theta, \delta, \lambda)$, is obtained by inverting (2) as

$$\alpha x_q^{\theta} + \gamma x_q^{\beta} + \ln\left\{1 - \sqrt[\delta]{\frac{(1+\lambda) - \sqrt{(1+\lambda)^2 - 4\lambda m}}{2\lambda}}\right\} = 0, \tag{12}$$

where

$$m = \left[1 - (1-q)^{\frac{1}{b}}\right]^{\frac{1}{a}}.$$

Since the above equation has no closed form solution in x_q, we have to use numerical methods to get the quantiles.

3.2 Moments

The rth moment, denoted by μ_r', of the Kw-TEAW$(\alpha, \beta, \gamma, \theta, \delta, \lambda, a, b, x)$ is given by the following theorem.

Theorem 3.1. *If X is a continuous random variable has the Kw-TEAW$(\alpha, \beta, \gamma, \theta, \delta, \lambda, a, b, x)$, then the rth non-central moment of X, is given by*

$$\mu_r' = \sum_{j,i,lw=0}^{\infty} \zeta_{j,i,l,w} \left\{ \sum_{k=0}^{\infty} \frac{(-1)^k \gamma^k \Gamma\left(\frac{r+\theta+k\beta}{\theta}\right)}{k! \alpha^{\frac{r+k\beta}{\theta}} (w+1)^{[k(\theta-\beta)-(\theta+r)]/\theta}} + \sum_{k=0}^{\infty} \frac{(-1)^k \alpha^k \Gamma\left(\frac{r+\beta+k\theta}{\beta}\right)}{k! \gamma^{\frac{r+k\theta}{\beta}} (w+1)^{[k(\beta-\theta)-(\beta+r)]/\beta}} \right\}. \tag{13}$$

Proof: *By definition*

$$\mu_r' = \int_0^{\infty} x^r f(x, \upsilon)\, dx = \sum_{j,i,lw=0}^{\infty} \zeta_{j,i,l,w} \int_0^{\infty} \left(\alpha\theta x^{r+\theta-1} + \gamma\beta x^{r+\beta-1}\right) e^{-(i+1)\left(\alpha x^{\theta} + \gamma x^{\beta}\right)} dx.$$

After some simplifications, we get

$$\mu_r' = \sum_{j,i,l,w,k=0}^{\infty} \upsilon_{j,i,l,w,k} \left\{ \frac{\gamma^k \Gamma\left(\frac{r+\theta+k\beta}{\theta}\right)}{\alpha^{\frac{r+k\beta}{\theta}} (w+1)^{[k(\theta-\beta)-(\theta+r)]/\theta}} + \frac{\alpha^k \Gamma\left(\frac{r+\beta+k\theta}{\beta}\right)}{\gamma^{\frac{r+k\theta}{\beta}} (w+1)^{[k(\beta-\theta)-(\beta+r)]/\beta}} \right\},$$

where

$$\upsilon_{j,i,l,w,k} = \frac{(-1)^{j+i+l+w}}{j!i!l!w!k!} \frac{ab\delta\Gamma(b)\Gamma(aj+a)\Gamma[\delta(aj+a+i+l)]2^l \lambda^{l+i}(1+\lambda)^{aj+a-i-l}}{\Gamma(b-j)\Gamma(aj+a-i)\Gamma(2+l)\Gamma[\delta(aj+a+i+l)-w]}.$$

□

The variation, skewness and kurtosis measures can be calculated from the ordinary moments using well-known relationships following.

Corollary 3.1. *Using the relation between the central moments and non-centeral moments, we can obtain the nth central moment, denoted by M_n, of a Kw-TEAW random variable as follows*

$$M_n = E(X-\mu)^n = \sum_{r=0}^{n} \binom{n}{r} (-\mu)^{n-r} E(X^r),$$

where $E(X^r)$ is the on-central moments of the Kw-TEAW$(\alpha, \beta, \gamma, \theta, \delta, \lambda, a, b, x)$. Therefore the nth central moments of the Kw-TEAW$(\alpha, \beta, \gamma, \theta, \delta, \lambda, a, b, x)$, is given by

$$M_n = \sum_{r=0}^{n} \binom{n}{r} (-\mu)^{n-r} \sum_{j,i,l,w,k=0}^{\infty} \upsilon_{j,i,l,w,k} \left\{ \frac{\gamma^k \Gamma\left(\frac{r+\theta+k\beta}{\theta}\right)}{\alpha^{\frac{r+k\beta}{\theta}} (w+1)^{[k(\theta-\beta)-(\theta+r)]/\theta}} + \frac{\alpha^k \Gamma\left(\frac{r+\beta+k\theta}{\beta}\right)}{\gamma^{\frac{r+k\theta}{\beta}} (w+1)^{[k(\beta-\theta)-(\beta+r)]/\beta}} \right\}.$$

3.3 Generating Function

The moment generating function of the Kw-TEAW is given by the following theorem

Theorem 3.2. *If X is a continuous random variable has the Kw-TEAW(υ, x), then the moment generating function of X, denoted by $M_X(t) = E\left(e^{tX}\right)$, is given as*

$$M_X(t) = \sum_{r=0}^{\infty}\frac{t^r}{r!}\sum_{j,i,l,w,k=0}^{\infty}\upsilon_{j,i,l,w,k}\left\{\frac{\gamma^k\Gamma\left(\frac{r+\theta+k\beta}{\theta}\right)}{\alpha^{\frac{r+k\beta}{\theta}}(w+1)^{\left[k(\theta-\beta)-(\theta+r)\right]/\theta}}+\frac{\alpha^k\Gamma\left(\frac{r+\beta+k\theta}{\beta}\right)}{\gamma^{\frac{r+k\theta}{\beta}}(w+1)^{\left[k(\beta-\theta)-(\beta+r)\right]/\beta}}\right\}. \quad (14)$$

Proof: *By definition*

$$\begin{aligned} M_X(t) &= \int_0^{\infty} e^{tx} f(x,\upsilon)\,dx \\ &= \sum_{r=0}^{\infty}\frac{t^r}{r!}\int_0^{\infty} x^r f(x,\upsilon)\,dx \\ &= \sum_{r=0}^{\infty}\frac{t^r}{r!}\mu_r'. \end{aligned} \quad (15)$$

By substituting from equation (14) into (11), we obtain the moment generating function as

$$M_X(t) = \sum_{r=0}^{\infty}\frac{t^r}{r!}\sum_{j,i,l,w,k=0}^{\infty}\upsilon_{j,i,l,w,k}\left\{\frac{\gamma^k\Gamma\left(\frac{r+\theta+k\beta}{\theta}\right)}{\alpha^{\frac{r+k\beta}{\theta}}(w+1)^{\left[k(\theta-\beta)-(\theta+r)\right]/\theta}}+\frac{\alpha^k\Gamma\left(\frac{r+\beta+k\theta}{\beta}\right)}{\gamma^{\frac{r+k\theta}{\beta}}(w+1)^{\left[k(\beta-\theta)-(\beta+r)\right]/\beta}}\right\}.$$

□

3.4 Incomplete Moments

The main application of the first incomplete moment refers to the Bonferroni and Lorenz curves. These curves are very useful in economics, reliability, demography, insurance and medicine. The answers to many important questions in economics require more than justknowing the mean of the distribution, but its shape as well. This is obvious not only in the study of econometrics but in other areas as well.

The *s-th* incomplete moments, denoted by $\varphi_s(t)$, of the Kw-TEAW random variable is given by

$$\varphi_s(t) = \int_0^t x^s f(x)\,dx,$$

Using equation (6) and the lower incomplete gamma function, we obtain

$$\varphi_s(t) = \sum_{j,i,l,w,k=0}^{\infty}\upsilon_{j,i,l,w,k}\left\{\frac{\gamma^k\Gamma\left(\frac{s+\theta+k\beta}{\theta}\right)}{\alpha^{\frac{s+k\beta}{\theta}}(w+1)^{\left[k(\theta-\beta)-(\theta+s)\right]/\theta}}+\frac{\alpha^k\Gamma\left(\frac{s+\beta+k\theta}{\beta}\right)}{\gamma^{\frac{s+k\theta}{\beta}}(w+1)^{\left[k(\beta-\theta)-(\beta+s)\right]/\beta}}\right\}. \quad (16)$$

The first incomplete moment of X, denoted by, $\varphi_1(t)$, is immediately calculated from equation (18) by setting $s = 1$ as

$$\varphi_1(t) = \sum_{j,i,l,w,k=0}^{\infty}\upsilon_{j,i,l,w,k}\left\{\frac{\gamma^k\Gamma\left(\frac{1+\theta+k\beta}{\theta}\right)}{\alpha^{\frac{1+k\beta}{\theta}}(w+1)(w+1)^{\left[k(\theta-\beta)-(\theta+1)\right]/\theta}}+\frac{\alpha^k\Gamma\left(\frac{1+\beta+k\theta}{\beta}\right)}{\gamma^{\frac{1+k\theta}{\beta}}(w+1)^{\left[k(\beta-\theta)-(\beta+1)\right]/\beta}}\right\}.$$

Another application of the first incomplete moment is related to the mean residual life and the mean waiting time (also known as mean inactivity time) given by

$$m_1(t;\theta) = (1-\varphi_1(t))R(t;\theta) - t$$

and

$$M_1(t;\theta) = t - (\varphi_1(t)F(t;\theta)),$$

respectively.

3.5 Mean Deviations

The amount of scatter in a population is evidently measured to some extent by the totality of deviations from the mean and median. The mean deviations about the mean $\delta_\mu(X) = E(|X - \mu_1'|)$ and about the median $\left(\delta_\mu(X) = E(|X - M|)\right)$ of X can be, used as measures of spread in a population, expressed by

$$\delta_\mu(X) = \int_0^\infty |X - \mu_1'| f(x)\,dx = 2\mu_1' F(\mu_1') - 2\varphi_1(\mu_1'), \tag{17}$$

and

$$\delta_M(X) = \int_0^\infty |X - M| f(x)\,dx = \mu_1' - 2\varphi_1(M), \tag{18}$$

respectively, where $\mu_1' = E(X)$ comes from (11), $F(\mu_1')$ is simply calculated from (1), $\varphi_1(\mu_1')$ is the first incomplete moments and M is the median of X.

The application of mean deviations refers to the Lorenz and Bonferroni curves defined by $L(p) = \varphi_1(q)/\mu_1'$ and $B(p) = \varphi_1(q)/p\mu_1'$, respectively, where $q = F^{-1}(p)$ can be computed fora given probability p by inverting (1) numerically. These curves are very useful in economics, reliability, demography, insurance and medicine.

3.6 Moments of the Residual Life

Several functions are defined related to the residual life. The failure rate function, mean residual life function and the left censored mean function, also called vitality function. It is well known that these three functions uniquely determine $F(x)$.

First, we present the nth moments of residual life, denoted by $m_n(t) = E((X - t)^n \mid X > t)$, $n = 1, 2, 3, \ldots$, uniquely determine $F(x)$. In a general way, the nth moments of the residual life random variable is given by

$$m_n(t) = \frac{1}{1 - F(t)} \int_t^\infty (x - t)^n\, dF(x). \tag{19}$$

Another interesting function is the mean residual life function (MRL) or the life expectancy at age t, defined by $m_1(x) = E((X - x) \mid X > x)$, and it represents the expected additional life length for a unit which is alive at age x. Definitions 3.1 and Result 3.1 present, respectively, the $m_n(t)$ and $m_1(x)$.

Definition 3.1. *The nth moments of the residual life of X is given by*

$$m_n(t) = \frac{1}{R(t)} \sum_{r=0}^{n} \frac{(-1)^{n-r}\Gamma(n+1)\, t^{n-r}}{r!\,\Gamma(n-r+1)} \sum_{j,i,l,w,k=0}^{\infty} \upsilon_{j,i,l,w,k} \left\{ \frac{\gamma^k \Gamma\left(\frac{r+\theta+k\beta}{\theta}, \alpha(w+1)t^\theta\right)}{\alpha^{\frac{r+k\beta}{\theta}} (w+1)^{[k(\theta-\beta)-(\theta+r)]/\theta}} + \frac{\alpha^k \Gamma\left(\frac{r+\beta+k\theta}{\beta}, \gamma(w+1)t^\beta\right)}{\gamma^{\frac{r+k\theta}{\beta}} (w+1)^{[k(\beta-\theta)-(\beta+r)]/\beta}} \right\}.$$

Here we can use the upper incomplete gamma function defined by

$$\Gamma(a, b) = \int_b^\infty y^{a-1} e^{-y} dy.$$

Result 3.1. *The mean residual life is obtained by setting $n = 1$ in equation (22) and it is given by*

$$m_1(t) = \frac{1}{R(t)} \sum_{j,i,l,w,k=0}^{\infty} \upsilon_{j,i,l,w,k} \left\{ \frac{\gamma^k \Gamma\left(\frac{1+\theta+k\beta}{\theta}\right)}{\alpha^{\frac{1+k\beta}{\theta}} (i+1)^{[k(\theta-\beta)-(\theta+1)]/\theta}} + \frac{\alpha^k \Gamma\left(\frac{1+\beta+k\theta}{\beta}\right)}{\gamma^{\frac{1+k\theta}{\beta}} (i+1)^{[k(\beta-\theta)-(\beta+1)]/\beta}} \right\} - t.$$

Guess and Proschan (1988) gave an extensive coverage of possible applications of the mean residual life. The MRL has many applications in survival analysis in biomedical sciences, life insurance, maintenance and product quality control, economics and social studies, Demography and product tecnology (see Lai and Xie (2006)).

3.7 Moments of the Reversed Residual Life

Definition 3.2. *Let X be a random variable, usually representing the life length for a certain unit at age t (where this unit can have multiple interpretations). Then, the nth moment of the reversed residual life of X, is given by*

$$M_n(t) = \frac{1}{F(t)} \sum_{r=0}^{n} \frac{(-1)^{r}\Gamma(n+1)\, t^{n-r}}{r!\,\Gamma(n-r+1)} \sum_{j,i,l,w,k=0}^{\infty} \upsilon_{j,i,l,w,k} \left\{ \frac{\gamma^k \Gamma\left(\frac{r+\theta+k\beta}{\theta}, \alpha(w+1)t^\theta\right)}{\alpha^{\frac{r+k\beta}{\theta}} (w+1)^{[k(\theta-\beta)-(\theta+r)]/\theta}} + \frac{\alpha^k \Gamma\left(\frac{r+\beta+k\theta}{\beta}, \gamma(w+1)t^\beta\right)}{\gamma^{\frac{r+k\theta}{\beta}} (w+1)^{[k(\beta-\theta)-(\beta+r)]/\beta}} \right\}.$$

Here we can use the lower incomplete gamma function defined by $\Gamma(a, b) = \int_0^b y^{a-1} e^{-y} dy$.

Note that, the nth moment of the reversed residual life, denoted by

$$M_n(t) = E((t-X)^n \mid X \le t), \quad t>0, \quad n=1,2,3,...,$$

uniquely determine $F(x)$. Then, the result presented in equation (20) is a result of the integral

$$M_n(t) = \frac{1}{F(t)} \int_0^t (t-x)^n \, dF(x). \tag{20}$$

Result 3.2. *The mean reversed residual life function, defined by* $M_1(t) = E((t-X) \mid X \le t)$, *is given by*

$$M_1(t) = t + \frac{1}{F(t)} \sum_{j,i,l,w,k=0}^{\infty} \upsilon_{j,i,l,w,k} \left\{ \frac{\gamma^k \Gamma\left(\frac{1+\theta+k\beta}{\theta}\right)}{\alpha^{\frac{s+k\beta}{\theta}} (w+1)^{[k(\theta-\beta)-(\theta+1)]/\theta}} + \frac{\alpha^k \Gamma\left(\frac{1+\beta+k\theta}{\beta}\right)}{\gamma^{\frac{s+k\theta}{\beta}} (w+1)^{[k(\beta-\theta)-(\beta+1)]/\beta}} \right\}.$$

Note that, the MRRL of the Kw-TEAW distribution can be obtained by setting $n=1$ in equation (20) presented in Definition 3.2

The mean inactivity time (MIT) or mean waiting time (MWT) also called mean reversed residual life function, presented in 3.2, represents the waiting time elapsed since the failure of an item on condition that this failure had occurred in $(0,x)$.

3.8 Rényi and q- Entropies

The Rényi entropy of a random variable X represents a measure of variation of the uncertainty and is defined as

$$I_\kappa(X) = \frac{1}{1-\kappa} \log \int_{-\infty}^{\infty} f^\kappa(x)\, dx, \ \kappa > 0 \text{ and } \kappa \ne 1.$$

Hence, the Rényi entropy reduces to

$$I_\kappa(X) = \frac{1}{1-\kappa} \log \left\{ (ab\delta)^\kappa \sum_{L=0}^{\kappa} \binom{\kappa}{L} (\gamma\beta)^L (\alpha\theta)^{\kappa-L} \sum_{j.i.k.w.h=0}^{\infty} \Upsilon_{j.i.k.w.h} \Gamma\left(\frac{\beta(L+h) + \theta(\kappa-L) - \kappa + 1}{\theta} \right) \right\}, \tag{21}$$

where

$$\Upsilon_{j.i.k.w.h} = \frac{(-1)^{j+i+k+w+h} \Gamma[\kappa(\beta-1)+1] \Gamma[\kappa+1]}{j!i!k!w!h!\Gamma[\kappa(\beta-1)-j+1]\Gamma[\kappa-i+1]} \frac{\Gamma[a(\kappa+j)-\kappa+1]\Gamma[H](1+\lambda)^{[a(\delta+j)-k-i]} \lambda^{k+i} 2^i \alpha^{s+\kappa-L} (\kappa+w)^{O+h}}{\Gamma[a(\kappa+j)-\kappa-k+1]\Gamma[H-w]},$$

$$s = \left(\frac{L\theta - h\beta + \kappa - 1 - L\beta - \kappa\theta}{\theta} \right),$$

and

$$H = \delta a(\kappa+j) + \delta(i+k) + 1 - \kappa.$$

The q-entropy, say $H_q(x)$, is defined by

$$H_q(x) = \frac{1}{q-1} \log\left(1 - \int_{-\infty}^{\infty} f^q(x)\, dx \right), \ q > 0 \text{ and } q \ne 1.$$

From equation (20), we can easily obtain

$$H_q(x) = \frac{1}{q-1} \log\left(1 - \left\{ (ab\delta)^q \sum_{L=0}^{q} \binom{q}{L} (\gamma\beta)^L (\alpha\theta)^{q-L} \sum_{j.i.k.w.h=0}^{\infty} \tau_{j.i.k.w.h} \Gamma\left(\frac{\beta(L+h) + \theta(q-L) - q + 1}{\theta} \right) \right\} \right), \tag{22}$$

where

$$\tau_{j.i.k.w.h} = \frac{(-1)^{j+i+k+w+h} \Gamma[q(\beta-1)+1] \Gamma[q+1]}{j!i!k!w!h!\Gamma[q(\beta-1)-j+1]\Gamma[q-i+1]} \frac{\Gamma[a(q+j)-q+1]\Gamma[d](1+\lambda)^{[a(\delta+j)-k-i]} \lambda^{k+i} 2^i \alpha^{\pi+q-L} (q+w)^{\pi+h}}{\Gamma[a(q+j)-q-k+1]\Gamma[d-w]},$$

$$\pi = \left(\frac{L\theta - h\beta + q - 1 - L\beta - q\theta}{\theta} \right),$$

and

$$d = \delta a(q+j) + \delta(i+k) + 1 - q.$$

4. Order Statistics

The order statistics and their moments have great importance in many statistical problems and they have many applications in reliability analysis and life testing. The order statistics arise in the study of reliability of a system. The order statistics can represent the lifetimes of units or components of a reliability system. Let $X_1, X_2, ..., X_n$ be a random sample of size n from the Kw-TEAW with cumulative distribution function, and the corresponding probability density function, as in (1) and (6), respectively. Let $X_{(1)}, X_{(2)}, ..., X_{(n)}$ be the corresponding order statistics. Then the pdf of jth order statistics, say $Y = X_{(j:n)},\ 1 \leq j \leq n,$ denoted by $f_Y(x)$, is given by

$$\begin{aligned} f_Y(x) &= ab\delta\binom{n}{i}\left(\alpha\theta x^{\theta-1}+\gamma\beta x^{\beta-1}\right)e^{-\left(\alpha x^{\theta}+\gamma x^{\beta}\right)}\left[1-e^{-\left(\alpha x^{\theta}+\gamma x^{\beta}\right)}\right]^{\delta a-1} \\ &\quad \left\{1+\lambda-\lambda\left[1-e^{-\left(\alpha x^{\theta}+\gamma x^{\beta}\right)}\right]^{\delta}\right\}^{a-1}\left\{1+\lambda-2\lambda\left[1-e^{-\left(\alpha x^{\theta}+\gamma x^{\beta}\right)}\right]^{\delta}\right\} \\ &\quad \left\{1-\left\{1-\left(1-e^{-\left(\alpha x^{\theta}+\gamma x^{\beta}\right)}\right)^{a\delta}\left[1+\lambda-\lambda\left(1-e^{-\left(\alpha x^{\theta}+\gamma x^{\beta}\right)}\right)^{\delta}\right]^{a}\right\}^{b}\right\}^{i-1} \\ &\quad \left[\left\{1-\left(1-e^{-\left(\alpha x^{\theta}+\gamma x^{\beta}\right)}\right)^{a\delta}\left[1+\lambda-\lambda\left(1-e^{-\left(\alpha x^{\theta}+\gamma x^{\beta}\right)}\right)^{\delta}\right]^{a}\right\}^{b}\right]^{b(n-i+1)-1}. \end{aligned} \tag{23}$$

Then, the pdf of Y can be expressed in a mixture form as

$$f_{i:n}(x) = \sum_{j,l,w,m,h=0}^{\infty} \varsigma_{j,l,w,m,h}\, g\left(x;\theta,\beta,\alpha(h+1),\gamma(h+1)\right), \tag{24}$$

where

$$\varsigma_{j,l,w,m,h} = ab\delta\binom{n}{i}\frac{(-1)^{j+l+w+m+h}(2)^{w}(\lambda)^{w+m}}{(h+1)(1+\lambda)^{-a(l+1)+w+m}}\binom{i-1}{j}\binom{al+1}{w}\binom{b(j+n-i+1)-1}{l}\binom{a-1}{m}\binom{\delta\left[a(l+1)+w+m\right]-1}{h},$$

and $g(x;\theta,\beta,\alpha(h+1),\gamma(h+1))$ is the additive Weibull density function with parameters $\theta, \beta, \alpha(h+1), \gamma(h+1), a$ and b. So, the density function of the Kw-TEAW order statistics is a mixture of AW densities. Based on equation (24), we can obtain some structural properties of Y from those AW properties.

The qth moment of the jth order statistics, $Y = X_{(j:n)}$, is given by

$$E\left(X^{q}_{(j:n)}\right) = \sum_{j,l,w,m,h=0}^{\infty} \varsigma_{(j,k,w,h)} E\left(Y^{q}_{\theta,\beta,\alpha(h+1),\gamma(h+1),a,b}\right), \tag{25}$$

where $Y_{\theta,\beta,\alpha(h+1),\gamma(h+1)} \sim AW(\theta,\beta,\alpha(h+1),\gamma(h+1))$.

The L-moments are analogous to the ordinary moments but can be estimated by linear combinations of order statistics. They exist whenever the mean of the distribution exists, even though some higher moments may not exist, and are relatively robust to the effects of outliers. Based upon the moments in equation (25), we can derive explicit expressions for the L-moments of X as infinite weighted linear combinations of the means of suitable AW distributions. They are linear functions of expected order statistics defined by

$$\lambda_r = \frac{1}{r}\sum_{d=0}^{r-1}(-1)^{d}\binom{r-1}{d}E\left(X_{r-d:d}\right),\ r \geq 1.$$

The first four L-moments are given by: $\lambda_1 = E(X_{1:1})$, $\lambda_2 = \frac{1}{2}E(X_{2:2} - X_{1:2})$, $\lambda_3 = \frac{1}{3}E(X_{3:3} - 2X_{2:3} + X_{1:3})$ and $\lambda_4 = \frac{1}{4}E(X_{4:4} - 3X_{3:4} + 3X_{2:4} - X_{1:4})$. One simply can obtain the λ's for X from equation (25) with $q = 1$.

5. Parameter Estimation

This section provides a system of equations that can be utilized to determine the maximum likelihood estimates of the parameters of the Kw-TEAW distribution. Let $\mathbf{X} = (X_1, ..., X_n)$ be a random sample of the Kw-TEAW distribution with unknown parameter vector $\upsilon = (\alpha, \beta, \gamma, \delta, \theta, \lambda, a, b)^T$.

Then, the log-likelihood function $\ell(\upsilon)$, is given by

$$\begin{aligned}\ell(\upsilon) &= n\ln a + n\ln b + n\ln\delta + (a\delta - 1)\sum_{i=1}^{n}\ln\left(1 - e^{S_i}\right) + \sum_{i=1}^{n}\ln Z_i \\ &- \sum_{i=1}^{n} S_i + \sum_{i=1}^{n}\ln K_i + (a-1)\sum_{i=1}^{n}\ln\{Q_i\}(b-1)\sum_{i=1}^{n}\ln\left\{1 - \left[1 - e^{S_i}\right]^{a\delta}\{Q_i\}^a\right\}.\end{aligned} \quad (26)$$

Therefore the score vector is

$$\mathbf{U}(\upsilon) = \frac{\partial\ell}{\partial\upsilon} = \left(\frac{\partial\ell}{\partial\alpha}, \frac{\partial\ell}{\partial\beta}, \frac{\partial\ell}{\partial\gamma}, \frac{\partial\ell}{\partial\delta}, \frac{\partial\ell}{\partial\theta}, \frac{\partial\ell}{\partial\lambda}, \frac{\partial\ell}{\partial a}, \frac{\partial\ell}{\partial b}\right)^T.$$

Let $Z_i = \alpha\theta x_i^{\theta-1} + \gamma\beta x_i^{\beta-1}$, $Q_i = 1 + \lambda - \lambda\left[1 - e^{-S_i}\right]^{\delta}$, $K_i = 1 + \lambda - 2\lambda\left[1 - e^{-S_i}\right]^{\delta}$ and $S_i = \alpha x_i^{\theta} + \gamma x_i^{\beta}$,

$$\begin{aligned}\frac{\partial\ell(\upsilon)}{\partial\alpha} &= (a\delta - 1)\sum_{i=1}^{n}\frac{e^{-S_i}x_i^{\theta}}{1 - e^{-S_i}} + \sum_{i=1}^{n}\frac{\theta x_i^{\theta-1}}{Z_i} - 2\lambda\delta\sum_{i=1}^{n}\frac{x_i^{\theta}e^{S_i}\left(1 - e^{-S_i}\right)^{\delta-1}}{K_i} - (a-1)\sum_{i=1}^{n}\frac{\lambda\delta e^{-S_i}x_i^{\theta}\left(1 - e^{-S_i}\right)^{\delta-1}}{Q_i} \\ &- \sum_{i=1}^{n} x_i^{\theta} + (a-1)\sum_{i=1}^{n}\frac{a\delta e^{S_i}x_i^{\theta}\left(1 - e^{-S_i}\right)^{a\delta-1}[Q_i]^{a-1}\left[2\lambda\left(1 - e^{-S_i}\right)^{\delta} - (1+\lambda)\right]}{1 - (1 - e^{-S_i})^{a\delta}[Q_i]^a},\end{aligned} \quad (27)$$

$$\begin{aligned}\frac{\partial\ell(\upsilon)}{\partial\beta} &= \gamma(a\delta - 1)\sum_{i=1}^{n}\frac{x_i^{\beta}e^{-S_i}\ln(x_i)}{1 - e^{-S_i}} + \gamma\sum_{i=1}^{n}\frac{x_i^{\beta-1}(\beta\ln x_i + 1)}{Z_i} - \gamma\delta\lambda(a-1)\sum_{i=1}^{n}\frac{x_i^{\beta}e^{-S_i}\ln\left[x_i\left(1 - e^{-S_i}\right)^{\delta-1}\right]}{Q_i} \\ &+ a\gamma\delta(b-1)\sum_{i=1}^{n}\frac{x_i^{\theta}e^{-S_i}(Q_i)^{a-1}\left[2\lambda\left(1 - e^{-S_i}\right)^{\delta} - (1+\lambda)\right]}{1 - (1 - e^{-S_i})^{a\delta}(Q_i)^a}\ln\left[x_i\left(1 - e^{-S_i}\right)^{a\delta-1}\right] \\ &- \gamma\sum_{i=1}^{n} x_i^{\beta}\ln(x_i) - 2\gamma\delta\lambda\sum_{i=1}^{n}\frac{x_i^{\beta}e^{-S_i}\ln\left[x_i\left(1 - e^{-S_i}\right)^{\delta-1}\right]}{K_i},\end{aligned} \quad (28)$$

$$\begin{aligned}\frac{\partial\ell(\upsilon)}{\partial\gamma} &= (a\delta - 1)\sum_{i=1}^{n}\frac{e^{-S_i}x_i^{\beta}}{1 - e^{-S_i}} + \sum_{i=1}^{n}\frac{\beta x_i^{\beta-1}}{Z_i} - \sum_{i=1}^{n} x_i^{\beta} - \delta\lambda(a-1)\sum_{i=1}^{n}\frac{e^{-S_i}x_i^{\theta}\left(1 - e^{-S_i}\right)^{\delta-1}}{Q_i} \\ &+ a\delta(b-1)\sum_{i=1}^{n}\frac{x_i^{\beta}e^{-S_i}(Q_i)^{a-1}\left(1 - e^{-S_i}\right)^{a\delta-1}\left[2\lambda\left(1 - e^{-S_i}\right)^{\delta} - (1+\lambda)\right]}{1 - (1 - e^{-S_i})^{a\delta}(Q_i)^a} + 2\delta\lambda\sum_{i=1}^{n}\frac{x_i^{\theta}e^{-S_i}\left(1 - e^{-S_i}\right)^{\delta-1}}{K_i},\end{aligned} \quad (29)$$

$$\begin{aligned}\frac{\partial\ell(\upsilon)}{\partial\delta} &= \frac{n}{\delta} + a\sum_{i=1}^{n}\ln\left(1 - e^{-S_i}\right) - 2\lambda\sum_{i=1}^{n}\frac{\left(1 - e^{-S_i}\right)^{\delta}\ln\left(1 - e^{-S_i}\right)}{K_i} - \lambda(a-1)\sum_{i=1}^{n}\frac{\left(1 - e^{-S_i}\right)^{\delta}\ln\left(1 - e^{-S_i}\right)}{Q_i} \\ &+ a(b-1)\sum_{i=1}^{n}\frac{\left(1 - e^{-S_i}\right)^{a\delta}(Q_i)^{a-1}\left[2\lambda\left(1 - e^{-S_i}\right)^{\delta} - (1+\lambda)\right]\ln\left(1 - e^{-S_i}\right)}{1 - (1 - e^{-S_i})^{a\delta}(Q_i)^a},\end{aligned} \quad (30)$$

$$\begin{aligned}\frac{\partial\ell(\upsilon)}{\partial\theta} &= (a\delta - 1)\sum_{i=1}^{n}\frac{\alpha e^{-S_i}x_i^{\theta}}{1 - e^{-S_i}} - \alpha\lambda\delta(a-1)\sum_{i=1}^{n}\frac{x_i^{\theta}e^{S_i}\ln\left[x_i\left(1 - e^{-S_i}\right)^{\delta-1}\right]}{K_i} + \alpha\sum_{i=1}^{n}\frac{x_i^{\theta-1}(\theta\ln x_i + 1)}{Z_i} \\ &- \alpha\sum_{i=1}^{n} x_i^{\theta}\ln x_i - 2\alpha\lambda\delta\sum_{i=1}^{n}\frac{x_i^{\theta}e^{S_i}\ln\left[x_i\left(1 - e^{-S_i}\right)^{\delta-1}\right]}{K_i} \\ &+ a\alpha\delta(b-1)\sum_{i=1}^{n}\frac{e^{S_i}\left[1 + \lambda - \lambda\left(1 - e^{-S_i}\right)^{\delta}\right]^{a-1}\left[2\lambda\left(1 - e^{-S_i}\right)^{\delta} - (1+\lambda)\right]}{\left\{\ln\left[x_i(1 - e^{-S_i})^{a\delta-1}\right]\right\}^{-1}\left\{1 - (1 - e^{-S_i})^{a\delta}\left[1 + \lambda - \lambda(1 - e^{-S_i})^{\delta}\right]^a\right\}},\end{aligned} \quad (31)$$

$$\frac{\partial \ell(\upsilon)}{\partial \lambda} = (a-1)\sum_{i=1}^{n} \frac{1-\left(1-e^{-S_i}\right)^{\delta}}{1+\lambda-\lambda(1-e^{-S_i})^{\delta}} + a(b-1)\sum_{i=1}^{n} \frac{(Q_i)^{a-1}\left(1-e^{-S_i}\right)^{a\delta}\left[1-\left(1-e^{-S_i}\right)\right]}{(Q_i)^{a}(1-e^{-S_i})^{a\delta}} + \sum_{i=1}^{n} \frac{1-2\left(1-e^{-S_i}\right)^{\delta}}{K_i}, \tag{32}$$

$$\frac{\partial \ell(\upsilon)}{\partial a} = \frac{n}{a} + \delta\sum_{i=1}^{n} \ln\left(1-e^{-S_i}\right) + \sum_{i=1}^{n} \ln(Q_i) - \delta(b-1)\sum_{i=1}^{n} \frac{Q_i\left(1-e^{-S_i}\right)^{\delta}\left[\ln\left(1-e^{-S_i}\right)Q_i\right]}{1-(1-e^{-S_i})^{a\delta}(Q_i)^{a}}, \tag{33}$$

and

$$\frac{\partial \ell(\upsilon)}{\partial b} = \frac{n}{b}\sum_{i=1}^{n} \ln\left[1-\left(1-e^{-S_i}\right)^{a\delta} Q_i\right]. \tag{34}$$

We can find the estimates of the unknown parameters by setting the score vector to zero, $\mathbf{U}(\widehat{\upsilon}) = 0$, and solving them simultaneously yields the ML estimators $\widehat{\alpha}, \widehat{\beta}, \widehat{\gamma}, \widehat{\delta}, \widehat{\theta}, \widehat{\lambda}, \widehat{a}$ and $\widehat{b}$. These equations cannot be solved analytically and statistical software can be used to solve them numerically by means of iterative techniques such as the Newton-Raphson algorithm.

In order to compute the standard error and asymptotic confidence interval we use the usual large sample approximation (Migon *et al.*, 2014) in which the maximum likelihood estimators of $\boldsymbol{\upsilon}$ can be treated as being approximately hepta-variate normal. For example, as $n \to \infty$ the asymptotic distribution of the MLE $(\widehat{\alpha}, \widehat{\beta}, \widehat{\gamma}, \widehat{\delta}, \widehat{\theta}, \widehat{\lambda}, \widehat{a}, \widehat{b})$, is given by,

$$\begin{pmatrix} \widehat{\alpha} \\ \widehat{\beta} \\ \widehat{\gamma} \\ \widehat{\delta} \\ \widehat{\theta} \\ \widehat{\lambda} \\ \widehat{a} \\ \widehat{b} \end{pmatrix} \sim N\left[\begin{pmatrix} \widehat{\alpha} \\ \widehat{\beta} \\ \widehat{\gamma} \\ \widehat{\delta} \\ \widehat{\theta} \\ \widehat{\lambda} \\ \widehat{a} \\ \widehat{b} \end{pmatrix}, \begin{pmatrix} \hat{V}_{11} & \dots & \hat{V}_{18} \\ \vdots & \ddots & \vdots \\ \hat{V}_{81} & \dots & \hat{V}_{88} \end{pmatrix}\right], \tag{35}$$

with, $\hat{V}_{ij} = V_{ij}\,|_{\theta=\hat{\theta}}$ and it is determined by the inverse of Fisher information that can be easily obtained since the second order derivatives of the log-likelihood function exist for all the eigth parameters of Kw-TEAW distribution. Thus, the approximate $100(1-\phi)\%$ confidence intervals for $\alpha, \beta, \gamma, \delta, \theta, \lambda, a$ and b can be determined as:

$$\widehat{\alpha} \pm Z_{\frac{\phi}{2}}\sqrt{\widehat{V}_{11}}, \quad \widehat{\beta} \pm Z_{\frac{\phi}{2}}\sqrt{\widehat{V}_{22}}, \quad \widehat{\gamma} \pm Z_{\frac{\phi}{2}}\sqrt{\widehat{V}_{33}}, \quad \widehat{\delta} \pm Z_{\frac{\phi}{2}}\sqrt{\widehat{V}_{44}},$$

$$\widehat{\theta} \pm Z_{\frac{\phi}{2}}\sqrt{\widehat{V}_{55}}, \quad \widehat{\lambda} \pm Z_{\frac{\phi}{2}}\sqrt{\widehat{V}_{66}}, \quad \widehat{a} \pm Z_{\frac{\phi}{2}}\sqrt{\widehat{V}_{77}}, \quad \widehat{b} \pm Z_{\frac{\phi}{2}}\sqrt{\widehat{V}_{88}},$$

where $Z_{\frac{\phi}{2}}$ is the upper ϕth percentile of the standard normal distribution. The applicability of the estimation method and the asymptotic confidence intervals are as follows, in Section 6.

6. Application

In this section, we provide an application of the Kw-TEAW distribution to show the importance and usefulness of the new model. For that, we use the data works with nicotine measurements, made from several brands of cigarettes in 1998, collected by the Federal Trade Commission which is an independent agency of the US government, whose main mission is the promotion of consumer protection.

The report entitled tar, nicotine, and carbon monoxide of the smoke of 1206 varieties of domestic cigarettes for the year of 1998 consists of the data sets and some information about the source of the data, smokers behavior and beliefs about nicotine, tar and carbon monoxide contents in cigarettes. The free form data set can be found at *http://pw1.netcom.com/rdavis2/smoke.html.* We analyzed data on nicotine, measured in milligrams per cigarette, from several cigarette brands and the TTT Plot of the rimes can be seen in Figure **??**-(a).

Table 2 shows the numerical values of the MLEs, the estimated standard error and four different selection criterias: -2 log, AIC, AICC and BIC. The adjustment of the model Kw-TEAW can be seen in Figure **??**-(b). Observe that all model are nested and the lowest value of -2 log likelihood observed was for the Kw-TEAW model, as expected.

6.1 Global and Local Influence to Kw-TEAW Estimated Model

In this section we will make an analysis of global and local influence for the data set given, using the Kw-TEAW model.

The first tool to assess the sensitivity analysis are measures of global influence. Starting with the case-deletion, that we study the effect of withdrawal of the ith element sampled. The first measure of global influence analysis is known as generalized Cook's distance, which is defined as the standard norm of $\boldsymbol{\psi}_i = (\alpha_i, \beta_i, \gamma_i, \delta_i, \theta_i, \lambda_i, a_i, b_i)$ and $\hat{\boldsymbol{\psi}} = (\hat{\alpha}, \hat{\beta}, \hat{\gamma}, \hat{\delta}, \hat{\theta}, \hat{\lambda}, \hat{a}, \hat{b})$ and is given by

$$CD_i(\boldsymbol{\psi}) = \left[\ \boldsymbol{\psi}_i - \hat{\boldsymbol{\psi}}\ \right]^T \left[-\ddot{L}(\boldsymbol{\psi})\right]\left[\ \boldsymbol{\psi}_i - \hat{\boldsymbol{\psi}}\ \right], \tag{36}$$

where $\ddot{L}(\boldsymbol{\psi})$ can be approximated by the estimated covariance and variance matrix.

Another way to measure the global influence is through the difference in likelihoods given by

$$LD_i(\boldsymbol{\psi}) = 2\left\{l(\hat{\boldsymbol{\psi}}) - l(\boldsymbol{\psi}_i)\right\}. \tag{37}$$

Figures 5-a and b show us, respectively, the Cook's generalized and likelihood distances where we could see possible influence points.

Furthermore, we know that the main objective of the local influence method is to evaluate changes in the results from the analysis when small perturbations are incorporated in the model and/or in the data. If such perturbations provoke disproportionate effects, it can be an indication that the model is fitted inadequately or serious departures from the assumptions of the model may exist.

In order to analyse the local influence, here we consider the response variable perturbation, ie, we will consider that each t_i is peturbed as $t_{im} = t_i + m_i V_t$, where V_t is a scale factor that may be the estimated standard deviation of T and $m_i \in \mathbb{R}$. Then, the perturbed log-likelihood function becomes expressed as

$$\begin{aligned}\ell(\psi|\mathbf{t}, \mathbf{m}) \quad = \quad & n\ln a + n\ln b + n\ln\delta + (a\delta - 1)\sum\nolimits_{i=1}^{n} \ln\left(1 - e^{S_{im}}\right) + \sum\nolimits_{i=1}^{n} \ln Z_{im} \\ & - \sum\nolimits_{i=1}^{n} S_{im} + \sum\nolimits_{i=1}^{n} \ln K_{im} + (a-1)\sum\nolimits_{i=1}^{n} \ln\{Q_{im}\} \\ & .(b-1)\sum\nolimits_{i=1}^{n} \ln\left\{1 - \left[1 - e^{S_{im}}\right]^{a\delta}\{Q_{im}\}^{a}\right\},\end{aligned} \tag{38}$$

where $Z_{im} = \alpha\theta t_{im}^{\theta-1} + \gamma\beta t_{im}^{\beta-1}$, $Q_{im} = 1 + \lambda - \lambda\left[1 - e^{-S_{im}}\right]^{\delta}$, $K_{im} = 1 + \lambda - 2\lambda\left[1 - e^{-S_{im}}\right]^{\delta}$ and $S_{im} = \alpha t_{im}^{\theta} + \gamma t_{im}^{\beta}$.
Figures 5-a and b, show us, respectively, the the Cook's generalized and likelihood distances and it is possible to see that the pertubation provoke some disproportionate effects.

After analyse the Figures 4 and 5, we can see the distinction of two observations in relation to others. Furthermore, we made a residual analyse by using the Martingale-type and deviance, see for example McCullagh and Nelder (1989), Barlow and Prentice (1988) and Therneau *et al.* (1990).

The first one, martingale-type residual, was introduced by Therneau *et al.* (1990) and was firstly used in a counting processes and that ones are skewed and have a maximum value at +1 and a minimum value at $-\infty$. By considering the Kw-TEAW model, the martingale-type residual can be written as

$$r_{M_i} = 1 + \ln\left[\left\{1 - \left(1 - e^{-(\alpha x^{\theta} + \gamma x^{\beta})}\right)^{a\delta}\left[1 + \lambda - \lambda\left(1 - e^{-(\alpha x^{\theta} + \gamma x^{\beta})}\right)^{\delta}\right]^{a}\right\}^{b}\right], \tag{39}$$

where $i = 1, \ldots, n$.

In addition, it is possible to use the deviance residual that has been widely applied in GLMs (generalized linear models). This one was proposed by the same authors (Therneau *et al.* (1990)) and it is a transformation of the martingale residual to attenuate the skewness. In our case, the deviance residuals is given by

$$r_{D_i} = \text{sign}(\hat{r}_{M_i})\left[-2\left(\hat{r}_{M_i} + \log\left(1 - \hat{r}_{M_i}\right)\right)\right], \tag{40}$$

$i = 1, \ldots, n$. Figures 6 show, respectively, the Martingale and deviance residuals.

In our case is clear that the observation $I = 30, 127$ can be influential points. In order to reveal the impact of these, the relative changes was measured as

$$\mathrm{RC}_{\zeta_j} = \left| \frac{\hat{\zeta}_j - \hat{\zeta}_{j(I)}}{\hat{\zeta}_j} \right| \times 100\%, \qquad j = 1, \ldots, p+1,$$

where $\hat{\zeta}_{j(I)}$ denotes the MLE of ζ_j after the set I of observations has been removed. Suggested by Lee *et al.* (2006), we use the total and maximum relative changes and the likelihood displacement given by

To reveal the impact of the detected influential observations, we use three measures defined by Lee *et al.* (2006),

$$\mathrm{TRC} = \sum_{i=1}^{n_p} \left| \frac{\hat{\zeta}_i - \hat{\zeta}_i^0}{\hat{\zeta}_i} \right|, \quad \mathrm{MRC} = \max_i \left| \frac{\hat{\zeta}_i - \hat{\zeta}_i^0}{\hat{\zeta}_i} \right| \quad \text{and} \quad LD_{(I)}(\zeta) = 2\{l(\hat{\zeta} - l(\hat{\zeta}^0)\},$$

where TRC is the total relative changes, MRC the maximum relative changes and LD the likelihood displacement, with n_p (the number of parameters) and $\hat{\zeta}^0$ denotes MLE of ζ after the set I of observations has been removed. Table 3 show us the impact of these two observations.

Note that, when we withdrew the 10 most influential points, β parameter was the most affected. Then, the Kw-TEAW was re-fitted to the data, Table 4 shows the MLEs and Figure 7 shows the empirical and re-adjusted curves.

7. Conclusions

In this paper, we propose a new model, called the Kumaraswamy transmuted exponentiated additive Weibull (Kw-TEAW) distribution, which extends the transmuted exponentiated additive Weibull (TEAW) distribution and some other well known distributions in the literature. An obvious reason for generalizing a standard distribution is the fact that the generalization provides more flexibility to analyze real life data.

In fact, the Kw-TEAW distribution is motivated by the wide use of the Weibull distribution in practice and also its hazard rate function very flexible in accommodating all forms of the hazard. Some of its mathematical and statistical properties was presented beside the explicit expressions for the ordinary and generating function, moments residual life, moments of the reversed residual life and Rényi and q entropies. Finally, we illustrate the usefulness of the model showing that this one provides consistently better fit than the other nested models mentioned above. We hope that the proposed model will attract wider application in areas such as engineering, survival and lifetime data, hydrology, economics (income inequality) and others.

There are a large number of possible extensions of the current work. The presence of covariates, as well as of long-term survivals, is very common in practice. Our approach should be investigate in both contexts. A possible approach is to consider the regression schemes adopted by Achcar and Louzada-Neto (1992) and Perdona and Louzada (2011), respectively. Other generalisation can be obtained as in Flores *et al.* (2013), which proposes a complementary exponential power series distribution, which arises on latent complementary risks scenarios, where the lifetime associated with a particular risk is not observable, rather we observe only the maximum lifetime value among all hazards.

References

Achcar, J. A., & Louzada-Neto, F. (1992). A Bayesian approach for accelerated life tests considering the Weibull distribution. *Computational Statistics Quarterly, 7,* 355-368.

Afify, A. Z., Nofal, Z. M., & Ebrahim, A. N. (2015). Exponentiated transmuted generalized Rayleigh distribution. *Pakistan Journal of Statistics and Operation Research, 11*(1), 113-132. http://dx.doi.org/10.18187/pjsor.v11i1.873

Al-Babtain, A., Fattah, A. A., Ahmed, A. N., & Merovci, F. (2015). The Kumaraswamy-transmuted exponentiated modified Weibull distribution. Accepted to be published in the Communications in Statistics-Simulation and Computation.

Aryal, G. R., & Tsokos, C. P. (2011). Transmuted Weibull distribution: a generalization of the Weibull probability distribution. *European Journal of Pure and Applied Mathematics, 4*(2), 89-102.

Barlow, W. E., & Prentice, R. L. (1988). Residuals for relative risk regression. Biometrika. http://dx.doi.org/10.1093/biomet/75.1.65

Bebbington, M., Lai, C., & Zitikis, R. (2007). A flexible Weibull extension. *Reliability Engineering & System Safety, 92*(6), 719-726. http://dx.doi.org/10.1016/j.ress.2006.03.004

Cordeiro, G. M., & de Castro, M. (2011). A new family of generalized distributions. *Journal of statistical computation and simulation, 81*(7), 883-898. http://dx.doi.org/10.1080/00949650903530745

Cordeiro, G. M., Hashimoto, E. M., & Ortega, E. M. (2014). McDonald Weibull model. *Statistics: A Journal of Theoretical and Applied Statistics, 48*, 256-278. http://dx.doi.org/10.1080/02331888.2012.748769

Elbatal, I. (2011). Exponentiated Modified Weibull Distribution. *Economic Quality Control, 26*(2), 189-200. http://dx.doi.org/10.1515/EQC.2011.018

Elbatal, I., & Aryal, G. (2013). On the transmuted additive Weibull distribution. *Austrian Journal of Statistics, 42*(2), 117-132.

Eltehiwy, M., & Ashour, S. (2013). Transmuted Exponentiated Modified Weibull Distribution. *International Journal of Basic and Applied Sciences, 2*(3), 258-269. http://dx.doi.org/10.14419/ijbas.v2i3.1074

Flores, J., Borges, P., Cancho, V. G., & Louzada, F. (2013). The complementary exponential power series distribution. *Brazilian Journal of Probability and Statistics, 27*, 565-584. http://dx.doi.org/10.1214/11-BJPS182

Gomes, A. E., da Silva, C. Q., Cordeiro, G. M., & Ortega, E. M. (2014). A new lifetime model: the Kumaraswamy generalized Rayleigh distribution. *Journal of Statistical Computation and Simulation, 84*(2), 290-309. http://dx.doi.org/10.1080/00949655.2012.706813

Guess, F., & Proschan, F. (1988). Mean residual life, theory and applications. In: Krishnaiah, P.R., Rao, C.R. (Eds.), Handbook of Statistics. Vol. 7. *Reliability and Quality Control,* pages 215-224. http://dx.doi.org/10.1016/s0169-7161(88)07014-2

Gupta, R. C., Gupta, P. L., & Gupta, R. D. (1998). Modeling failure time data by Lehman alternatives. *Communications in Statistics-Theory and methods, 27*(4), 887-904. http://dx.doi.org/10.1080/03610929808832134

Khan, M. S., & King, R. (2013). Transmuted modified Weibull distribution: A generalization of the modified Weibull probability distribution. *European Journal of Pure and Applied Mathematics, 6*, 66-86.

Kundu, D., & Raqab, M. Z. (2005). Generalized Rayleigh distribution: different methods of estimations. *Computational statistics & data analysis, 49*(1), 187-200. http://dx.doi.org/10.1016/j.csda.2004.05.008

Lai, C. D., & Xie, M. (2006). Stochastic ageing and dependence for reliability. New York.

Lee, S. Y., Lu, B., & Song, X. Y. (2006). Assessing local influence for nonlinear structural equation models with ignorable missing data. *Computational Statistics & Data Analysis.* http://dx.doi.org/10.1016/j.csda.2004.11.012

M., S. A., & Zaindin, M. (2013). Modified Weibull distribution. *Applied Sciences, 11,* 123-136.

Marshall, A., & Olkin, I. (1997). A new method for adding a parameter to a family of distributions with application to the exponential and Weibull families. *Biometrika, 84*(3), 641-652. http://dx.doi.org/10.1093/biomet/84.3.641

McCullagh, P., & Nelder, J. A. (1989). Generalized Linear Models. McGraw Hill, London. http://dx.doi.org/10.1007/978-1-4899-3242-6

Merovci, F. (2013a). Transmuted exponentiated exponential distribution. *Mathematical Sciences and Applications ENotes, 1*(2), 112-122.

Merovci, F. (2013b). Transmuted Rayleigh distribution. *Austrian Journal of Statistics, 42*(1), 21-31.

Migon, H. S., Gamerman, D., & Louzada, F. (2014). Statistical inference: an integrated approach. London.

Perdona, G. S. C., & Louzada, F. (2011). A General Hazard Model for Lifetime Data in the Presence of Cure Rate. *Journal of Applied Statistics, 38*, 1395-1405. http://dx.doi.org/10.1080/02664763.2010.505948

Pham, H., & Lai, C. D. (2007). On recent generalizations of the Weibull distribution. *IEEE Transactions on Reliability, 56*(3), 454-458. http://dx.doi.org/10.1109/TR.2007.903352

Rayleigh, J. W. S. (1880). On the resultant of a largenumber of vibration of the same pitch and arbitrary phase. *Philosophical magazine 5th Series, 10*, 73-78. http://dx.doi.org/10.1080/14786448008626893

Mudholkar, G. S., & Srivastava, D. K. (1993). Exponentiated Weibull family for analysing bathtub failure rate data. *IEEE Transactions on Reliability, 42*(2), 299-302. http://dx.doi.org/10.1109/24.229504

Shaw, W. T., & Buckley, I. R. C. (2007). The alchemy of probability distributions: beyond Gram-Charlier expansions, and a skew-kurtotic-normal distribution from a rank transmutation map. UCL Discovery Repository, pages 1-16.

Therneau, T. M., Grambsch, P. M., & Fleming, T. R. (1990). Martingale-based residuals for survival models. *Biometrika.* http://dx.doi.org/10.1093/biomet/77.1.147

Weibull, W. (1951). A statistical distribution function of wide applicability. *J. Appl. Mech.-Trans, 18*(3), 293297.

Xie, M., & Lai, C. D. (1995). Reliability analysis using an additive Weibull model with bathtub-shaped failure rate function. *Reliability Engineering and System Safety, 52,* 87-93. http://dx.doi.org/10.1016/0951-8320(95)00149-2

Xie, M., Tang, Y., & Goh, T. N. (2002). A modified Weibull extension with bathtub-shaped failure rate function. *Reliability Engineering and System Safety, 76*(3), 279C285. http://dx.doi.org/10.1016/S0951-8320(02)00022-4

Appendix

A. Tables

Table 1. Sub-models of the Kw-TEAW $(\alpha, \beta, \gamma, \theta, \delta, \lambda, a, b)$.

No.	Distribution	α	β	γ	θ	δ	λ	a	b	Author
1	kw-EAW	α	β	γ	θ	δ	0	a	b	–
2	kw-TAW	α	β	γ	θ	1	λ	a	b	–
3	Kw-AW	α	β	γ	θ	1	0	a	b	–
4	Kw-TEME	α	1	γ	1	δ	λ	a	b	–
5	Kw-TME	α	1	γ	1	1	λ	a	b	–
6	kw-EME	α	1	γ	1	δ	0	a	b	–
7	kw-ME	α	1	γ	1	1	0	a	b	–
8	GTEAW	α	β	γ	θ	δ	λ	a	1	–
9	GEAW	α	β	γ	θ	δ	0	a	1	–
10	GTAW	α	β	γ	θ	1	λ	a	1	–
11	GAW	α	β	γ	θ	1	0	a	1	–
12	GTEME	α	1	γ	1	δ	λ	a	1	–
13	GTME	α	1	γ	1	1	λ	a	1	–
14	GEME	α	1	γ	1	δ	0	a	1	–
15	GME	α	1	γ	1	1	0	a	1	–
16	TEAW	α	β	γ	θ	δ	λ	1	1	–
17	TAW	α	β	γ	θ	1	λ	1	1	Elbatal and Aryal (2013)
18	EAW	α	β	γ	θ	δ	0	1	1	–
19	AW	α	β	γ	θ	1	0	1	1	Xie and Lai (1995)
20	TEME	α	1	γ	1	δ	λ	1	1	–
21	TME	α	1	γ	1	1	λ	1	1	Elbatal and Aryal (2013)
22	EME	α	1	γ	1	δ	0	1	1	–
23	ME	α	1	γ	1	1	0	1	1	Elbatal and Aryal (2013)
24	Kw-TEMW	α	β	γ	1	δ	λ	a	b	Al-Babtain *et al.* (2015)
25	ETR	α	2	γ	1	δ	0	a	1	–
26	New-ER	0	2	γ	1	δ	0	a	1	–
27	New-EE	α	β	0	1	δ	0	a	1	–
28	Kw-TELFR	α	2	γ	1	δ	λ	a	b	–
29	Kw-TEW	0	β	γ	1	δ	λ	a	b	–
30	Kw-TER	0	2	γ	1	δ	λ	a	b	–
31	Kw-TEE	α	β	0	1	δ	λ	a	b	–
32	Kw-TMW	α	β	γ	1	1	λ	a	b	–
33	Kw-TLFR	α	2	γ	1	1	λ	a	b	–
34	Kw-TW	0	β	γ	1	1	λ	a	b	–
35	Kw-TR	0	2	γ	1	1	λ	a	b	–
36	Kw-TE	α	β	0	1	δ	λ	a	b	–
37	Kw-EMW	α	β	γ	1	δ	0	a	b	–
38	Kw-MW	α	β	γ	1	1	0	a	b	Cordeiro *et al.* (2014)
39	Kw-EW	0	β	γ	1	δ	0	a	b	–
40	ETEMW	α	β	γ	1	δ	λ	a	1	–
41	ETELFR	α	β	γ	1	δ	1	a	2	–
42	ETEW	0	β	γ	1	δ	λ	a	1	–
43	ETGR	0	2	γ	1	δ	λ	a	1	Afify *et al.* (2015)
44	ETEE	α	β	0	1	δ	λ	a	1	–
45	ETMW	α	β	γ	1	1	λ	a	1	–
46	ETLFR	α	2	γ	1	1	λ	a	1	–
47	ETW	0	β	γ	1	1	λ	a	1	–
48	ETR	0	2	γ	1	1	λ	a	1	–
49	ETE	α	β	0	1	1	λ	a	1	–
50	New-EMW	α	β	γ	1	δ	0	a	1	–

Continuing...

No.	Distribution	α	β	γ	θ	δ	λ	a	b	Author
51	New-ELFR	α	2	γ	1	δ	0	a	1	–
52	New-EW	0	β	γ	1	δ	0	a	1	–
53	Kw-EE	0	β	0	1	δ	λ	a	b	Gomes *et al.* (2014)
54	Kw-LFR	α	2	γ	1	1	0	a	b	Al-Babtain *et al.* (2015)
55	Kw-ELFR	α	2	γ	1	δ	0	a	b	Elbatal (2011)
56	Kw-ER	0	2	γ	1	δ	0	a	b	Gomes *et al.* (2014)
57	EMW	α	β	γ	1	1	0	a	1	Elbatal (2011)
58	EW	0	β	γ	1	δ	0	a	1	S. and K. (1993)
59	Kw-W	0	β	γ	1	1	0	a	b	Cordeiro and de Castro (2011)
60	Kw-R	0	2	γ	1	1	0	a	b	–
61	Kw-E	α	β	0	1	1	0	a	b	–
62	EE	α	β	0	1	1	λ	a	1	Gupta *et al.* (1998)
63	TEMW	α	β	γ	1	δ	λ	1	1	Eltehiwy and Ashour (2013)
64	TELFR	α	2	γ	1	δ	λ	1	1	–
65	TEW	0	β	γ	1	δ	λ	1	1	–
66	TER	0	2	0	1	δ	λ	1	1	Merovci (2013a)
67	TEE	α	β	γ	1	δ	λ	1	1	Merovci (2013b)
68	TMW	α	β	γ	1	1	λ	1	1	Khan and King (2013)
69	TLFR	α	2	γ	1	1	λ	1	1	–
70	TW	0	β	γ	1	1	λ	1	1	Aryal and Tsokos (2011)
71	TR	0	2	γ	1	δ	λ	1	1	Khan and King (2013)
72	TE	α	β	0	1	1	λ	1	1	Shaw and Buckley (2007)
73	ELFR	α	2	γ	1	δ	0	1	1	M. and Zaindin (2013)
74	ER	0	2	γ	1	δ	0	1	1	Kundu and Raqab (2005)
75	MW	α	β	γ	1	1	0	1	1	M. and Zaindin (2013)
76	LFR	α	2	γ	1	1	0	1	1	–
77	W	0	β	γ	1	1	0	1	1	Weibull (1951)
78	R	0	2	γ	1	1	0	1	1	Rayleigh (1880)
79	E	0	1	γ	1	1	0	1	1	–

Table 2. Estimates of parameters and confidence interval of Kw-TEAW model and some nested ones.

Model	Parameters	Estimate	Standard Error	Selection Criteria −2 log	AIC	AICC	BIC
Kw-TEAW	α	1.027	1.637	216.0	232.0	232.4	262.7
	β	2.033	1.989				
	λ	−0.606	0.513				
	θ	0.369	0.355				
	γ	1.218	2.470				
	a	1.385	0.944				
	b	1.649	7.173				
	δ	2.896	4.736				
TAW	α	0.388	0.354	218.1	228.1	228.3	247.4
	β	2.664	0.314				
	λ	−0.708	0.213				
	θ	1.216	0.517				
	γ	1.172	0.347				
AW	α	0.426	0.167	217.6	225.6	225.8	241.0
	β	2.652	0.235				
	θ	0.700	0.221				
	γ	1.245	0.187				
TEMW	α	0.722	0.501	217.1	227.1	227.2	246.3
	β	2.599	0.272				
	λ	−0.629	0.230				
	γ	1.177	0.265				
	δ	1.525	0.494				
EMW	α	0.453	0.305	218.9	226.9	227.0	242.3
	β	2.841	0.216				
	γ	1.068	0.133				
	δ	1.626	0.424				
EW	β	3.063	0.354	226.3	232.3	232.4	243.9
	γ	0.947	0.173				
	δ	0.812	0.152				
W	β	2.719	0.114	227.6	231.6	231.6	239.2
	γ	1.047	0.022				

Table 3. Some influence measures of set I.

Parameters	RC/100	TRC	MRC	LD
μ	0.484	11.370	6.012	12.1
β	6.012			
λ	1.635			
θ	0.856			
σ	0.367			
a	0.458			
b	1.454			
δ	0.103			

Table 4. MLEs considering the Kw-TEAW model after been removed the set I.

Parameters	Estimate	Standard Error	Confidence Interval (95%) Lower	Upper
μ	1.9903	1.1536	−0.2787	4.2593
β	0.2899	0.2959	−0.2922	0.872
λ	−0.2299	1.0181	−2.2324	1.7727
θ	2.5705	0.4928	1.6013	3.5397
σ	1.9237	1.7946	−1.6062	5.4535
a	2.5579	5.8507	−8.9498	14.0655
b	0.6719	0.5704	−0.4501	1.7938
δ	3.2272	5.8322	−8.2441	14.6984

B. Graphical Representations

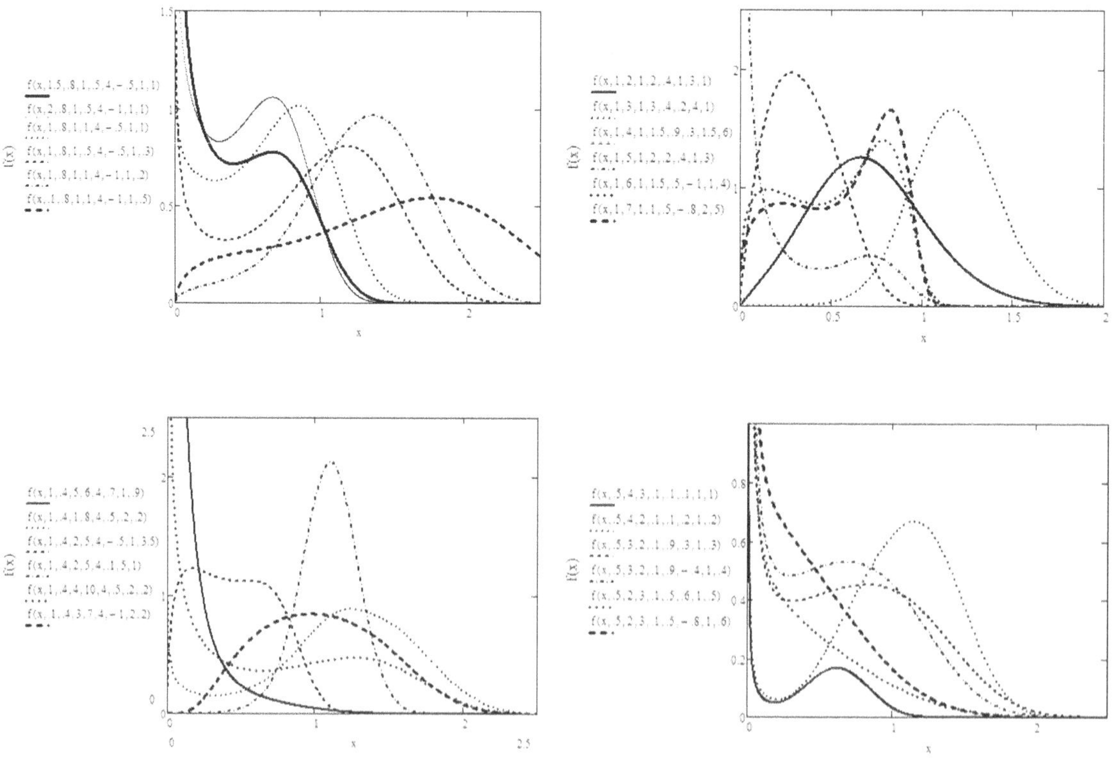

Figure 1. Density curves of Kw-TEAW model for different values of parameters.

Figure 2, lower and upper panels, provide some plotsof the Kw-TEAW hazard rate function, showing that it is quite flexible for modelling survival data.

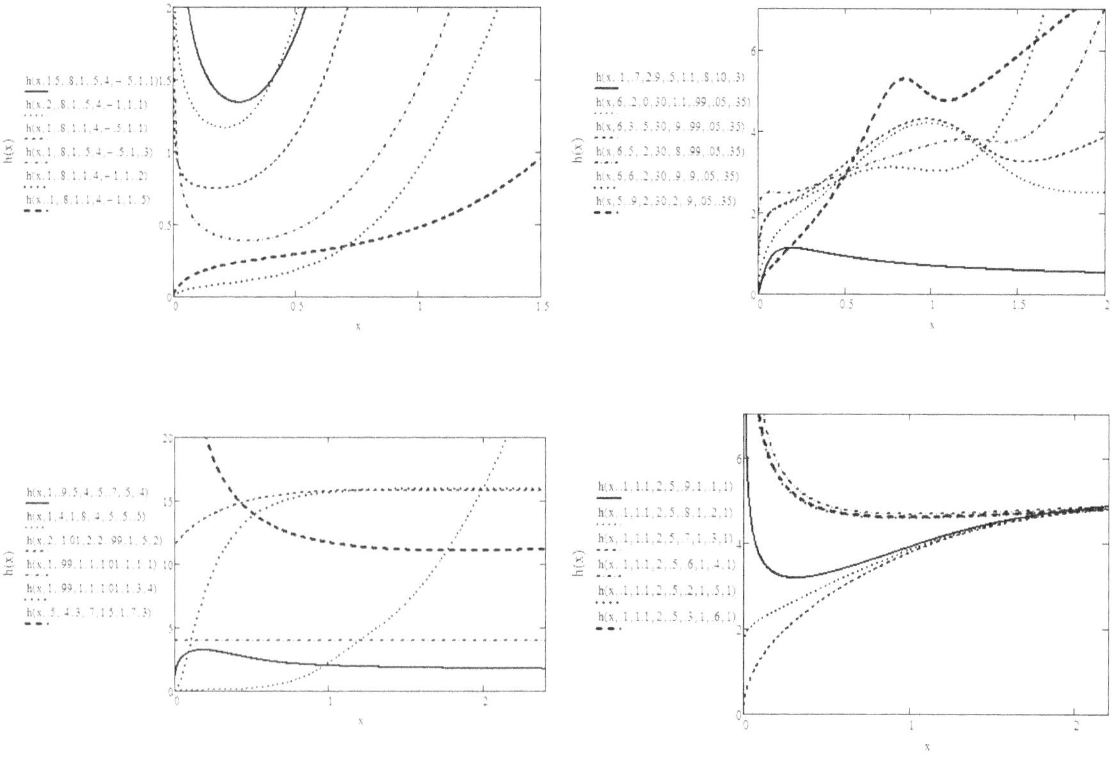

Figure 2. Hazard rate curves of Kw-TEAW model for different values of parameters.

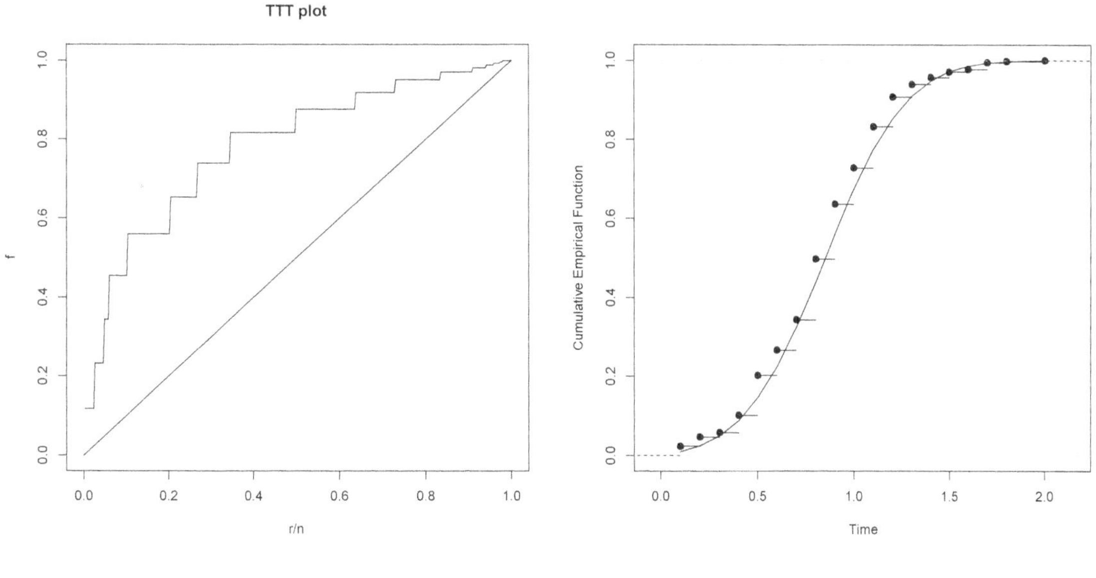

Figure 3. TTT-Plot and cumulative curves, empirical and estimated by Kw-TEAW model.

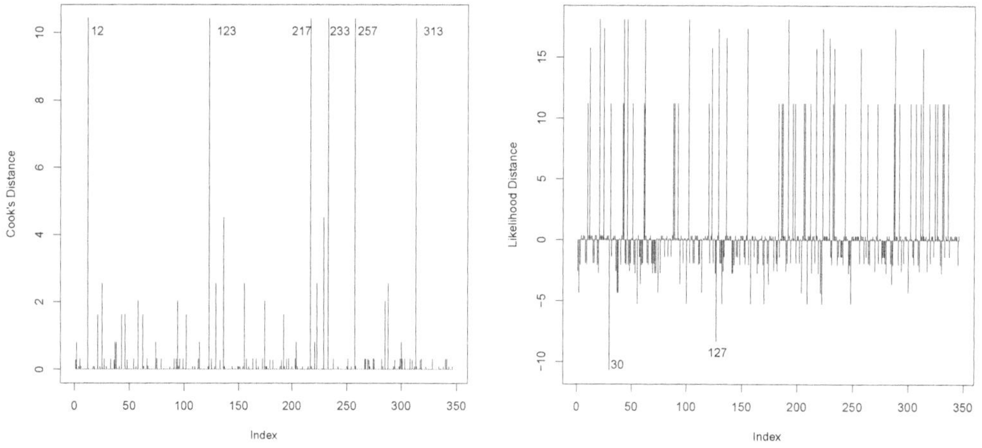

Figure 4. (a) Cooks's distance and (b) Likelihood distance.

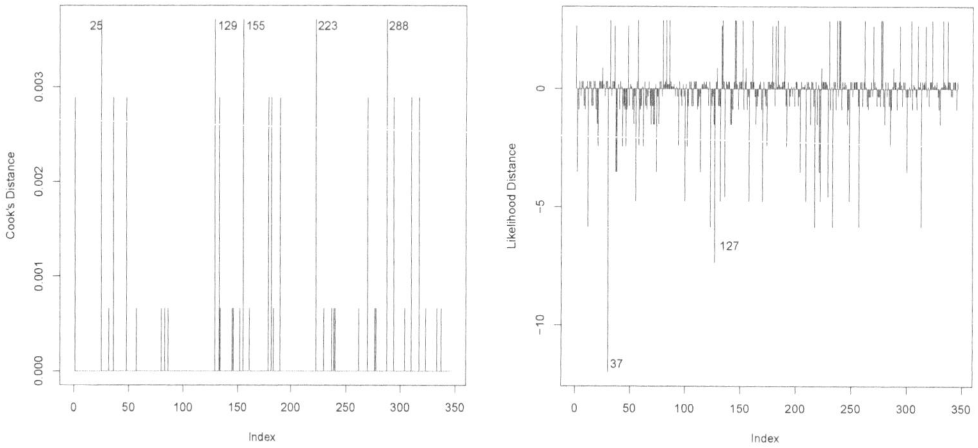

Figure 5. (a) Cooks's distance and (b) Likelihood distance after response perturbation.

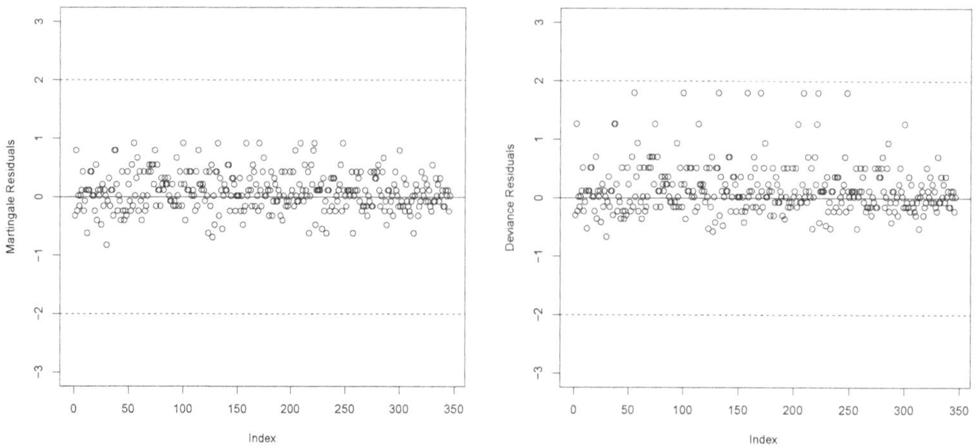

Figure 6. (a) Martingale residuals; (b) Deviance residuals.

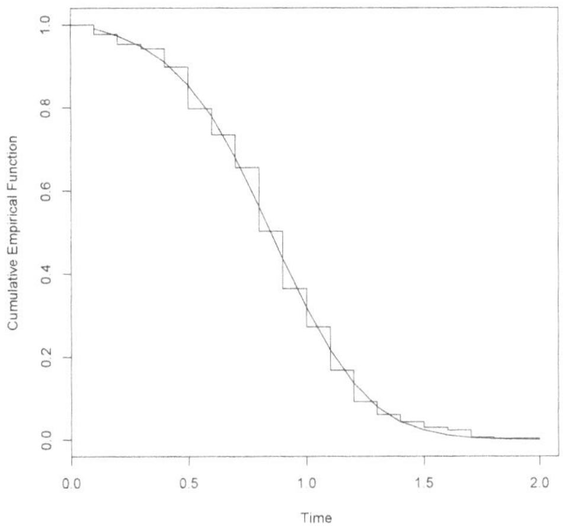

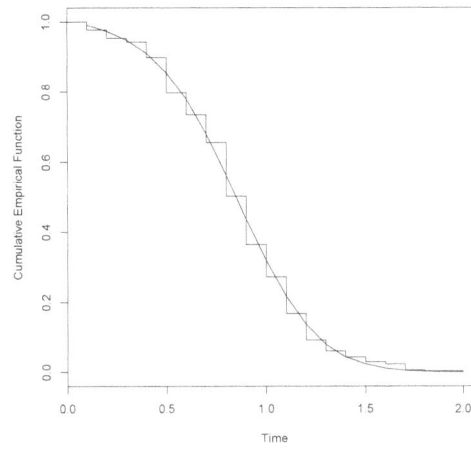

Figure 7. Kaplan-Meier empirical survival curve vs the re-adjusted model Kw-TEAW.

2

Limit Distribution of a Generalized Ornstein – Uhlenbeck Process

Andriy Yurachkivsky[1]

[1] Taras Shevchenko National University, Kyiv, Ukraine

Correspondence: Correspondence: Andriy Yurachkivsky, Taras Shevchenko National University, 64 Volodymyrska St., Kyiv, 01601, Ukraine. E-mail: andriy.yurachkivsky@gmail.com

Abstract

Let an $\mathbb{R}^d$-valued random process ξ be the solution of an equation of the kind $\xi(t) = \xi(0) + \int_0^t A(u)\xi(u)\mathrm{d}\iota(u) + S(t)$, where $\xi(0)$ is a random variable measurable w. r. t. some σ-algebra $\mathcal{F}(0)$, S is a random process with $\mathcal{F}(0)$-conditionally independent increments, ι is a continuous numeral random process of locally bounded variation, and A is a matrix-valued random process such that for any $t > 0$ $\int_0^t \|A(s)\|\,|\mathrm{d}\iota(s)| < \infty$. Conditions guaranteing existence of the limiting, as $t \to \infty$, distribution of $\xi(t)$ are found. The characteristic function of this distribution is written explicitly.

Keywords: limit distribution, generalized Ornstein – Uhlenbeck process

Mathematics Subject Classification: 60F05, 60H10

1. Introduction

The random processes under consideration are assumed given on a common probability space $(\Omega, \mathcal{F}, \mathsf{P})$. It is assumed that a a filtration $\mathbb{F} = (\mathcal{F}(t),\ t \in \mathbb{R}_+)$ on $\mathcal{F}$ is given. We consider, without loss generality, that it is right-continuous and each $\mathcal{F}(t)$ contains all P-negligible sets from $\mathcal{F}$ (these are so called usual conditions – see (Gikhman & Skorokhod, 1982; Jacod & Shiryaev, 1987; Liptser & Shiryaev, 1989)). We will consider also the trivial filtration $\mathbb{F}^0 = (\mathcal{F}^0(t),\ t \in \mathbb{R}_+)$, where $\mathcal{F}^0(t) = \mathcal{F}(0)$ for all t. Thus a random process is $\mathbb{F}^0$-adapted iff its value at any nonrandom time is an $\mathcal{F}(0)$-measurable random variable. We introduce the notation: $\mathsf{E}^0 = \mathsf{E}(\cdots|\mathcal{F}(0))$; $\mathcal{V}_0^c$ is the class of all starting from zero $\mathbb{F}^0$-adapted continuous random processes; $\mathfrak{S}_+$ is the class of all nonnegative (in the spectral sense) symmetric $d\times d$ matrices with real entries; $\mathbb{M}_+(\mathfrak{C})$ is the class of all σ-finite measures on a σ-algebra $\mathfrak{C}$; l.i.p. signifies the limit in probability. In integrals with a discontinuous integrator, $\int_a^b$ means $\int_{]a,b]}$. The indicator of a set $\{\cdots\}$ is denoted by $I\{\cdots\}$.

All vectors are thought of, unless otherwise stated, as columns; all matrices are meant of size $d \times d$, with real entries. The unit matrix is denoted by $\mathbf{I}$, and the space of all d-dimensional row vectors with real components by $\mathbb{R}^{d*}$. We use the Euclidean norm $|\cdot|$ of vectors and the operator norm $\|\cdot\|$ of matrices. For symmetric matrices A and B, the inequality $A \le B$ means that $B - A \in \mathfrak{S}_+$ (so that one may speak about increasing $\mathfrak{S}_+$-valued functions).

The quadratic characteristic of a locally square-integrable $\mathbb{R}^d$-valued martingale Z will be denoted by $\langle Z\rangle$. This is an increasing $\mathfrak{S}_+$-valued random process.

The words "almost surely" are tacitly implied in relations between random variables, including the convergence relation unless it is explicitly written as the convergence in probability (denoted by $\xrightarrow{\mathsf{P}}$) or in distribution (denoted by $\xrightarrow{\mathrm{d}}$).

The reference books for the notions and results of stochastic analysis used in this paper are (Gikhman & Skorokhod, 1982, 2009; Jacod & Shiryaev, 1987; Liptser & Shiryaev, 1989). A number of more specific definitions and statements relevant to the topic can be found in (Yurachkivsky, 2013).

The goal of the article is to find the limit, as $t \to \infty$ (which will be tacitly meant in all asymptotic relations), of the one-dimensional distribution of the solution of the stochastic equation

$$\xi(t) = \xi(0) + \int_0^t A(u)\xi(u)\mathrm{d}\iota(u) + S(t), \tag{1}$$

where S is a random process with $\mathcal{F}(0)$-conditionally independent increments, $\iota \in \mathcal{V}_0^c$, and A is a matrix-valued $\mathbb{F}^0$-adapted random process such that for any $t > 0$

$$\int_0^t \|A(s)\|\,|\mathrm{d}\iota(s)| < \infty. \tag{2}$$

We call thus defined ξ a *generalized Ornstein – Uhlenbeck process*. Recall that a classical Ornstein – Uhlenbeck process is the solution of (1) with $d = 1, U = 0, \iota(t) = t$, constant A and a homogeneous Wiener process as S. For them, the problem is easy and solved long ago (the limit distribution exists iff $A < 0$). But even in a seemingly simple case when S is a homogeneous generalized (i. e., with random jumps) Poisson process the problem is nontrivial. It was solved first by Zakusilo in 1981 (this result is contained in (Anisimov, Zakusilo & Donchenko, 1987)). A more general theorem, but also only for the case of homogeneous S and $\iota(t) = t$, was proved in (Sato & Yamazato, 1984). Unlike these authors, we focus on the described above model. This level of generality requires quite different technique that will be demonstrated below. The only known to the author work with $\iota(t)$ possibly other than t and nonhomogeneous S is (Ivanenko, 2009). But the proof therein relies on very specific assumptions (for example, $A = -c\mathbf{I}$) and does not carry over to a more general situation.

The structure of the article is clear from the titles of its sections: General preliminaries, Special preliminaries and The main result.

2. General Preliminaries

The following statement is immediate from Lebesgue's dominated convergence theorem and Lévy's continuity theorem.

Proposition 2.1. *Let ξ be an $\mathbb{R}^d$-valued random process. Suppose that there exists a $\mathbb{C}$-valued random function J on $\mathbb{R}^{d*}$ such that:* $\operatorname{l.i.p.}_{z\to 0} J(z) = 0$; *for any $z \in \mathbb{R}^{d*}$ $\mathsf{E}^0 e^{iz\xi(t)} \xrightarrow{\text{d}} e^{J(z)}$. Then there exists a random variable ξ_∞ such that $\mathsf{E}e^{iz\xi_\infty} = e^{J(z)}$ and $\xi(t) \xrightarrow{\text{d}} \xi_\infty$.*

Let us consider the equation

$$K(t, s) = \mathbf{I} + \int_s^t A(\tau)K(\tau, s)\mathrm{d}\iota(\tau). \tag{3}$$

The standard convention $\int_s^t = -\int_t^s$ entitles us to consider that the variables s and t independently range over $\mathbb{R}_+$. The integral on the r. h. s. of (22) being pathwise, the variable $\omega \in \Omega$ performs in it as a parameter. Thus we may consider this equation deterministic. Its solution will be an important tool in our study.

Lemma 2.2 (Yurachkivsky, 2013; Lemma 3.10)**.** *Let ι be a numeral function of locally bounded variation, and A be a Borel matrix-valued function satisfying condition* (2). *Then the solution of equation* (3) *exists, is unique and has locally bounded variation in each argument.*

In what follows, $\mathfrak{C}$ is a σ-algebra of subsets of some set Θ.

We denote by $\mathcal{K}$ the class of $\mathbb{F}$-adapted $\mathbb{R}^d$-valued random processes M such that, firstly, $\mathsf{E}^0|M(t)|^2 < \infty$ for all $t \in \mathbb{R}_+$, secondly, $\mathsf{E}(M(t) - M(s)|\mathcal{F}(s)) = M(s)$ for all $t > s \geq 0$ and, thirdly, the process $\mathsf{E}^0|M|^2$ is continuous (here E^0 is the extended conditional expectation, so we need not assume that $\mathsf{E}|M(t)| < \infty$). This class is contained in the class of locally square integrable martingales (see, e. g., (Yurachkivsky, 2013, 2014)).

From this time on we deal with the following particular case of (1):

$$\xi(t) = \xi(0) + \int_0^t A(s)\xi(s)\mathrm{d}\iota(s) + U(t) + \int_0^t\int_\Theta f(s,\theta)\nu(\mathrm{d}s\,\mathrm{d}\theta) + \int_0^t\int_\Theta g(s,\theta)\,(\upsilon - \widetilde{\upsilon})\,(\mathrm{d}s\,\mathrm{d}\theta) + W(t) \tag{4}$$

which may be written shortly in the notation of stochastic analysis (see (Gikhman & Skorokhod, 1982; Jacod & Shiryaev, 1987; Liptser & Shiryaev, 1989)) as

$$\xi = \xi(0) + (A\xi) \circ \iota + U + f * \nu + g * (\upsilon - \widetilde{\upsilon}) + W.$$

To make it quite definite we impose the assumptions:

1. $\xi(0)$ is an $\mathbb{R}^d$-valued $\mathcal{F}(0)$-measurable random variable.

2. ι is an $\mathbb{R}$-valued random process of class $\mathcal{V}_0^c$.

3. A is a matrix-valued $\mathcal{F}(0) \otimes \mathcal{B}_+$-measurable in (ω, t) random process satisfying, for all t, condition (2).

4. U is an $\mathbb{R}^d$-valued random process of class $\mathcal{V}_0^c$.

5. ν and υ are $\mathbb{F}$-adapted quasicontinuous integer-valued random measures on $\mathcal{B}_+ \otimes \mathfrak{C}$ with $\mathbb{F}^0$-adapted compensators $\widetilde{\nu}$ and $\widetilde{\upsilon}$, respectively.

6. For any random $\mathcal{F}(0)\otimes\mathcal{B}_+\otimes\mathfrak{C}$-measurable functions q_1 and q_2 on $\mathbb{R}_+\times\Theta$ such that $|q_1|*\widetilde{\nu}(t)+|q_2|^2*\widetilde{\upsilon}(t)<\infty$ for all t, the processes $q_1*\nu$ and $q_2*(\upsilon-\widetilde{\upsilon})$ have no simultaneous jumps.

7. W is a starting from zero $\mathbb{R}^d$-valued continuous random process of class $\mathcal{K}$ with $\mathbb{F}^0$-adapted quadratic characteristic.

8. f and g are $\mathbb{R}^d$-valued $\mathcal{F}(0)\otimes\mathcal{B}_+\otimes\mathfrak{C}$-measurable random functions on $\mathbb{R}_+\times\Theta$.

9. $(|g|^2\wedge|g|)*\widetilde{\upsilon}(t)<\infty$ for all t.

10. $(|f|\wedge 1)*\widetilde{\nu}(t)<\infty$ for all t.

11. $|f|*\nu(t)<\infty$ for all t.

Theorem 2.3 (Yurachkivsky, 2013; Corollary 6.2). *Let assumptions **1 – 11** be satisfied. Then for any $t\in\mathbb{R}_+$ and $z\in\mathbb{R}^{d*}$*

$$\mathsf{E}^0 e^{iz\xi(t)}=e^{izK(t,0)\xi(0)+izQ_t-zR_tz^\top/2+F_t(z)+G_t(z)},$$

where

$$Q_t=\int_0^t K(t,s)\mathrm{d}U(s), \tag{5}$$

$$R_t=\int_0^t K(t,s)\mathrm{d}\langle W\rangle(s)K(t,s)^\top, \tag{6}$$

$$F_t(z)=\int_0^t\int_\Theta\left(e^{izK(t,s)f(s,\theta)}-1\right)\widetilde{\nu}(\mathrm{d}s\,\mathrm{d}\theta), \tag{7}$$

$$G_t(z)=\int_0^t\int_\Theta\left(e^{izK(t,s)g(s,\theta)}-1-izK(t,s)g(s,\theta)\right)\widetilde{\upsilon}(\mathrm{d}s\,\mathrm{d}\theta). \tag{8}$$

Corollary 2.4. *Let assumptions **1 – 11** be satisfied, and $Q_t, R_t, F_t(z)$ and $G_t(z)$ be defined by* (5) – (8). *Suppose also that*

$$K(t,0)\xrightarrow{\mathsf{P}}0 \tag{9}$$

and there exist an $\mathbb{R}^d$-valued random variable ϑ, a matrix-valued random variable Υ and a $\mathbb{C}$-valued random function Φ and Γ on $\mathbb{R}^d$, all given on a common probability space and such that

$$\underset{z\to 0}{\mathrm{l.i.p.}}\,\Phi(z)=0, \tag{10}$$

$$\underset{z\to 0}{\mathrm{l.i.p.}}\,\Gamma(z)=0 \tag{11}$$

and for all $z\in\mathbb{R}^{d*}$

$$(Q_t,R_t,F_t(z),G_t(z))\xrightarrow{\mathrm{d}}(\vartheta,\Upsilon,\Phi(z),\Gamma(z)). \tag{12}$$

Then the distribution of $\xi(t)$ weakly converges, as $t\to\infty$, to the distribution with characteristic function $\mathsf{E}e^{iz\vartheta-z\Upsilon z^\top/2+\Phi(z)+\Gamma(z)}$.

Lemma 2.5. *Let μ be a finite measure on some σ-algebra $\mathcal{X}$ of subsets of a set X, and let for each $t>0$ ζ_t be a nonnegative measurable random function on X. Suppose the following: for any $x\in X$*

$$\zeta_t(x)\xrightarrow{\mathsf{P}}0; \tag{13}$$

there exists a measurable random function Z on X such that

$$\int_X Z\mathrm{d}\mu<\infty \tag{14}$$

and for all $t>0$ and $x\in X$

$$\zeta_t(x)\le Z(x). \tag{15}$$

Then

$$\int_X \zeta_t\mathrm{d}\mu\xrightarrow{\mathsf{P}}0. \tag{16}$$

Proof. In this and the subsequent proofs, $\int$ stands for $\int_X$.

Denote $Z^L = ZI\{Z > L\}$. The relations

$$\zeta_t = \zeta_t I\{\zeta_t > L\} + \zeta_t I\{\zeta_t \le L\} \le Z^L + \zeta_t \wedge L,$$

of which the inequality follows from (15), imply that for any positive ε and L

$$\mathsf{P}\left\{\int \zeta_t \mathrm{d}\mu > 2\varepsilon\right\} \le \mathsf{P}\left\{\int Z^L \mathrm{d}\mu > \varepsilon\right\} + \mathsf{P}\left\{\int \zeta_t(u) \wedge L\mathrm{d}\mu > \varepsilon\right\}. \tag{17}$$

By construction $Z^L \to 0$ as $L \to$ for all $x \in X$ and $\omega \in \Omega$. Hence and from (14) we get by the dominated convergence theorem (DCT) applied at those $\omega \in \Omega$ where (14) holds

$$\lim_{L\to\infty} \int Z^L \mathrm{d}\mu = 0. \tag{18}$$

Condition (13) implies by the DCT that $\mathsf{E}(\zeta_t(x) \wedge L) \to 0$ for all $L > 0$ and $x \in X$. Hence by the DCT and due to finiteness of μ

$$\int \mathsf{E}(\zeta_t \wedge L)\mathrm{d}\mu \to 0.$$

Herein

$$\int \mathsf{E}(\zeta_t \wedge L)\mathrm{d}\mu = \mathsf{E}\int \zeta_t \wedge L\mathrm{d}\mu$$

by the Fubini – Tonelli theorem. This together with the preceding relation and (17) implies that for any positive ε and L

$$\overline{\lim_{t\to\infty}} \mathsf{P}\left\{\int \zeta_t \mathrm{d}\mu > 2\varepsilon\right\} \le \mathsf{P}\left\{\int Z^L \mathrm{d}\mu > \varepsilon\right\}.$$

It remains to make use of (18). □

Lemma 2.6. *Let μ be a σ-finite measure on some σ-algebra $\mathcal{X}$ of subsets of a set X, and let for each $t > 0$ ζ_t be a nonnegative measurable random function on X. Suppose the following: for any $x \in X$ relation* (13) *holds; there exists a nonrandom measurable function Z on X satisfying conditions* (14), (15) *and, for any $r > 0$, the condition*

$$\mu\{x : Z(x) > r\} < \infty. \tag{19}$$

Then relation (16) *holds.*

Proof. The set $\{x \in X : Z(x) > r\}$ is nonrandom (since in this lemma so is Z), belongs to $\mathcal{X}$ (because Z is measurable) and has, due to (19), finite measure as $r > 0$. Then it follows from (13) – (15) by Lemma 2.5 that for any $r > 0$

$$\int \zeta_t I\{Z > r\}\mathrm{d}\mu \xrightarrow{\mathsf{P}} 0.$$

Consequently, for any positive r and ε

$$\overline{\lim_{t\to\infty}} \mathsf{P}\left\{\int \zeta_t \mathrm{d}\mu > 2\varepsilon\right\} \le \overline{\lim_{t\to\infty}} \mathsf{P}\left\{\int \zeta_t I\{Z \le r\}\mathrm{d}\mu > \varepsilon\right\}. \tag{20}$$

Condition (15) implies that Z is nonnegative, whence with account of (14) we get by the DCT

$$\lim_{r\to 0} \int ZI\{Z \le r\}\mathrm{d}\mu = 0,$$

which together with (15) yields

$$\lim_{r\to 0} \overline{\lim_{t\to\infty}} \mathsf{P}\left\{\int \zeta_t I\{Z \le r\}\mathrm{d}\mu > \varepsilon\right\} = 0$$

for any $\varepsilon > 0$. Now, (16) follows from (20). □

Let n and d be natural numbers, P and S be $n \times d$-matrices, and B be a $d \times d$ matrix. The identity

$$PBP^\top - SBS^\top = (P - S)BP^\top + SB(P - S)^\top$$

yields the estimate

$$\left\|PBP^\top - SBS^\top\right\| \leq \|P - S\|\,\|B\|\,\left\|P^\top\right\| + \|S\|\,\|B\|\,\left\|(P - S)^\top\right\|.$$

In particular, for any $B \in \mathfrak{S}_+$ (so that $\|B\| \leq \operatorname{tr} B$) and $p, q \in \mathbb{R}^{d*}$

$$\left\|pBp^\top - qBq^\top\right\| \leq (|p| + |q|)|p - q| \operatorname{tr} B.$$

Hence the following conclusion is immediate.

Lemma 2.7. *Let V be an increasing $\mathfrak{S}_+$-valued function on $[a, b]$. Then for any continuous $\mathbb{R}^{d*}$-valued functions φ and ψ on $[a, b]$*

$$\left|\int_a^b \varphi(s)\mathrm{d}V(s)\varphi(s)^\top - \int_a^b \psi(s)\mathrm{d}V(s)\psi(s)^\top\right| \leq \int_a^b (|\varphi(s)| + |\psi(s)|)|\varphi(s) - \psi(s)|\,\mathrm{d}\operatorname{tr} V(s).$$

Lemma 2.8. *For arbitrary $z \in \mathbb{R}^{d*}$, $x, y \in \mathbb{R}^d$* $|e^{izx} - e^{izy}| \leq 2(|z| \vee 2)\frac{|x-y|}{1+|x-y|}$.

Proof. For $y = 0$ this is Lemma 7.1 in (Yurachkivsky, 2013). It remains to note that $|e^{izx} - e^{izy}| = |e^{iz(x-y)} - 1|$. □

Lemma 2.9. *Let K be the solution of equation (3) on $\mathbb{R}_+^2$, where ι is a continuous numeral function of locally bounded variation, and A is a matrix-valued Borel function such that for any $t > 0$ (2) holds. Then for every Borel function $L(\cdot, \cdot)$ that has locally bounded variation in the first argument the solution of the equation*

$$\phi(t, s) = L(t, s) + \int_s^t A(u)\phi(u, s)\mathrm{d}\iota(u) \tag{21}$$

is given by the formula

$$\phi(t, s) = K(t, s)L(s, s) + \int_s^t K(t, \tau)\mathrm{d}_\tau L(\tau, s). \tag{22}$$

Proof. Without loss of generality $s = 0$. Then this statement is a particular case of Corollary 3.19 in (Yurachkivsky, 2013). □

Throughout below, the prime does not signify differentiation.

Corollary 2.10. *Let ι, A, K be as in Lemma 2.9, and K' be the solution of the equation*

$$K'(t, s) = \mathbf{I} + \int_s^t A'(\tau)K'(\tau, s)\mathrm{d}\iota(\tau), \tag{23}$$

on $\mathbb{R}_+^2$, where A' is a matrix-valued measurable function such that for any $t > 0$

$$\int_0^t \|A'(u)\|\,|\mathrm{d}\iota(u)| < \infty. \tag{24}$$

Then for all $t \geq s \geq 0$

$$K(t, s) - K'(t, s) = \int_s^t K(t, \tau)(A(\tau) - A'(\tau))K'(\tau, s)\mathrm{d}\iota(\tau).$$

Proof. To deduce this statement from Lemma 2.9 it suffices to write, on the basis of (3) and (23),

$$K(t, s) - K'(t, s) = \int_s^t (A(\tau) - A'(\tau))K'(\tau, s)\mathrm{d}\iota(\tau) + \int_s^t (A(\tau)(K(\tau, s) - K'(\tau, s))\mathrm{d}\iota(\tau)$$

□

Corollary 2.11. *Let ι, A, A', K and K' be as in Corollary 2.10. Suppose also that*

$$\|K(t, s)\| \vee \|K'(t, s)\| \leq e^{q(t-s)}$$

for some $q \in \mathbb{R}$ and all $t, s \in \mathbb{R}_+$. Then for all $t \geq s \geq 0$

$$\|K(t, s) - K'(t, s)\| \leq e^{q(t-s)} \int_s^t \|A(\tau) - A'(\tau)\|\,|\mathrm{d}\iota(\tau)|. \tag{25}$$

3. Special Preliminaries

From now on, $\int_\Theta$ will be written shortly as $\int$.

We impose two more assumptions:

12. There exists an $\mathbb{M}_+(\mathfrak{C})$-valued random process π such that the equality

$$\int_0^t \int \chi(s,\theta)\widetilde{\nu}(\mathrm{d}s\mathrm{d}\theta) = \int_0^t \mathrm{d}s \int \chi(s,\theta)\pi(s,\mathrm{d}\theta) \tag{26}$$

holds for every nonnegative $\mathcal{F}(0)\otimes\mathcal{B}_+\otimes\mathfrak{C}$-measurable random function χ on $\mathbb{R}_+\times\Theta$ and all $t>0$.

13. There exists an $\mathbb{M}_+(\mathfrak{C})$-valued random processes ϖ such that the equality

$$\int_0^t \int \chi(s,\theta)\widetilde{\upsilon}(\mathrm{d}s\mathrm{d}\theta) = \int_0^t \mathrm{d}s \int \chi(s,\theta)\varpi(s,\mathrm{d}\theta) \tag{27}$$

holds for every nonnegative $\mathcal{F}(0)\otimes\mathcal{B}_+\otimes\mathfrak{C}$-measurable random function χ on $\mathbb{R}_+\times\Theta$ and all $t>0$.

For an $\mathbb{M}_+(\mathfrak{C})$-valued random process κ we denote by $\mathcal{E}^\kappa$ the class of all $\mathbb{C}$-valued $\mathcal{F}(0)\otimes\mathcal{B}_+\otimes\mathfrak{C}$-measurable random functions χ on $\mathbb{R}_+\times\Theta$ such that

$$\int_0^t \mathrm{d}s \int |\chi(s,\theta)|\kappa(s,\mathrm{d}\theta) < \infty$$

for all $t>0$.

The next statement is obvious.

Lemma 3.1. *If assumption **12** is satisfied, then equality* (26) *holds for all* $\chi\in\mathcal{E}^\pi$ *and* $t>0$*; under assumption **13**, equality* (27) *holds for all* $\chi\in\mathcal{E}^\varpi$ *and* $t>0$.

Lemma 3.2. *Let assumptions **2**, **3**, **8** – **10**, **12** and **13** be satisfied. Then*

$$F_t(z) = \int_0^t \mathrm{d}s \int \left(e^{izK(t,s)f(s,\theta)} - 1\right)\pi(s,\mathrm{d}\theta), \tag{28}$$

$$G_t(z) = \int_0^t \mathrm{d}s \int \left(e^{izK(t,s)g(s,\theta)} - 1 - izK(t,s)g(s,\theta)\right)\varpi(s,\mathrm{d}\theta). \tag{29}$$

Proof. Let us fix $v>0$, $z\in\mathbb{R}^{d*}$ and denote $\varphi(s,\theta) = e^{izK(v,s)f(s,\theta)} - 1$. By construction and assumptions **2** and **3** the random process $zK(v,\cdot)$ is $\mathcal{F}(0)\otimes\mathcal{B}_+$-measurable, so by assumption **8** φ is $\mathcal{F}(0)\otimes\mathcal{B}_+\otimes\mathfrak{C}$-measurable. Herein

$$|\varphi(s,\theta)| \le (|zK(v,s)f(s,\theta)|\,|f(s,\theta)|)\wedge 2 \le (|z|\,\|K(v,s)\|\vee 2)(|f(s,\theta)|\wedge 1),$$

whence

$$\int_0^t \mathrm{d}s \int |\varphi(s,\theta)|\pi(s,\mathrm{d}\theta) \le \left(|z|\max_{s\le t}\|K(v,s)\| + 2\right)\int_0^t \mathrm{d}s \int (|f(s,\theta)|\wedge 1)\pi(s,\mathrm{d}\theta). \tag{30}$$

The random function $|f|\wedge 1$ is by assumption **8** $\mathcal{F}(0)\otimes\mathcal{B}_+\otimes\mathfrak{C}$-measurable, so by assumption **12**

$$\int_0^t \mathrm{d}s \int (|f(s,\theta)|\wedge 1)\pi(s,\mathrm{d}\theta) = \int_0^t \int (|f(s,\theta)|\wedge 1)\widetilde{\nu}(\mathrm{d}s,\mathrm{d}\theta),$$

whence in view of **10** and (30) we get $\int_0^t \mathrm{d}s \int_\Theta |\varphi(s,\theta)|\pi(s,\mathrm{d}\theta) < \infty$. Now, Lemma 3.1 asserts equality (26). Setting $v=t$, we turn it into

$$\int_0^t \int \left(e^{izK(t,s)f(s,\theta)} - 1\right)\widetilde{\nu}(\mathrm{d}s\mathrm{d}\theta) = \int_0^t \mathrm{d}s \int \left(e^{izK(t,s)f(s,\theta)} - 1\right)\pi(s,\mathrm{d}\theta).$$

And this is, in view of (7), none other than (28).

Likewise from **2**, **3**, **8**, **9**, (27) and the evident inequality

$$|e^{ip} - 1 - ip| \le p^2 \wedge 2|p|, \qquad p\in\mathbb{R}, \tag{31}$$

we get by Lemma 3.1

$$\int_0^t \int \left(e^{izK(t,s)g(s,\theta)} - 1 - izK(t,s)g(s,\theta)\right)\widetilde{\upsilon}(\mathrm{d}s\mathrm{d}\theta) = \int_0^t \int \left(e^{izK(t,s)g(s,\theta)} - 1 - izK(t,s)g(s,\theta)\right)\varpi(s,\mathrm{d}\theta),$$

which together with (8) proves (9). □

We continue the list of assumptions:

14. There exist a nonrandom σ-finite measure Π on $\mathfrak{C}$ and a positive random variable Ξ such that for all $t \in \mathbb{R}_+$ and $B \in \mathfrak{C}$ $\quad \pi(t, B) \leq \Xi\Pi(B)$.

15. There exist a nonrandom σ-finite measure Σ on $\mathfrak{C}$ and a positive random variable Ξ such that for all $t \in \mathbb{R}_+$ and $B \in \mathfrak{C}$ $\quad \varpi(t, B) \leq \Xi\Sigma(B)$.

16. There exists a nonrandom measurable function $\mathfrak{f}$ on Θ such that (i) $|f(s,\theta)| \leq \mathfrak{f}(\theta)$ and (ii) $\int \ln(1+\mathfrak{f})d\Pi < \infty$.

17. There exists a nonrandom measurable function $\mathfrak{g}$ on Θ such that (i) $|g(s,\theta)| \leq \mathfrak{g}(\theta)$ and (ii) $\int \left(\mathfrak{g}^2 \wedge \mathfrak{g}\right) d\Sigma < \infty$.

In the next two assumptions and four statements, K_0 and K are continuous matrix-valued random function on $\mathbb{R}^2_+$; the assumption that K satisfies equation (3) is not used. We assume the following:

18. There exists a positive random variable $\varkappa$ such that for all $t > s$

$$\|K_0(t,s)\| \vee \|K(t,s)\| \leq e^{\varkappa(s-t)}.$$

19. There exists an increasing random process Λ such that for all $t > s$

$$\|K(t,s) - K_0(t,s)\| \leq e^{\varkappa(s-t)}(\Lambda(t) - \Lambda(s)),$$

where $\varkappa$ is the same as in **18**.

If assumption **19** is imposed and **18** is not, then $\varkappa$ will signify simply an $\mathcal{F}(0)$-measurable positive random variable.

Throughout below, we tacitly use the fact that the function $x \mapsto \frac{x}{1+x}$ increases on $\mathbb{R}_+$.

Lemma 3.3. *Let assumptions **14**, **16** and **19** be satisfied, and let*

$$\lim_{t\to\infty} \Lambda(t) < \infty. \tag{32}$$

Then for any $z \in \mathbb{R}^{d*}$

$$\int_0^t ds \int \left(e^{izK(t,s)f(s,\theta)} - e^{izK_0(t,s)f(s,\theta)}\right)\pi(s, d\theta) \to 0.$$

Proof. Denote $H(t,s) = \|K(t,s) - K_0(t,s)\|$, $\epsilon = \lim_{t\to\infty}(\Lambda(t) - \Lambda(0))$ ($< \infty$ by condition (32)). It suffices, in view of Lemma 2.8, the definition of H, the Fubini – Tonelli theorem and assumptions **14** and **16**(i), to show that

$$\int \Pi(d\theta) \int_0^t \frac{\mathfrak{f}(\theta)H(t,s)}{1+\mathfrak{f}(\theta)H(t,s)} ds \to 0. \tag{33}$$

Assumption **19** and the definition of ϵ yield

$$\frac{\mathfrak{f}(\theta)H(t,s)}{1+\mathfrak{f}(\theta)H(t,s)} \leq \frac{\epsilon\mathfrak{f}(\theta)e^{\varkappa(s-t)}}{1+\epsilon\mathfrak{f}(\theta)e^{\varkappa(s-t)}},$$

whence for any $c \in [0, 1]$

$$\int_0^{ct} \frac{\mathfrak{f}(\theta)H(t,s)}{1+\mathfrak{f}(\theta)H(t,s)} ds \leq \ln\left(1 + \epsilon e^{-(1-c)\varkappa t}\mathfrak{f}(\theta)\right) - \ln\left(1 + \epsilon e^{-\varkappa t}\mathfrak{f}(\theta)\right).$$

And this together with **16**(ii) implies by the DCT that for any $c \in [0, 1[$

$$\int \Pi(d\theta) \int_0^{ct} \frac{\mathfrak{f}(\theta)H(t,s)}{1+\mathfrak{f}(\theta)H(t,s)} ds \to 0. \tag{34}$$

Denote $\rho(t,c) = \Lambda(t) - \Lambda(ct)$. It follows from (32) that for any $c > 0$

$$\lim_{t\to\infty} \rho(t,c) = 0. \tag{35}$$

By assumption **19** $H(t,s) \le e^{\varkappa(s-t)}\rho(t,c)$ as $s \ge ct$, so

$$\int_{ct}^{t} \frac{\mathfrak{f}(\theta)H(t,s)}{1+\mathfrak{f}(\theta)H(t,s)}\mathrm{d}s \le \ln\left(1+\rho(t,c)\mathfrak{f}(\theta)\right) - \ln\left(1+\rho(t,c)e^{-(1-c)\varkappa t}\mathfrak{f}(\theta)\right).$$

And this together with **16**(ii) and (35) implies by the DCT that for any $c \in]0,\ 1]$

$$\int \Pi(\mathrm{d}\theta)\int_{ct}^{t} \frac{\mathfrak{f}(\theta)H(t,s)}{1+\mathfrak{f}(\theta)H(t,s)}\mathrm{d}s \to 0,$$

whence with account of (34) relation (33) follows. □

Denote

$$\gamma(s,\theta,z) = e^{izg(s,\theta)} - izg(s,\theta) - 1.$$

Lemma 3.4. *Let assumptions* **15**, **17** – **19** *be satisfied and condition* (32) *be fulfilled. Then for any* $z \in \mathbb{R}^{d*}$

$$\int_0^t \mathrm{d}s \int (\gamma(s,\theta,zK(t,s)) - \gamma(s,\theta,zK_0(t,s)))\varpi(s,\mathrm{d}\theta) \to 0.$$

Proof. It suffices, in view of assumptions **15** and the Fubini – Tonelli theorem, to show that

$$\int_0^t \mathrm{d}s \int |\gamma(s,\theta,zK(t,s)) - \gamma(s,\theta,zK_0(t,s))|\,\Sigma(\mathrm{d}\theta) \to 0. \tag{36}$$

Denote $B_\varepsilon = \{\theta : \mathfrak{g}(\theta) \le \varepsilon\}$, $B^\varepsilon = \{\theta : \mathfrak{g}(\theta) > \varepsilon\}$. Assumption **17**(ii) implies, obviously, that for any $\varepsilon > 0$

$$\int_{B_\varepsilon} \mathfrak{g}^2 \mathrm{d}\Sigma < \infty \tag{37}$$

and

$$\int_{B^\varepsilon} \mathfrak{g}\mathrm{d}\Sigma < \infty. \tag{38}$$

Inequality (31) implies that

$$|\gamma(s,\theta,zK(t,s)) - \gamma(s,\theta,zK_0(t,s))| \le 2|z|^2|g(s,\theta)|^2(||K_0(t,s)||^2 + ||K(t,s)||^2),$$

whence in view of **17**(i) and **18**

$$\int_0^t \mathrm{d}s \int_{B_\varepsilon} |\gamma(s,\theta,zK(t,s)) - \gamma(s,\theta,zK_0(t,s))|\,\Sigma(\mathrm{d}\theta) \le 4|z|^2 \int_0^t e^{2\varkappa(s-t)}\mathrm{d}s \int_{B_\varepsilon} \mathfrak{g}^2\mathrm{d}\Sigma.$$

Hence, writing

$$\int_0^t e^{2\varkappa(s-t)}\mathrm{d}s = \frac{1-e^{-2\varkappa t}}{2\varkappa}$$

and noting that

$$\lim_{\varepsilon\to 0}\int_{B_\varepsilon} \mathfrak{g}^2\,\mathrm{d}\Sigma = 0$$

because of (37), we get

$$\lim_{\varepsilon\to 0}\overline{\lim_{t\to\infty}}\int_0^t \mathrm{d}s \int_{B_\varepsilon} |\gamma(s,\theta,zK(t,s)) - \gamma(s,\theta,zK_0(t,s))|\,\Sigma(\mathrm{d}\theta) = 0. \tag{39}$$

On the other hand, $|e^{ia} - ia - (e^{ib} - ib)| \le 2|a-b|$ for any $a, b \in \mathbb{R}$, so

$$|\gamma(s,\theta,zK(t,s)) - \gamma(s,\theta,zK_0(t,s))| \le 2|z|\,|g(s,\theta)|\,||K_0(t,s) - K(t,s)||,$$

which together with **17**(i) and **19** yields

$$\int_0^t \mathrm{d}s \int_{B^\varepsilon} |\gamma(s,\theta,zK(t,s)) - \gamma(s,\theta,zK_0(t,s))|\,\Sigma(\mathrm{d}\theta) \le 2|z| \int_0^t e^{\varkappa(s-t)}(\Lambda(t) - \Lambda(s))\mathrm{d}s \int_{B^\varepsilon} \mathfrak{g}\mathrm{d}\Sigma. \tag{40}$$

Writing, for an arbitrary $c \in]0, 1[$,

$$\int_0^t e^{\varkappa(s-t)}(\Lambda(t) - \Lambda(s))\mathrm{d}s = \int_0^{ct} e^{\varkappa(s-t)}(\Lambda(t) - \Lambda(s))\mathrm{d}s + \int_{ct}^t e^{\varkappa(s-t)}(\Lambda(t) - \Lambda(s))\mathrm{d}s$$

and arguing as in the previous proof, we get from (32) and the assumption that Λ increases

$$\lim_{t\to\infty} \int_0^t e^{\varkappa(s-t)}(\Lambda(t) - \Lambda(s))\mathrm{d}s = 0. \tag{41}$$

And this jointly with (40) and (38) implies that for all $z \in \mathbb{R}^{d*}$ and $\varepsilon > 0$

$$\lim_{t\to\infty} \int_0^t \mathrm{d}s \int_{B^\varepsilon} |\gamma(s, \theta, zK(t, s)) - \gamma(s, \theta, zK_0(t, s))|\, \Sigma(\mathrm{d}\theta) = 0.$$

Hence and from (39) relation (36) follows. □

Lemma 3.5. *Let assumptions **7**, **18** and **19** be satisfied and condition* (32) *be fulfilled. Suppose also that there exists a positive random variable β such that for all $t > s \geq 0$*

$$\mathrm{tr}\langle W\rangle(t) - \mathrm{tr}\langle W\rangle(s) \leq \beta(t - s). \tag{42}$$

Then for any $t \in \mathbb{R}_+$

$$\int_0^t K(t, s)\mathrm{d}\langle W\rangle(s)K(t, s)^\top - \int_0^t K_0(t, s)\mathrm{d}\langle W\rangle(s)K_0(t, s)^\top \to 0.$$

Proof. By Lemma 2.7

$$\left|\int_0^t zK(t, s)\mathrm{d}\langle W\rangle(s)(zK(t, s))^\top - \int_0^t zK_0(t, s)\mathrm{d}\langle W\rangle(s)(zK_0(t, s))^\top\right| \leq |z|^2 \int_0^t (\|K_0(t, s)\| + \|K(t, s)\|)H(t, s)\,\mathrm{d}\,\mathrm{tr}\langle W\rangle(s). \tag{43}$$

Assumptions **18**, **19**, condition (42) and the inequality $e^{2p} \leq e^p$ for $p \leq 0$ imply that

$$\int_0^t (\|K_0(t, s)\| + \|K(t, s)\|)H(t, s)\,\mathrm{d}\,\mathrm{tr}\langle W\rangle(s) \leq 2\beta \int_0^t e^{\varkappa(s-t)}(\Lambda(t) - \Lambda(s))\,\mathrm{d}s,$$

which together with (41) (emerging from (32)) and (43) proves the lemma. □

Lemma 3.6. *Let assumptions **4** and **19** be satisfied and condition* (32) *be fulfilled. Suppose also that there exists a positive random variable β such that for all $t > s \geq 0$*

$$\operatorname*{var}_{[s,t]} U \leq \beta(t - s). \tag{44}$$

Then

$$\int_0^t K(t, s)\mathrm{d}U(s) - \int_0^t K_0(t, s)\mathrm{d}U(s) \to 0.$$

Proof. Writing on the basis of (44) and **19**

$$\left|\int_0^t K(t, s)\mathrm{d}U(s) - \int_0^t K_0(t, s)\mathrm{d}U(s)\right| \leq \beta \int_0^t e^{\varkappa(s-t)}(\Lambda(t) - \Lambda(s))\mathrm{d}s,$$

we deduce the desired conclusion from (41) (emerging from (32)). □

4. The Main Result

Theorem 4.1. *Let assumptions **1** – **8** and **11** – **17** be satisfied. Suppose also that there exist:*

- *an $\mathbb{R}^d$-valued random variable ϑ,*
- *random matrices A_0 and Υ,*
- *random σ-finite measures π_0 and ϖ_0 on $\mathfrak{C}$,*
- *positive random variables $\varkappa$, α and β,*
- *$\mathbb{R}^d$-valued measurable random functions f_0 and g_0 on Θ,*
- *a positive number a,*

• *an increasing random process* Λ *with property* (32)
— *such that:*

$$e^{tA_0}\int_0^t e^{-sA_0}\mathrm{d}U(s)\xrightarrow{\mathsf{P}}\vartheta, \tag{45}$$

$$\int_0^t e^{(t-s)A_0}\mathrm{d}\langle W\rangle(s)e^{(t-s)A_0^\top}\xrightarrow{\mathsf{P}}\Upsilon; \tag{46}$$

for any $B\in\mathfrak{C}$

$$\pi_0(B)\le\Xi\Pi(B), \tag{47}$$

$$\varpi_0(B)\le\Xi\Pi(B), \tag{48}$$

where Π, Σ *and* Ξ *are from assumptions* ***14*** *and* ***15****; for all* $u\in\mathbb{R}_+$

$$\left\|e^{uA_0}\right\|\le e^{-au}; \tag{49}$$

for all $t>s\ge 0$

$$\|K(t,s)\|\le e^{\varkappa(s-t)}, \tag{50}$$

$$\left\|K(t,s)-e^{(t-s)A_0}\right\|\le e^{\alpha(s-t)}(\Lambda(t)-\Lambda(s)) \tag{51}$$

and inequalities (42) *and* (44) *hold; for any* $\theta\in\Theta$

$$f(t,\theta)\xrightarrow{\mathsf{P}}f_0(\theta), \tag{52}$$

$$g(t,\theta)\xrightarrow{\mathsf{P}}g_0(\theta); \tag{53}$$

for any $z\in\mathbb{R}^{d*}$

$$\int(e^{izf_0(\theta)}-1)\pi(t,\mathrm{d}\theta)\xrightarrow{\mathsf{P}}\int(e^{izf_0(\theta)}-1)\pi_0(\mathrm{d}\theta), \tag{54}$$

$$\int(e^{izg_0(\theta)}-1-izg(\theta))\varpi(t,\mathrm{d}\theta)\xrightarrow{\mathsf{P}}\int(e^{izg_0(\theta)}-1-izg_0(\theta))\varpi_0(\mathrm{d}\theta). \tag{55}$$

Then the distribution of $\xi(t)$ *weakly converges, as* $t\to\infty$*, to the distribution with characteristic function*

$$\mathsf{E}e^{iz\vartheta-z\Upsilon z^\top/2+\Phi(z)+\Gamma(z)},$$

where

$$\Phi(z)=\int_0^\infty\mathrm{d}u\int\left(e^{ize^{uA_0}f_0(\theta)}-1\right)\pi_0(\mathrm{d}\theta), \tag{56}$$

$$\Gamma(z)=\int_0^\infty\mathrm{d}u\int\left(e^{ize^{uA_0}g_0(\theta)}-1-ize^{uA_0}g_0(\theta)\right)\varpi_0(\mathrm{d}\theta). \tag{57}$$

Proof. 1°. Denote $D(u)=e^{uA_0}$, $K_0(t,s)=D(t-s)$,

$$\eta_t(u,z)=\int\left(e^{izD(u)f(t-u,\theta)}-1\right)\pi(t-u,\mathrm{d}\theta), \tag{58}$$

$$\psi_t(u,z)=\int\left(e^{izD(u)g(t-u,\theta)}-1-izD(u)g(t-u,\theta)\right)\varpi(t-u,\mathrm{d}\theta). \tag{59}$$

Rewriting condition (49) in the form

$$\|D(u)\|\le e^{-au}, \tag{60}$$

and taking to account (50), (51) and the assumed properties of Λ, we see that K and K_0 satisfy assumptions **18** and **19** (with $\varkappa\wedge\alpha$ as the former $\varkappa$). Hence and from assumptions **14** – **17** we get by Lemmas 3.3 and 3.4

$$F_t(z)-\int_0^t\eta_t(u,z)\mathrm{d}u\to 0, \tag{61}$$

$$G_t(z)-\int_0^t\psi_t(u,z)\mathrm{d}u\to 0, \tag{62}$$

where $F_t(z)$ and $G_t(z)$ are defined by equalities (28) and (29), respectively.

2°. Let us show that

$$\int_0^t \eta_t(u,z)\mathrm{d}u - \int_0^t \mathrm{d}u \int \left(e^{izD(u)f_0(\theta)} - 1\right)\pi(t-u,\mathrm{d}\theta) \xrightarrow{\mathsf{P}} 0. \tag{63}$$

Denote $\delta_t(u,\theta) = |f(t-u,\theta) - f_0(\theta)|$. From (58) we have by Lemma 2.8, assumption **14** and the Fubini – Tonelli theorem

$$\left|\int_0^t \eta_t(u,z)\mathrm{d}u - \int_0^t \mathrm{d}u \int \left(e^{izD(u)f_0(\theta)} - 1\right)\pi(t-u,\mathrm{d}\theta)\right| \le (|z| \vee 2)\Xi \int \Pi(\mathrm{d}\theta) \int_0^t \frac{\|D(u)\|\delta_t(u,\theta)}{1+\|D(u)\|\delta_t(u,\theta)}\mathrm{d}u. \tag{64}$$

On the strength of (60)

$$\frac{\|D(u)\|\delta_t(u,\theta)}{1+\|D(u)\|\delta_t(u,\theta)} \le \kappa_t(u,\theta), \tag{65}$$

where

$$\kappa_t(u,\theta) = \frac{e^{-au}\delta_t(u,\theta)}{1+e^{-au}\delta_t(u,\theta)}. \tag{66}$$

For convenience of the subsequent derivations, we set $\kappa_t(u,\theta) = 0$ as $u > t$.

By construction $0 \le \kappa_t(u,\theta) < 1$. By condition (52)

$$\kappa_t(u,\theta) \xrightarrow{\mathsf{P}} 0,$$

whence by Lemma 2.5 for any $T > 0$

$$\int_0^T \kappa_t(u,\theta)\mathrm{d}u \xrightarrow{\mathsf{P}} 0. \tag{67}$$

By assumption **16**(i) $|f(t-u,\theta)| \le \mathfrak{f}(\theta)$, which together with (52) yields

$$|f_0(\theta)| \le \mathfrak{f}(\theta). \tag{68}$$

Hence and (again) from **16**(i) we get $\delta_t(u,\theta) \le 2\mathfrak{f}(\theta)$, whence in view of (66)

$$\kappa_t(u,\theta) \le \frac{2\mathfrak{f}(\theta)e^{-au}}{1+2\mathfrak{f}(\theta)e^{-au}},$$

and therefore for any $[a,\ b] \subset \mathbb{R}_+$

$$\int_a^b \kappa_t(u,\theta)\mathrm{d}u \le \ln(1+2e^{-\alpha a}\mathfrak{f}(\theta)).$$

Herein, obviously, $\ln(1+2x) \le 2\ln(1+x)$. Thus

$$\int_0^T \kappa_t(u,\theta)\mathrm{d}u \le 2\ln(1+\mathfrak{f}(\theta)),$$

which together with (66) and finiteness (by assumption **16**(ii)) of $\int \ln(1+\mathfrak{f}(\theta))\Pi(\mathrm{d}\theta)$ implies by Lemma 2.6 (applied to $X = \Theta,\ \mathcal{X} = \mathfrak{C},\ \mu = \Pi,\ \zeta_t(\theta) = \int_0^T \kappa_t(u,\theta)\mathrm{d}u$) that

$$\operatorname*{l.i.p.}_{t\to\infty} \int \Pi(\mathrm{d}\theta) \int_0^T \kappa_t(u,\theta)\mathrm{d}u = 0. \tag{69}$$

Further, for any $t \ge T$

$$\int \Pi(\mathrm{d}\theta) \int_T^t \kappa_t(u,\theta)\mathrm{d}u \le \int \ln\left(1+2e^{-\alpha T}\mathfrak{f}(\theta)\right)\Pi(\mathrm{d}\theta),$$

which together with **16**(ii) and the DCT yields

$$\lim_{T\to\infty}\sup_{t\ge T} \int \Pi(\mathrm{d}\theta) \int_T^t \kappa_t(u,\theta)\mathrm{d}u = 0.$$

And this jointly with (69) entails (63).

3°. Let us denote

$$\gamma_t(u,\theta,z) = e^{izD(u)g(t-u,\theta)} - 1 - izD(u)g(t-u,\theta),$$
$$\gamma_0(u,\theta,z) = e^{izD(u)g_0(\theta)} - 1 - izD(u)g_0(\theta)$$

and show that

$$\int_0^t \psi_t(u,z)\mathrm{d}u - \int_0^t \mathrm{d}u \int \gamma_0(u,\theta,z)\varpi(t-u,\mathrm{d}\theta) \xrightarrow{\mathsf{P}} 0. \tag{70}$$

Recalling (59), we see that $\psi_t(u,z) = \int \gamma_t(u,\theta,z)\varpi(t-u,\mathrm{d}\theta)$, so it suffices, in view of **15**, to prove the relation

$$\int_0^t \mathrm{d}u \int |\gamma_t(u,\theta,z) - \gamma_0(u,\theta,z)|\,\Sigma(\mathrm{d}\theta) \to 0. \tag{71}$$

Inequalities (31) and (60) imply that

$$|\gamma_t(u,\theta,z) - \gamma_0(u,\theta,z)| \le 2|z|^2 e^{-2au}(|g(t-u,\theta)|^2 + |g_0(\theta)|^2).$$

By assumption **17**(i) $|g(t-u,\theta)| \le \mathfrak{g}(\theta)$, which together with (53) yields

$$|g_0(\theta)| \le \mathfrak{g}(\theta). \tag{72}$$

Thus

$$|\gamma_t(u,\theta,z) - \gamma_0(u,\theta,z)| \le 4|z|^2 e^{-2au}\mathfrak{g}(\theta)^2.$$

Hence and from (37) we get

$$\lim_{\varepsilon\to 0}\overline{\lim_{t\to\infty}} \int_0^t \mathrm{d}u \int_{B_\varepsilon} |\gamma_t(u,\theta,z) - \gamma_0(u,\theta,z)|\,\Sigma(\mathrm{d}\theta) = 0. \tag{73}$$

On the other hand, $|e^{ib} - ib - (e^{ic} - ic)| \le 2|b-c|$ for any $b, c \in \mathbb{R}$, so

$$|\gamma_t(u,\theta,z) - \gamma_0(u,\theta,z)| \le 2e^{-au}|z|h_t(u,\theta),$$

where

$$h_t(u,\theta) = |g(t-u,\theta) - g_0(\theta)|.$$

Thus

$$\int_0^t \mathrm{d}s \int_{B^\varepsilon} |\gamma_t(u,\theta,z) - \gamma_0(u,\theta,z)|\,\Sigma(\mathrm{d}\theta) \le 2|z| \int_0^t e^{-au} q_t(u,\varepsilon)\mathrm{d}u, \tag{74}$$

where

$$q_t(u,\varepsilon) = \int_{B^\varepsilon} h_t(u,\theta)\Sigma(\mathrm{d}\theta). \tag{75}$$

Condition (53) means that $h_t(u,\theta) \xrightarrow{\mathsf{P}} 0$ for all u and θ; inequality (38) and the definition of B^ε show that $\Pi(B^\varepsilon) < \infty$; assumption **17**(i) and inequality (72) imply that

$$h_t(u,\theta) \le 2\mathfrak{g}(\theta). \tag{76}$$

Hence and from (38) we get by Lemma 2.5

$$q_t(u,\varepsilon) \xrightarrow{\mathsf{P}} 0 \tag{77}$$

for all positive u and ε. Herein

$$q_t(u,\theta) \le C \equiv 2\int_{B^\varepsilon} \mathfrak{g}\mathrm{d}\Pi \tag{78}$$

in view of (75) and (76). Then again by Lemma 2.5

$$\int_0^T e^{-au} q_t(u,\varepsilon)\mathrm{d}u \xrightarrow{\mathsf{P}} 0 \tag{79}$$

for all positive T and ε. Besides, inequality (78) implies that for any $t > T$

$$\int_T^t e^{-au} q_t(u,\varepsilon)\mathrm{d}u \le C\alpha^{-1}e^{-aT}$$

and therefore

$$\lim_{T\to\infty}\sup_{t>T}\int_T^t e^{-au}q_t(u,\varepsilon)\mathrm{d}u = 0,$$

which together with (79) implies

$$\int_0^t e^{-au}q_t(u,\varepsilon)\mathrm{d}u \xrightarrow{\mathsf{P}} 0.$$

This relation jointly with (74) and (73) proves (70).

4°. Lemma 2.8 and inequality (60) yield

$$\left|e^{izD(u)f_0(\theta)}-1\right| \le 2(|z|\vee 2)\frac{e^{-au}|f_0(\theta)|}{1+e^{-au}|f_0(\theta)|},$$

whence, using the Fubini – Tonelli theorem, we get

$$\int_0^\infty \mathrm{d}u\int\left|e^{izD(u)f_0(\theta)}-1\right|\pi_0(\mathrm{d}\theta) \le 2(|z|\vee 2)\int\pi_0(\mathrm{d}\theta)\int_0^\infty\frac{e^{-au}|f_0(\theta)|}{1+e^{-au}|f_0(\theta)|}\mathrm{d}u,$$

which together with (47), (68) and **16**(ii) results in

$$\int_0^\infty \mathrm{d}u\int\left|e^{izD(u)f_0(\theta)}-1\right|\pi_0(\mathrm{d}\theta) < \infty.$$

Thus the definition $\Phi(z)$ by formula (56) is correct and there holds the estimate

$$|\Phi(z)| \le 2\Xi(|z|\vee 2)\int\ln(1+\mathfrak{f})\mathrm{d}\Pi$$

showing that Φ satisfies condition (10).

Let us show that

$$\int_0^t \mathrm{d}u\int\left(e^{izD(u)f_0(\theta)}-1\right)\pi(t-u,\mathrm{d}\theta) \xrightarrow{\mathsf{P}} \Phi(z). \qquad (80)$$

This will be done if we establish the relation

$$\int_0^\infty \zeta_t(u,z)\mathrm{d}u \xrightarrow{\mathsf{P}} 0, \qquad (81)$$

where

$$\zeta_t(u,z) = \frac{1}{\Xi}\left|\int\left(e^{izD(u)f_0(\theta)}-1\right)\pi(t-u,\mathrm{d}\theta)-\int\left(e^{izD(u)f_0(\theta)}-1\right)\pi_0(\mathrm{d}\theta)\right|,$$

and π is defined for negative values of the temporal argument by zero. (The divider Ξ will enable us to construct a nonrandom majorant for ζ_t and thereon to apply Lemma 2.6.)

By construction, Lemma 2.8, assumption **14**, condition (47) and formula (68)

$$0 \le \zeta_t(u,z) \le Z(u,z), \qquad (82)$$

where

$$Z(u,z) = 4(|z|\vee 2)\int\frac{e^{-au}\mathfrak{f}(\theta)}{1+e^{-au}\mathfrak{f}(\theta)}\Pi(\mathrm{d}\theta). \qquad (83)$$

By condition (54)

$$\zeta_t(u,z) \xrightarrow{\mathsf{P}} 0 \qquad (84)$$

for all $u\in\mathbb{R}_+$ and $z\in\mathbb{R}^{d*}$. By construction $Z(u,z)$ is nonrandom. Formula (83) and assumption **16**(ii) imply that

$$\int_0^\infty Z(u,z)\mathrm{d}u < \infty \qquad (85)$$

and (by the DCT) $Z(u,z)\to 0$ as $u\to\infty$, so that for any $r>0$ the Lebesgue measure of the set $\{u : Z(u,z) > r\}$ is finite. Now, (81) follows from (82) and (84) by Lemma 2.6.

5°. Inequalities (31) and (60) yield

$$\left|e^{izD(u)g_0(\theta)}-1-izD(u)g_0(\theta)\right| \le 2(|z|^2\vee|z|)e^{-au}(|g_0(\theta)|^2\wedge|g_0(\theta)|),$$

whence, using the Fubini – Tonelli theorem, we get

$$\int_0^\infty \mathrm{d}u \int \left|e^{izD(u)g_0(\theta)} - 1 - izD(u)g_0(\theta)\right| \varpi_0(\mathrm{d}\theta) \leq 2(|z|^2 \vee |z|)a^{-1} \int (|g_0|^2 \wedge |g_0|)\mathrm{d}\varpi_0,$$

which together with (48), (72) and **17**(ii) results in

$$\int_0^\infty \mathrm{d}u \int \left|e^{izD(u)g_0(\theta)} - 1 - izD(u)g_0(\theta)\right| \varpi_0(\mathrm{d}\theta) < \infty. \tag{86}$$

Thus the definition $\Gamma(z)$ by formula (57) is correct and there holds the estimate

$$|\Gamma(z)| \leq 2(|z|^2 \vee |z|)a^{-1} \int (\mathfrak{g}_0^2 \wedge \mathfrak{g}_0)\mathrm{d}\Sigma$$

showing that Γ satisfies condition (11).

Let us show that

$$\int_0^t \mathrm{d}u \int \left(e^{izD(u)g_0(\theta)} - 1 - izD(u)g_0(\theta)\right) \varpi(t-u, \mathrm{d}\theta) \xrightarrow{\mathsf{P}} \Gamma(z). \tag{87}$$

This will be done if we establish relation (81), where this time (unlike item 4°)

$$\zeta_t(u,z) = \frac{1}{\Xi}\left|\int \left(e^{izD(u)g_0(\theta)} - 1 - izD(u)g_0(\theta)\right) \varpi(t-u, \mathrm{d}\theta) - \int \left(e^{izD(u)g_0(\theta)} - 1 - izD(u)g_0(\theta)\right) \varpi_0(\mathrm{d}\theta)\right|,$$

and ϖ is defined for negative values of the temporal argument by zero.

The above expression, assumption **15**, condition (48) and formulas (72), (31) imply (82), where this time (unlike item 4°)

$$Z(u,z) = 4(|z|^2 \vee |z|)e^{-au} \int (\mathfrak{g}^2 \wedge \mathfrak{g})\mathrm{d}\Sigma. \tag{88}$$

By condition (55) relation (84) holds for all $u \in \mathbb{R}_+$ and $z \in \mathbb{R}^{d*}$. Obviously, $Z(u,z) \to 0$ as $u \to \infty$, so that for any $r > 0$ the Lebesgue measure of the set $\{u : Z(u,z) > r\}$ is finite. Now, (81) follows from (82) and (84) by Lemma 2.6.

6°. Obviously, condition (50) entails (9).

Conditions (51) amounts to assumption **19** for the K_0 defined in item 1°. So Lemma 3.6 whose other conditions are retained in this theorem asserts that

$$\int_0^t K(t,s)\mathrm{d}U(s) - e^{tA_0}\int_0^t e^{-sA_0}\mathrm{d}U(s) \xrightarrow{\mathsf{P}} 0,$$

hereon (45) implies that

$$\int_0^t K(t,s)\mathrm{d}U(s) \xrightarrow{\mathsf{P}} \vartheta. \tag{89}$$

Likewise from (51) and (44) we deduce by Lemma 3.5 that

$$\int_0^t K(t,s)\mathrm{d}\langle W\rangle(s)K(t,s)^\top - \int_0^t e^{(t-s)A_0}\mathrm{d}\langle W\rangle(s)e^{(t-s)A_0^\top} \xrightarrow{\mathsf{P}} 0,$$

hereon (46) implies that

$$\int_0^t K(t,s)\mathrm{d}\langle W\rangle(s)K(t,s)^\top \xrightarrow{\mathsf{P}} \Upsilon. \tag{90}$$

From (61), (63) and (80) we have

$$F_t(z) \xrightarrow{\mathsf{P}} \Phi(z), \tag{91}$$

where $\Phi(z)$ is defined by (56). It follows from (62), (70) and (87) that

$$G_t(z) \xrightarrow{\mathsf{P}} \Gamma(z), \tag{92}$$

where $\Gamma(z)$ is defined by (57). Properties (10) and (11) of these Φ and Γ were verified in items 4° and 5°.

Relations (89) – (92) entail (12) (even in a stronger form — with $\xrightarrow{\mathsf{P}}$ instead of $\xrightarrow{\mathrm{d}}$). Now, noting that assumptions **9** and **10** are satisfied because of **14** – **16** and the evident inequality $p \wedge 1 \leq 2\ln(1+p) \quad (p > 0)$, we deduce the conclusion of the theorem from Corollary 2.4. □

Conditions (50) and (51) (the latter accompanied by (32)) of the theorem, unlike the others, involve the function K that is not pre-specified. Its explicit expression is available only in very special cases, so to finalize the main result we must give more elementary conditions guaranteing the fulfilment of (51), (32) and (50). In the next statement, (50) is still an assumption.

Proposition 4.2. *Let condition* (50) *be fulfilled. Suppose also that there exists random variables* ϱ, ς *and an increasing random process* H *such that for all* $t > s \geq 0$

$$\|e^{(\iota(t)-\iota(s))A_0}\| \leq e^{\varrho(s-t)}, \tag{93}$$

$$\|e^{(\iota(t)-\iota(s))A_0} - e^{(t-s)A_0}\| \leq e^{\varsigma(s-t)}(H(t) - H(s)). \tag{94}$$

Then: (i) inequality (51) *holds with* $\alpha = \varkappa \wedge \varrho \wedge \varsigma$, $\Lambda = H + V$,

$$V(t) = \int_0^t \|A(u) - A_0\| \, |\mathrm{d}\iota(u)|; \tag{95}$$

(ii) if, moreover, $\lim_{t\to\infty} H(t) < \infty$ *and*

$$\int_0^\infty \|A(u) - A_0\| \, |\mathrm{d}\iota(u)| < \infty,$$

then $\lim_{t\to\infty} \Lambda(t) < \infty$.

Proof. Denote $K'(t, s) = e^{(\iota(t)-\iota(s))A_0}$. From (50), and (93) and (95) we get by Corollary 2.11

$$\|K(t, s) - K'(t, s)\| \leq e^{(\varkappa\wedge\varrho)(s-t)}(V(t) - V(s)),$$

which together with (94) completes the proof. □

It remains to find elementary sufficient conditions for (50).

Lemma 4.3. *Let* K *be the solution of equation* (3) *on* $\mathbb{R}^2_+$, *where* ι *is a continuous increasing function, and* A *is a matrix-valued Borel function such that for any* $t > 0$

$$\int_0^t \|A(u)\| \mathrm{d}\iota(u) < \infty.$$

Suppose that there exists a nonnegative Borel function $\mathfrak{a}$ *such that*

$$x^\top A(t)x \leq -\mathfrak{a}(t)|x|^2 \tag{96}$$

for all $t > 0$ *and* $x \in \mathbb{R}^d$. *Then for all* $t > s \geq 0$

$$\|K(t, s)\| \leq \exp\left\{-\int_s^t \mathfrak{a}(u)\mathrm{d}\iota(u)\right\}. \tag{97}$$

Proof. Fix x and denote $X(t, s) = K(t, s)x$. By construction X satisfies the equation

$$X(t, s) = x + \int_s^t A(s)X(u, s)\mathrm{d}\iota(u).$$

Then it follows from (96) by Theorem 3.1 in (Yurachkivsky, 2014) (actually, by its very special particular case) that

$$|X(t, s)| \leq |x| \exp\left\{-\int_s^t \mathfrak{a}(u)\mathrm{d}\iota(u)\right\},$$

which together with the definition of the operator norm implies (97). □

References

Anisimov, V. V., Zakusilo, O. K., & Donchenko, V. S. (1987). *Elements of Queueing Theory and Asymptotic Analysis of Systems* (in Russian). Kiev: Vyshcha Shkola.

Gikhman, I. I. & Skorokhod, A. V. (1982). *Stochastic Differential Equations and Their Applications* (in Russian). Kiev: Naukova Dumka.

Gikhman, I. I. & Skorokhod, A. V. (2009). *Theory of Stochastic Processes* (Vol.3, 2nd ed.). Berlin: Springer.

Ivanenko, D. O. (2007). Existence of a limit distribution of a solution of a linear inhomogeneous stochastic differential equation. *Theory of Probability and Mathematical Statistics, 78*, 49-60.

Jacod, J. & Shiryayev, A. N. (1987). *Limit Theorems for Stochastic Processes*. Berlin: Springer.

Liptser, R. Sh. & Shiryayev, A. N. (1989). *Theory of Martingales*. Dordrecht: Kluwer.

Sato, K. & Yamazato, M. (1984) Operator-self-decomposable distributions as limit distributions of processes of Ornstein–Uhlenbeck type. *Stochastic Processes and Their Applications, 17*, 73-100.

Yurachkivsky, A. (2013) An ergodic-type theorem for Ornstein–Uhlenbeck processes. *Random Operators and Stochastic Equations, 21*, 217-269. http://dx.doi.org/10.1515/rose-2013-0011

Yurachkivsky, A. (2014) A conditional mean square estimate for the solution of a SDE. *International Journal of Statistics and Probability, 3*, 1-10. http://dx.doi.org/10.5539/ijsp.v3n4p1

3

Installation and Dispatch of the Traffic Patrol Service Platform

Tang Jiahui[1], Zhang Yuanbiao[2], Peng Churu[3], & Huang Xinxin[3]

[1] Mathematical Modeling Innovative Practice Base, JiNan University, ZhuHai GuangDong, China

Correspondence: Mathematical Modeling Innovative Practice Base, JiNan University, ZhuHai GuangDong,China

Abstract

In this paper, we construct three mathematical models to install and dispatch the traffic patrol service platform properly based on the real data of a certain city. Firstly, we build the shortest path model based on the Floyd algorithm to determine the jurisdictional of each platform. Then, we designed the dispatch model combined with 0-1 integer programming and the Hungarian algorithm to find the dispatching schemes when coming across large-scale emergencies. Lastly, we build the multiple-objective location model to optimize the present distribution situation of the traffic patrol platforms considering the workload differences among these existing platforms and overlong response time in some places.

Keywords: Floyd algorithm, shortest path model, dispatch model, multiple-objective location model

1. Introduction

The police undertake four major functions, including the criminal enforcement, public security management, traffic management and servicing the masses. In order to implement these functions effectively, we need to set up traffic patrol service platforms in some city traffic arteries and important positions. The function and police resources of each traffic patrol service platform are basically the same. However, the police resources are not infinite. As a result, how to install and dispatch the traffic patrol service platform reasonably has been a practical task faced by the police department according to the actual situation such as the distribution intersection nodes and crime probability difference of an area (Yu, Song, Zhao & Li, 2012). The shortest path model based on the Floyd algorithm and dispatch model can solve such kinds of problems effectively and properly.

The shortest path problem is a basic but very important problem in network optimization, and the Floyd algorithm is a common method to solve the shortest path problem. The Floyd algorithm begins with the weighted adjacency matrix, which represents the distance between any two intersection nodes v_i and v_j. Then it inserts an extra vertex to be the transfer station, and compares the shortest path's distance before and after inserting the vertex v_k, choosing the minimum to get a new distance matrix. Keeping loop iteration till all the vertices are as transfer stations between $v_i$and v_j. Next, the final weighted adjacency matrix will be worked out, which reflects the shortest distance information between any two vertices, becoming a distance matrix of the original image. Lastly, we can get the shortest path between any two points and the shortest path distance from the final distance matrix (Zhu & Zhang, 2012)

The Vehicle Scheduling Problem (VSP) has been a hot topic since it was put forward by Dantzing and Ramser(1959) for the first time (Dantzing & Ramser, 1959), which attracted many experts to research, including experts in operational research, combinatorial mathematics, graph theory and network analysis, and so on. Savelsbergh(1985) proved VSP is a NP-Complete problem(Savelsbergh, 1985), so how to get the answer is the key point in VSP when in large scale. Laporte and Mercure(1992) came up with a kind of vehicle routing problem with stochastic travel time. They also developed the chance constrained model and solved the problem successfully by the branch cut algorithm(Laporte & Mercure, 1992). Recently, some intelligent algorithms are widely applied into VSP solving, such as genetic algorithm, ant colony algorithm, particle swarm optimization and so on, which achieved much progress in VSP(Baker 2003; Berger,2003; Blanton,1993; Bullnheimer,1999; Salman,2002; Ellabib, 2002). In China, there are also some scholars studied the vehicle scheduling problem. Huo and Wang(2005) solved the vehicle scheduling problem based on constraint programming(Huo & Wang,2005). Xie, Li and Guo (2000) has carried on the thorough research concerning the logistics vehicle scheduling problem, and put forward a variety of algorithm (Xie, Li & Guo,2000).

In this paper, we develop three mathematical models to install and dispatch the traffic patrol service platform properly based on the real date of a certain city. Firstly, we distribute the jurisdictional of each platform according to the shortest path model. Then, we designed the dispatch model to find the reasonable dispatching schemes when come across large-scale emergencies. We lastly proposed the multiple-objective location model to optimize present distribution situation of the traffic patrol service platforms considering the workload differences and the overlong response time in some places.

2. Problem-Descriptions

This paper develops several mathematical models to install and dispatch the traffic patrol service platform reasonably, based on the data and assumption of Problem B in China Undergraduate Mathematical Contest in Modeling (China Undergraduate Mathematical Contest in Modeling[CUMCM],2011). The specific backgrounds of this problem are as following:

(1) The traffic network and the distribution of the existing 20 traffic patrol service platforms in Area A are shown in Fig1. We are required to determine the jurisdictional scope of each platform, making sure that the police can arrive at the scene within 3 minutes when emergency happens in its jurisdictional scope. And the speed of police car is 60Km/h.

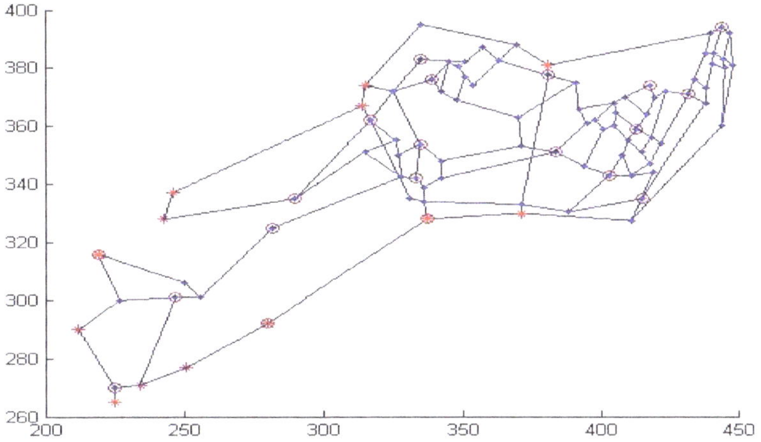

Figure 1. The traffic network and platforms location of city A

(2) When coming across large-scale emergencies, it is required to blockade all the 13 roads in city A as quickly as possible by dispatching the 20 traffic patrol service platforms. Actually, the police resource of each platform can only blockade one intersection node. What we are supposed to do is figuring out the proper methods to dispatch the traffic patrol service platforms.

(3) Considering the workload differences among these existing platforms and overlong response time in some places, the government is going to add 2 to 5 police platforms to improve the present situation. We are required to determine the specific quantity and position of these traffic patrol service platforms that are going to be set.

3. Development of Models

3.1 Model for Calculating the Shortest Path

In order to make sure that police can arrive at the scene as quickly as possible when there are emergencies in its jurisdictional scope, we find out the nearest traffic patrol service platform of each intersection and nominate the nearest one to be the responsible one for this intersection node. In other words, we determine the jurisdiction of each platform by finding the traffic patrol service platform that is responsible for each intersection node. When all the intersection nodes are allocated to the nearest platform, the jurisdiction is determined. Floyd algorithm is an ideal method to solve such a problem.

3.1.1 The Principle of Floyd Algorithm

Floyd algorithm was put up in 1962, which can be used to solve any kind of network shortest path problem when arc power is real, as well as figure out the shortest path distance between any two points. Floyd algorithm can be understood as a successive approximation approach, whose basic methods are as follows(Mao Yuan-jie,1942):

The Floyd algorithm begins with the weighted adjacency matrixD^0 and then generates a sequent of recursive matrixes $D^1,\cdots,D^k,\cdots,D^n$ through iteration. The formula of calculating $D^0$and D^k are as follows:

$$D^0 = \begin{bmatrix} d_{11} & d_{12} & \cdots & d_{1n} \\ d_{21} & d_{22} & \cdots & d_{2n} \\ \vdots & \vdots & \vdots & \vdots \\ d_{n1} & d_{n2} & \cdots & d_{nn} \end{bmatrix} \quad (1)$$

$$D^k = (d_{ij}^k)_{n\times n} \quad (2)$$

While d_{ij} is the distance between any two intersection nodes; k is the number of iterations; d_{ij}^k is the shortest path's distance between the two intersection nodes v_i and v_j when the number of intermediate nodes isn't greater than k. We can calculate d_{ij}^k by the following iterative formula:

$$d_{ij}^k = min\{d_{ij}^{k-1}, d_{ik}^{k-1} + d_{kj}^{k-1}\} \quad (3)$$

When $k=n$, the value of d_{ij}^n is the shortest distance between v_i and v_j , which is $d(v_i,v_j)$.

3.1.2 Result— the Jurisdictional Scope

In order to confirm the jurisdictional scope of each traffic patrol service platform, we firstly figure out the length of each path, which is the distance between every two intersection nodes. Secondly, we arrange these numbers into a weighted adjacency matrix D_0 of 20×72 dimensions. Then, based on the Floyd algorithm and MATLAB, we find out the nearest traffic patrol service platform of each intersection and nominate the nearest one to be the responsible one for the corresponding intersection. That is to say, when all the intersection nodes are assigned, the jurisdiction is determined. The specific results are shown in Table 1.

Table 1. jurisdictional scope of the traffic patrol service platform in city A

Platforms	Jurisdictional scope	Platforms	Jurisdictional scope
1	1,67,68,69,71,73,74,75,76,78	11	11, 26, 27
2	2, 39, 40, 43, 44, 70, 72	12	12, 25
3	3, 54, 55, 65, 66	13	13, 21, 22, 23, 24
4	4, 57, 60, 62, 63, 64	14	14
5	5, 49, 50, 51, 52, 53, 56, 58, 59	15	15, 28, 29
6	6	16	16, 36, 37, 38
7	7, 30, 32, 47, 48, 61	17	17, 41, 42
8	8, 33, 46	18	18, 80, 81, 82, 83
9	9, 31, 34, 35, 45	19	19, 77, 79
10	10	20	20, 84, 85, 86, 87, 88, 89, 90, 91, 92

We can see from table 1.that the workload differences among traffic patrol service platforms in Area A is obvious. For example, the 6th、 the 10th and the 14th platforms, they only need to be responsible for the intersection node where they are, because they are so far away from other road crosses that they can't rush to the scene in 3 minutes according to the shortest distance calculated by the Floyd algorithm. Oppositely, the 1th、 the 5th、 and the 20th platforms, their jurisdictional scope are wide and the workload are especially more. Considering the existing problems, the government is going to add 2 to 5 police platforms, and we will discuss the specific number and location of them later.

3.2 Model for Dispatching Traffic Patrol Service Platforms

We know that there are total 20 traffic patrol service platforms in Area A, and if it comes across large-scale emergencies, we need blockade all the 13 roads in Area A as quickly as possible by dispatching the 20 traffic patrol service platforms. Now, we are going to build a dispatch model based on 0-1 integer programming and the Hungarian algorithm to solve such a dispatching problem reasonably. Firstly, we introduce the 0-1 variables x_{ij} （$i=1,2,...,13; j=1,2,...,20$ ）: When the jth traffic patrol service platform was assigned to blockade the ith traffic artery, $x_{ij} = 1$. Otherwise, $x_{ij} = 0$.

3.2.1 The Objective Functions

We know that there are total 20 traffic patrol service platforms in Area A, and when large-scale emergencies happens, we need dispatch all the 20 traffic patrol service platforms reasonably to all the 13 roads. Reasonably dispatching means that blockade all the 13 roads base on the principal of time priority. Considering the speeds of police car are same in all conditions, we transfer the time priority to distance priority, which is:

$$\min\ z = \sum_{i=1}^{m}\sum_{j=1}^{n} x_{ij} l_{ij} \tag{4}$$

Where l_{ij} is the shortest path's length between the jth traffic patrol service platform and the ith traffic artery; x_{ij} represents whether the jth traffic patrol service platform was assigned to blockade the ith traffic artery, if yes, then $x_{ij} = 1$, or $x_{ij} = 0$.

3.2.2 The Constraints

(1) In order to guarantee that the whole Area A is blockaded, every traffic artery requires at least one traffic patrol service platform, which means:

$$\sum_{i=1}^{n} x_{ij} = 1, \qquad j = 1, 2, \cdots, n; n = 20 \tag{5}$$

(2) Each traffic patrol service platform can only be assigned to manage a traffic artery, which means:

$$\sum_{j=1}^{m} x_{ij} = 1, \qquad i = 1, 2, \cdots, m; m = 13 \tag{6}$$

3.2.3 The Methods

Firstly, we figure out the shortest path from the jth traffic patrol service platform to the ith traffic artery and its length l_{ij}. And we arrange the date of distance into the coefficient matrix of x_{ij}, matrix X_0:

$$X_0 = \begin{bmatrix} l_{11} & l_{12} & \cdots & l_{1n} \\ l_{21} & l_{22} & \cdots & l_{2n} \\ \cdots & \cdots & \cdots & \cdots \\ l_{m1} & l_{m2} & \cdots & l_{mn} \end{bmatrix} \tag{7}$$

Then, in order to meet the requirements of the Hungarian algorithm to matrix, we add *n-m* imaginary missions to convert the original matrix X_0 of $m \times n$ dimensions into *n* phalanxX_1:

$$X_1 = \begin{bmatrix} l_{11} & l_{12} & \cdots & l_{1n} \\ \cdots & \cdots & \cdots & \cdots \\ l_{m1} & l_{m2} & \cdots & l_{mn} \\ 0 & 0 & \cdots & 0 \\ \cdots & \cdots & \cdots & \cdots \\ 0 & 0 & \cdots & 0 \end{bmatrix} \tag{8}$$

Next, we use the Hungarian algorithm to transfer phalanxX_1 into the identity matrixX_2:

$$X_2 = \begin{bmatrix} 0 & 0 & 0 & \cdots & 0 \\ 0 & 1 & 0 & \cdots & 0 \\ 0 & 0 & 1 & \cdots & 0 \\ \vdots & \vdots & \vdots & \vdots & \vdots \\ 0 & 0 & 1 & \cdots & 0 \end{bmatrix} \tag{9}$$

Finally, we can figure out the proper Scheduling Scheme of the traffic patrol service platforms when large-scale emergencies happen according to the matrixX_2.

3.2.4 Result— Scheduling Scheme to Blockade Area A

The specific results are shown in Table 2.

Table 2. Scheduling Scheme of the traffic patrol service platform

Platforms	A12	A14	A16	A9	A10	A13	A11	A15	A8	A7	A2	A5	A4
Traffic arteries	A12	A14	A16	A21	A22	A23	A24	A28	A29	A30	A38	A48	A62

We can see from the table 2 that the specific Scheduling Scheme is: the 12th, 14th, and 16th platforms stay at their original positions to stand by; the 9th platform is appointed to block the 21th traffic artery; the 10th platform is appointed to block the 22th traffic artery; the 13th platform is appointed to block the 23th traffic artery; the 11th platform is appointed to block the 24th traffic artery; the 15th platform is appointed to blockade the 28th traffic artery; the 8th platform is appointed to block the 29th traffic artery; the 17th platform is appointed to block the 30th traffic artery; the 2th platform is appointed to block the 38th traffic artery; the 5th platform is appointed to block the 48th traffic artery; the 4th platform is appointed to block the 62th traffic artery.

3.3 Model of Multiple-Objective Location

3.3.1 The Preliminary Location of New Traffic Patrol Service Platforms

Concerning that the added traffic patrol service platforms are mainly used to optimize the imbalanced workload situation and decrease the response time, and the existing 20 traffic patrol service platforms are not to make any changes, we consider decreasing the response time firstly. The specific methods are as follows:

We firstly find out the intersection nodes, whose guardian platforms can't arrive in 3 minutes when accident happens. Then we find out the traffic patrol service platforms that can arrive in these non-effective guarded platforms within 3 minutes, making them the alternative new platforms' location. We can see the preliminary location of these new platforms in table 3:

Table 3. The preliminary location of these new platforms in table 3

Non-effective guarded platforms	Preliminary location of new platforms
A28	A26, A27, A28, A29
A29	A28, A29
A38	A38, A39, A40
A39	A38, A39; A40
A61	A48; A61
A92	A87; A88; A89; A90; A91; A92

3.3.2 Location Decision of New Platforms

Objective1: Minimize the total response time, which means decreasing the response time of platforms as much as possible. Thus we obtain the first objective function V_1:

$$V_1 = \min\sum_{i=1}^{n} t_i \qquad (i = 1, 2, \cdots, 92) \tag{10}$$

With V_1 representing the cumulative time spent by each traffic patrol service platform to arrive at the designated area; t_i represents response time, which is the amount of time spent by the jth traffic patrol service platform to the ith intersection node.

According to the equation 'time = distance/ speed', we can generate the value of t_i

$$t_i = s_i / v \qquad (i = 1, 2, \cdots, 92) \tag{11}$$

With s_i representing the distance between the jth traffic patrol service platform and the ith intersection node; v represents the value of police car speed, and we assume v is 60Km/h in this paper.

At the same time, in order to make sure that the police can timely tackle the accidents in its jurisdictional area, the response time is supposed to less than 3minutes:

$$t_i < 3\min = 0.05\text{h} \quad (i = 1, 2, \cdots, 92) \tag{12}$$

Objective2: Maximize the coverage of crime probability, which can increase the efficiency of police resources usage. Thus we obtain the second objective function V_2:

$$V_2 = \max \sum_{i=1}^{n} a_i A_i \quad (i = 1, 2, \cdots, 92) \tag{13}$$

With V_2 representing the crime probability in areas managed by all the traffic patrol service platforms; a_i representing the crime probability in the ith intersection node; A_i representing whether the ith intersection node is properly guarded by a certain traffic patrol service platform, if yes, then A_i=1, or A_i=0.

Objective3: Minimize the number of total traffic patrol service platforms, which can decrease the cost of setting platforms. Thus we obtain the third objective function V_3:

$$V_3 = \min Z = \min \sum_{j=1}^{n} X_j \tag{14}$$

With Z representing the number of total traffic patrol service platforms in Area A; X_j representing whether the jth traffic artery is equipped with traffic patrol service platform, if yes, then X_j=1, or X_j=0.

Considering that changing the location of these existing 20 traffic patrol service platforms can cost a lot, which goes against the cost-benefit principle, so the government will only set new traffic patrol service platforms in the remaining intersection nodes and the existing 20 platforms remain their present positions. And these can be concluded through the following two equations:

$$\text{When } j = 1, 2, \cdots, 20, \quad X_j = 1; \tag{15}$$

$$\text{When } j = 21, 22, \cdots, 92, \quad X_j \in \{0, 1\}; \tag{16}$$

After adding 2 to 5 new platforms, the number of total traffic patrol service platforms in Area A should be from 22 to 25:

$$22 \le Z \le 25 \tag{17}$$

Objective4: Minimize the workload difference among traffic patrol service platforms. In this paper, we use the mean difference of workload to measure the degree of workload imbalance, and the smaller of the mean difference, the more balanced of the workload distribution. Thus we obtain the forth objective function V_4:

$$V_4 = \min \left(\frac{\sum_{j=1}^{Z} \left| \sum_{i=1}^{n} a_i A_{i,j} - \bar{X} \right|}{Z} \right) \tag{18}$$

With V_4 representing the mean difference of workload; a_i representing the case happening probability in the ith intersection node; A_{ij} representing whether the jth traffic patrol service platform is arranged to the ith intersection node. If yes, then $A_{i,j}$=1, or $A_{i,j}$=0.

3.3.3 Result—the final location of new traffic patrol service platforms

On the basis of the preliminary scheme, we combine the multi-objective location decision model, which aims at minimizing the total response time, maximizing the coverage of the crime probability, minimizing the number of total traffic patrol service platforms, and minimizing the workload difference, to figure out the number of traffic patrol service platform and their location eventually. And we find that adding 4 traffic patrol service platforms is OK, which located in A28, A39, A61, A92 intersection nodes.

4. Concluding Remarks

Firstly, we use the Floyd algorithm to confirm the jurisdictional scope of each traffic patrol service platform, making sure that police can arrive at the scene as soon as possible when there are emergencies in its jurisdictional scope, and the specific results can be seen in table1. However, we are surprised to find that the workload differences among some traffic patrol service platforms in Area A is obvious, for example, the 6th only needs to be responsible for the intersection where it is and the 1th platform needs manage 10 platforms.

Then, we develop a dispatch model based on 0-1 integer programming and the Hungarian algorithm to dispatch the 20 traffic patrol service platforms in Area A, guaranteeing that the police can blockade all the 13 traffic arteries as quickly as possible when large-scale emergencies occur. The specific scheduling scheme is shown in Table 2.

Lastly, on the basis of multi-objective location decision model, we find that adding 4 traffic patrol service platforms in

A28, A39, A61, A92 intersection nodes can decrease the workload disequilibrium phenomenon and cut down the long response time.

In this paper, we find that the mathematical models combined with Floyd algorithm can solve the problem about the installation and dispatch of the patrol service platform effectively and feasibly, including the shortest path model, the dispatch model and the multiple-objective location model, which provides a good reference for the practical situations.

References

Baker, B. M., & Ayechew, M. A. (2003). A genetic algorithm for the vehicle routing problem. *Computers and operations research, 30*(5), 787-800. https://doi.org/10.1016/S0305-0548(02)00051-5

Berger, J., & Barkaoui, M. (2003). A hybrid genetic algorithm for the capacitated vehicle routing problem.*Geneticand evolutionary computation conference*. Springer Berlin Heidelberg, 646-656. http://dx.doi.org/10.1007/3-540-45105-6_80

Blanton, J. L., & Wainwright, R. L. Multiple vehicle routing with time and capacity constraints using geneticalgorithms. International conference on genetic algorithms. 1993,452-459.

Bullnheimer, B., Hartl, R. F., & Strauss, C. (1999). Applying the ANT system to the vehicle routing problem. *Meta-heuristics*. Springer US, 285-296. http://dx.doi.org/10.1007/978-1-4615-5775-3

China Undergraduate Mathematical Contest in Modeling, The competition problems in 2011. http://www.mcm.edu.cn.

Dantzing, G., & Ramser, J. (1959). The truck dispatching problem. *Management Science, 10*(6), 80-91. https://doi.org/10.1287/mnsc.6.1.80

Ellabib, I., Basir, O. A., & Calamai, P. (2002). An experimental study of a simple ant colony system for the vehicle routing problem with time windows. *Lecture notes in computer science, 2463*, 53-64. http://dx.doi.org/10.1007/3-540-45724-0_5

Huo, J. Z., & Wang, X. H. (2005). Solving vehicle scheduling problem based on constraint programming. *Logistics technology, 9,* 110-112.

Laporte, G., & Mercure, H. (1992). The vehicle routing problem with stochastic travel times. *Transportation Science, 26*(3), 161-170. http://dx.doi.org/10.1287/trsc.26.3.161

Mao Yuan-jie. (2013). Floyd Algorithm and matlab program realization of shortest path problem. *Journal of Hebei north university(natural science edition), 29*(5), 1673-1942. http://dx.doi.org/10.3969/j.issn.1673-1492.2013.05.004

Salman, A., Ahmad, I., & Al-Madani S. (2002). Particle swarm optimization for task assignment problem. *Microprocessors & Microsystems, 26*(8), 363-371. http://dx.doi.org/10.1016/S0141-9331(02)00053-4

Savelsbergh, M. W. P. (1985). Local search in routing problems with time windows. *Annals of Operations Research, 4*(1), 285-305. http://dx.doi.org/10.2498/cit.1002276

Xie, B. L., Li, J., & Guo, Y. H. (2000). Genetic algorithm for vehicle scheduling problem of non-full loads with time windows. *Journal of Systems Engineering, 9(*03), 235-239.

Yu, J. X., Song, D. C., Zhao, X. Y., & Li, J. Q. (2012). Research on the reasonable dispatch of service platform of traffic and patrol police. *Science Technology and Engineering, 12*(1), 1671-1815.

Zhu, H., & Zhang, Y. (2011). Finding shortest path between nodes based on improved floyd algorithm. *Network and Multimedia, 12*(3), 1002-8684. http://dx.doi.org/10.16311/j.audioe.2011.12.025

4

Computation of the Survival Probability of Brownian Motion with Drift When the Absorbing Boundary is a Piecewise Affine or Piecewise Exponential Function of Time

Tristan Guillaume [1]

[1] Université de Cergy-Pontoise, Laboratoire Thema, Cergy, France

Correspondence: Tristan Guillaume, Université de Cergy-Pontoise, Laboratoire Thema, 33 boulevard du port, F-95011 Cergy-Pontoise Cedex, France. E-mail: tristan.guillaume@u-cergy.fr

Abstract

A closed form formula is provided for the probability, in a closed time interval, that an arithmetic Brownian motion remains under or above a sequence of three affine, one-sided boundaries (equivalently, for the probability that a geometric Brownian motion remains under or above a sequence of three exponential, one-sided boundaries). The numerical evaluation of this formula can be done instantly and with the accuracy required for all practical purposes. The method followed can be extended to sequences of absorbing boundaries of higher dimension. It is also applied to sequences of two-sided boundaries.

Keywords: boundary crossing probability; survival probability; probability of absorption; first passage time; hitting time; Brownian motion; affine boundary; exponential boundary

1. Introduction

The question of the crossing of a non-constant boundary by a diffusion process is of central importance in many mathematical sciences. As mentioned in Wang and Pötzelberger (2007), it arises in biology, economics, engineering reliability, epidemiology, finance, genetics, seismology and sequential statistical analysis. The probability that a diffusion process will remain under or above some critical threshold over a given time interval can be referred to as a survival probability or probability of non-absorption. The vast majority of the research articles published on this topic either focus on numerical algorithms for general classes of processes or boundaries, usually involving recursive multidimensional quadrature, or they seek to obtain approximate solutions, typically substituting the initial boundary with another one for which computations are easier and then deriving a bound for the error entailed by using the approximating boundary. Much attention has also been paid to asymptotic estimates. However, known closed form results are scarce. By closed form results, we mean fully explicit formulae involving functions whose numerical evaluation can be carried out with the accuracy and the efficiency required for all practical purposes, in contrast to approximate analytical solutions that are quickly computed but inaccurate, and to numerical algorithms that can only produce the required standard of precision through heavy computational burden. The most classical of these closed form results is the so-called Bachelier-Levy formula (Levy, 1948), which provides the first-passage time density of Brownian motion to a linear boundary. This result is extended to a two-sided linear boundary by Doob (1949), but only in infinite time. The generalisation to a closed time interval is given by Anderson (1960), who is also able to integrate the density. The first passage time density of Brownian motion to a quadratic boundary is obtained independently by Salminen (1988) and Groeneboom (1989), while Novikov et al. (1999) manage to derive the hitting time density of Brownian motion to a square root boundary, but the numerical evaluation is quite involved in both cases, requiring infinite series of roots of combinations of Airy functions or confluent hypergeometric functions. By integrating these first passage time densities, the corresponding survival probabilities can be derived, though the integration is not actually performed by the mentioned authors and is far from trivial. Scheike (1992) provides a closed form solution for the survival probability of Brownian motion in infinite time when the boundary consists of two successive linear functions of time but cannot explicitly compute the corresponding integral in finite time. There are also a few closed form results for a Brownian motion (Daniels, 1996; Wang and Pötzelberger, 2007), an Ornstein-Uhlenbeck process (Choi and Nam, 2003; Wang and Pötzelberger, 2007) and a growth process (Wang and Pötzelberger, 2007), that involve very specific forms of the boundary and thus have limited use in practice, although they are quite valuable to test numerical algorithms.

This paper provides new results for the survival probability of Brownian motion. The problem raised by Scheike (1992)

is reformulated, extended and analytically solved. The extension with regard to the existing literature can be summarized as follows :

- cumulative distribution functions are provided, i.e. the integration of the first passage time densities is performed
- results are provided for generalised Brownian motion (whether arithmetic or geometric Brownian motion), i.e. the underlying stochastic dynamics include drift and volatility coefficients
- sequences of up to three general affine boundaries (in the case of arithmetic Brownian motion) or exponential boundaries (in the case of geometric Brownian motion) are handled
- sequences of two-sided piecewise affine or exponential boundaries are also tackled, under the assumption that the growth rate of the boundary is identical on the downside and on the upside, i.e. the upper and the lower sides of the boundary are parallel curves

Only distributions in finite time are considered, as they are the ones used in practice in the various mathematical sciences. The choice of affine and exponential boundaries is because they allow to model a reasonably large variety of time-dependent conditions for real life problems, while preserving analytical tractability. There are potentially many applications, for example in the valuation and risk management of various path dependent financial options or insurance contracts as well as in structural models of credit risk (see, e.g., Jeanblanc et al., 2009).

Section 2 of this article states a closed form formula for the survival probability of an arithmetic or a geometric Brownian under or above a sequence of three different one-sided affine or exponential boundaries over a finite time interval and provides a few numerical results, then outlines a proof omitting cumbersome computations, and finally discusses generalization to higher-dimensional boundaries. Section 3 of this article states a closed form formula for the survival probability of an arithmetic or a geometric Brownian motion under and above a sequence of two different two-sided, parallel, affine or exponential boundaries over a finite time interval, provides a few numerical results and outlines the proof.

2. Survival Probability of an Arithmetic or a Geometric Brownian Motion under or Above a Sequence of One-sided Affine or Exponential Boundaries over a Finite Time Interval

2.1 Definitions

Let μ be a real constant, σ be a positive real constant, and $\{B(t), t \geq 0\}$ be a standard Brownian motion defined on a probability space with measure $\mathbb{P}$. Let $\{X_1(t), t \geq 0\}$ be an arithmetic Brownian motion driven, under $\mathbb{P}$, by :

$$dX_1(t) = \mu\, dt + \sigma\, dB(t) \tag{2.1}$$

Let $\{X_2(t), t \geq 0\}$ be a geometric Brownian motion driven, under $\mathbb{P}$, by :

$$dX_2(t) = \mu X_2(t) dt + \sigma X_2(t) dB(t) \tag{2.2}$$

A finite time interval $[0, T]$ is considered and divided into a partition Π of n subintervals $[t_0 = 0, t_1], [t_1, t_2]$, $\ldots [t_{n-1}, t_n = T]$, which are not necessarily of equal length, with $t_n \geq t_{n-1} \geq \ldots \geq t_1 \geq t_0$. Let $\mathbb{I}$ denote the indicator function. For a given $n \in \mathbb{N}$, two piecewise affine absorbing boundaries $g_1(t)$ and $g_2(t)$ are defined as follows :

$$g_1(t) = \sum_{i=1}^{n} \left(a_i + b_i(t - t_{i-1})\right) \mathbb{I}_{[t_{i-1}, t_i]}(t), a_i \in \mathbb{R}, b_i \in \mathbb{R},\ i \in \{1, 2, \ldots, n\} \tag{2.3}$$

$$g_2(t) = \sum_{i=1}^{n} (a_i + b_i t)\, \mathbb{I}_{[t_{i-1},t_i]}(t) \quad , \quad a_i \in \mathbb{R} \quad , \quad b_i \in \mathbb{R}, \quad i \in \{1,2,...,n\} \tag{2.4}$$

The difference between $g_1(t)$ and $g_2(t)$ is that $g_1(t)$ is time-homogeneous.

Similarly, we have the two following piecewise exponential boundaries :

$$h_1(t) = \sum_{i=1}^{n} X(0)\exp(a_i + b_i(t - t_{i-1}))\, \mathbb{I}_{[t_{i-1},t_i]}(t), \quad a_i \in \mathbb{R} \quad , \quad b_i \in \mathbb{R}, \quad i \in \{1,2,...,n\} \tag{2.5}$$

$$h_2(t) = \sum_{i=1}^{n} X(0)\exp(a_i + b_i t)\, \mathbb{I}_{[t_{i-1},t_i]}(t), \quad a_i \in \mathbb{R} \quad , \quad b_i \in \mathbb{R}, \quad i \in \{1,2,...,n\} \tag{2.6}$$

Consider the cumulative distribution function of a sequence of n maxima or n minima and n endpoints in II in the two following cases :

- the absorbing boundary is defined either by $g_1(t)$ or $g_2(t)$ and the process under consideration is X_1

- the absorbing boundary is defined either by $h_1(t)$ or $h_2(t)$ and the process under consideration is X_2

Such a function is often referred to as a survival probability. As shown by Wang and Pötzelberger (1997), its value can be approximated by a Monte Carlo simulation scheme drawing on the Markovian nature of X_1 and X_2 in the following manner : the endpoint values of X_1 and X_2 in each time subinterval $[t_{i-1}, t_i]$ are randomly drawn at each performed simulation; if the relevant conditions at each t_i are met, then a cumulative variable records the product of the conditional probabilities that the boundary has not been crossed in each (t_{i-1}, t_i), which admit simple analytical formulae (Siegmund, 1986). This is obviously much more efficient and accurate than discretizing the whole path of the process at each run. For $n > 1$, the survival probability under consideration does not admit any known closed form formula. Although it does not seem possible to come up with an explicit and compact formula for any $n \in \mathbb{N}$, one can actually solve the problem analytically in "moderate" dimension. In this paper, the case $n = 3$ is tackled. More specifically, let $\mathrm{P}_{[...]}(\mu, \sigma, a_1, a_2, a_3, b_1, b_2, b_3, k_1, k_2, k_3, t_1, t_2, t_3)$ be defined as one of the following eight cumulative distribution functions :

$$\mathrm{P}_{[AU1]}(\mu, \sigma, a_1, a_2, a_3, b_1, b_2, b_3, k_1, k_2, k_3, t_1, t_2, t_3), \; a_1 \in \mathbb{R}_+, (a_2, a_3, b_1, b_2, b_3, k_1, k_2, k_3) \in \mathbb{R}^8 \tag{2.7}$$

$$= \mathbb{P}\begin{pmatrix} (X_1(t) < a_1 + b_1 t, \forall 0 \le t \le t_1) \cap X_1(t_1) < k_1 \cap (X_1(t) < a_2 + b_2(t - t_1), \forall t_1 \le t \le t_2) \\ \cap X_1(t_2) < k_2 \cap (X_1(t) < a_3 + b_3(t - t_2), \forall t_2 \le t \le t_3) \cap X_1(t_3) < k_3 \end{pmatrix}$$

$$\mathrm{P}_{[AU2]}(\mu, \sigma, a_1, a_2, a_3, b_1, b_2, b_3, k_1, k_2, k_3, t_1, t_2, t_3), \; a_1 \in \mathbb{R}_+, (a_2, a_3, b_1, b_2, b_3, k_1, k_2, k_3) \in \mathbb{R}^8 \tag{2.8}$$

$$= \mathbb{P}\begin{pmatrix} (X_1(t) < a_1 + b_1 t, \forall 0 \le t \le t_1) \cap X_1(t_1) < k_1 \cap (X_1(t) < a_2 + b_2 t, \forall t_1 \le t \le t_2) \\ \cap X_1(t_2) < k_2 \cap (X_1(t) < a_3 + b_3 t, \forall t_2 \le t \le t_3) \cap X_1(t_3) < k_3 \end{pmatrix}$$

$$P_{[GU1]}\left(\mu,\sigma,a_1,a_2,a_3,b_1,b_2,b_3,k_1,k_2,k_3,t_1,t_2,t_3\right),\left(a_1,k_1,k_2,k_3\right)\in\mathbb{R}_+^4,\left(a_2,a_3,b_1,b_2,b_3\right)\in\mathbb{R}^5 \tag{2.9}$$

$$=\mathbb{P}\begin{pmatrix}\left(X_2(t)<X_2(0)\exp\left(a_1+b_1t\right),\forall 0\le t\le t_1\right)\cap X_2(t_1)<k_1\\ \cap\left(X_2(t)<X_2(0)\exp\left(a_2+b_2(t-t_1)\right),\forall t_1\le t\le t_2\right)\\ \cap X_2(t_2)<k_2\cap\left(X_2(t)<X_2(0)\exp\left(a_3+b_3(t-t_2)\right),\forall t_2\le t\le t_3\right)\cap X_2(t_3)<k_3\end{pmatrix}$$

$$P_{[GU2]}\left(\mu,\sigma,a_1,a_2,a_3,b_1,b_2,b_3,k_1,k_2,k_3,t_1,t_2,t_3\right),\left(a_1,k_1,k_2,k_3\right)\in\mathbb{R}_+^4,\left(a_2,a_3,b_1,b_2,b_3\right)\in\mathbb{R}^5 \tag{2.10}$$

$$=\mathbb{P}\begin{pmatrix}\left(X_2(t)<X_2(0)\exp\left(a_1+b_1t\right),\forall 0\le t\le t_1\right)\cap X_2(t_1)<k_1\\ \cap\left(X_2(t)<X_2(0)\exp\left(a_2+b_2t\right),\forall t_1\le t\le t_2\right)\\ \cap X_2(t_2)<k_2\cap\left(X_2(t)<X_2(0)\exp\left(a_3+b_3t\right),\forall t_2\le t\le t_3\right)\cap X_2(t_3)<k_3\end{pmatrix}$$

$$P_{[AL1]}\left(\mu,\sigma,a_1,a_2,a_3,b_1,b_2,b_3,k_1,k_2,k_3,t_1,t_2,t_3\right),\ a_1\in\mathbb{R}_-,\left(a_2,a_3,b_1,b_2,b_3,k_1,k_2,k_3\right)\in\mathbb{R}^8 \tag{2.11}$$

$$=\mathbb{P}\begin{pmatrix}\left(X_1(t)>a_1+b_1t,\forall 0\le t\le t_1\right)\cap X_1(t_1)>k_1\cap\left(X_1(t)>a_2+b_2(t-t_1),\forall t_1\le t\le t_2\right)\\ \cap X_1(t_2)>k_2\cap\left(X_1(t)>a_3+b_3(t-t_2),\forall t_2\le t\le t_3\right)\cap X_1(t_3)>k_3\end{pmatrix}$$

$$P_{[AL2]}\left(\mu,\sigma,a_1,a_2,a_3,b_1,b_2,b_3,k_1,k_2,k_3,t_1,t_2,t_3\right),\ a_1\in\mathbb{R}_-,\left(a_2,a_3,b_1,b_2,b_3,k_1,k_2,k_3\right)\in\mathbb{R}^8 \tag{2.12}$$

$$=\mathbb{P}\begin{pmatrix}\left(X_1(t)>a_1+b_1t,\forall 0\le t\le t_1\right)\cap X_1(t_1)>k_1\cap\left(X_1(t)>a_2+b_2t,\forall t_1\le t\le t_2\right)\\ \cap X_1(t_2)>k_2\cap\left(X_1(t)>a_3+b_3t,\forall t_2\le t\le t_3\right)\cap X_1(t_3)>k_3\end{pmatrix}$$

$$P_{[GL1]}\left(\mu,\sigma,a_1,a_2,a_3,b_1,b_2,b_3,k_1,k_2,k_3,t_1,t_2,t_3\right),\ a_1\in\mathbb{R}_-,\left(k_1,k_2,k_3\right)\in\mathbb{R}_+^3,\left(a_2,a_3,b_1,b_2,b_3\right)\in\mathbb{R}^5 \tag{2.13}$$

$$=\mathbb{P}\begin{pmatrix}\left(X_2(t)>X_2(0)\exp\left(a_1+b_1t\right),\forall 0\le t\le t_1\right)\cap X_2(t_1)>k_1\\ \cap\left(X_2(t)>X_2(0)\exp\left(a_2+b_2(t-t_1)\right),\forall t_1\le t\le t_2\right)\\ \cap X_2(t_2)>k_2\cap\left(X_2(t)>X_2(0)\exp\left(a_3+b_3(t-t_2)\right),\forall t_2\le t\le t_3\right)\cap X_2(t_3)>k_3\end{pmatrix}$$

$$P_{[GL2]}\left(\mu,\sigma,a_1,a_2,a_3,b_1,b_2,b_3,k_1,k_2,k_3,t_1,t_2,t_3\right),\ a_1\in\mathbb{R}_-,\left(k_1,k_2,k_3\right)\in\mathbb{R}_+^3,\left(a_2,a_3,b_1,b_2,b_3\right)\in\mathbb{R}^5 \tag{2.14}$$

$$=\mathbb{P}\begin{pmatrix}\left(X_2(t)>X_2(0)\exp\left(a_1+b_1t\right),\forall 0\le t\le t_1\right)\cap X_2(t_1)>k_1\\ \cap\left(X_2(t)>X_2(0)\exp\left(a_2+b_2t\right),\forall t_1\le t\le t_2\right)\\ \cap X_2(t_2)>k_2\cap\left(X_2(t)>X_2(0)\exp\left(a_3+b_3t\right),\forall t_2\le t\le t_3\right)\cap X_2(t_3)>k_3\end{pmatrix}$$

In other words, taking $n=3$,

- $P_{[AU1]}$ is the probability that an arithmetic Brownian motion will remain under the piecewise affine time-homogeneous boundary $g_1(t)$ defined by (2.3) and under the successive endpoints k_1,k_2,k_3

- $P_{[AU2]}$ is the probability that an arithmetic Brownian motion will remain under the piecewise affine time-inhomogeneous boundary $g_2(t)$ defined by (2.4) and under the successive endpoints k_1,k_2,k_3

- $P_{[GU1]}$ is the probability that a geometric Brownian motion will remain under the piecewise exponential time-homogeneous boundary $h_1(t)$ defined by (2.5) and under the successive endpoints k_1, k_2, k_3

- $P_{[GU2]}$ is the probability that a geometric Brownian motion will remain under the piecewise exponential time-inhomogeneous boundary $h_2(t)$ defined by (2.6) and under the successive endpoints k_1, k_2, k_3

- $P_{[AL1]}$ is the probability that an arithmetic Brownian motion will remain above the piecewise affine boundary $g_1(t)$ defined by (2.3) and above the successive endpoints k_1, k_2, k_3

- $P_{[AL2]}$ is the probability that an arithmetic Brownian motion will remain above the piecewise affine boundary $g_2(t)$ defined by (2.4) and above the successive endpoints k_1, k_2, k_3

- $P_{[GL1]}$ is the probability that a geometric Brownian motion will remain above the piecewise exponential boundary $h_1(t)$ defined by (2.5) and above the successive endpoints k_1, k_2, k_3

- $P_{[GL2]}$ is the probability that a geometric Brownian motion will remain above the piecewise exponential boundary $h_2(t)$ defined by (2.6) and above the successive endpoints k_1, k_2, k_3

2.2 Statement of Formula 1

Formula 1 *Let* $P_{[...]}(\mu, \sigma, a_1, a_2, a_3, b_1, b_2, b_3, k_1, k_2, k_3, t_1, t_2, t_3)$ *be defined as in Subsection 2.1. Then,*

$$P_{[...]}(\mu, \sigma, a_1, a_2, a_3, b_1, b_2, b_3, k_1, k_2, k_3, t_1, t_2, t_3) \tag{2.15}$$

$$= \Phi_3\left[\theta\frac{z_1 - b_1t_1 - \mu_1t_1}{\sigma\sqrt{t_1}}, \theta\frac{z_2 - b_2t_2 - \mu_1t_1 - \mu_2(t_2 - t_1)}{\sigma\sqrt{t_2}}, \theta\frac{z_3 - b_3t_3 - \mu_1t_1 - \mu_2(t_2 - t_1) - \mu_3(t_3 - t_2)}{\sigma\sqrt{t_3}}; \sqrt{t_1 / t_2}, \sqrt{t_2 / t_3}\right]$$

$$-\exp\left(\frac{\lambda_1}{\sigma^2}\right)\Phi_3\left[\begin{array}{l}\theta\frac{z_1 - b_1t_1 - 2a_1 - \mu_1t_1}{\sigma\sqrt{t_1}}, \theta\frac{z_2 - b_2t_2 - 2a_1 - \mu_1t_1 - \mu_2(t_2 - t_1)}{\sigma\sqrt{t_2}}, \\ \theta\frac{z_3 - b_3t_3 - 2a_1 - \mu_1t_1 - \mu_2(t_2 - t_1) - \mu_3(t_3 - t_2)}{\sigma\sqrt{t_3}}; \sqrt{t_1 / t_2}, \sqrt{t_2 / t_3}\end{array}\right]$$

$$-\exp\left(\frac{\lambda_2}{\sigma^2}\right)\Phi_3\left[\begin{array}{l}\theta\frac{z_1 - b_1t_1 - \mu_1t_1 + 2\mu_2t_1}{\sigma\sqrt{t_1}}, \theta\frac{z_2 - b_2t_2 - 2\alpha_2 + \mu_1t_1 - \mu_2(t_1 + t_2)}{\sigma\sqrt{t_2}}, \\ \theta\frac{z_3 - b_3t_3 - 2\alpha_2 + \mu_1t_1 - \mu_2(t_1 + t_2) - \mu_3(t_3 - t_2)}{\sigma\sqrt{t_3}}; -\sqrt{t_1 / t_2}, \sqrt{t_2 / t_3}\end{array}\right]$$

$$+\exp\left(\frac{\lambda_3}{\sigma^2}\right)\Phi_3\left[\begin{array}{l}\theta\frac{z_1-b_1t_1-2a_1-\mu_1t_1+2\mu_2t_1}{\sigma\sqrt{t_1}},\theta\frac{z_2-b_2t_2-2\alpha_2+2a_1+\mu_1t_1-\mu_2(t_1+t_2)}{\sigma\sqrt{t_2}}\\ \theta\frac{z_3-b_3t_3-2\alpha_2+2a_1+\mu_1t_1-\mu_2(t_1+t_2)-\mu_3(t_3-t_2)}{\sigma\sqrt{t_3}};-\sqrt{t_1/t_2},\sqrt{t_2/t_3}\end{array}\right]$$

$$-\exp\left(\frac{\lambda_4}{\sigma^2}\right)\Phi_3\left[\begin{array}{l}\theta\frac{z_1-b_1t_1-\mu_1t_1+2\mu_3t_1}{\sigma\sqrt{t_1}},\theta\frac{z_2-b_2t_2-\mu_1t_1-\mu_2(t_2-t_1)+2\mu_3t_2}{\sigma\sqrt{t_2}},\\ \theta\frac{z_3-b_3t_3-2\alpha_3+\mu_1t_1+\mu_2(t_2-t_1)-\mu_3(t_2+t_3)}{\sigma\sqrt{t_3}};\sqrt{t_1/t_2},-\sqrt{t_2/t_3}\end{array}\right]$$

$$+\exp\left(\frac{\lambda_5}{\sigma^2}\right)\Phi_3\left[\begin{array}{l}\theta\frac{z_1-b_1t_1-2a_1+2\mu_3t_1-\mu_1t_1}{\sigma\sqrt{t_1}},\theta\frac{z_2-b_2t_2-2a_1-\mu_1t_1-\mu_2(t_2-t_1)+2\mu_3t_2}{\sigma\sqrt{t_2}},\\ \theta\frac{z_3-b_3t_3-2\alpha_3+2a_1+\mu_1t_1+\mu_2(t_2-t_1)-\mu_3(t_2+t_3)}{\sigma\sqrt{t_3}};\sqrt{t_1/t_2},-\sqrt{t_2/t_3}\end{array}\right]$$

$$+\exp\left(\frac{\lambda_6}{\sigma^2}\right)\Phi_3\left[\begin{array}{l}\theta\frac{z_1-b_1t_1-2\mu_3t_1+2\mu_2t_1-\mu_1t_1}{\sigma\sqrt{t_1}},\theta\frac{z_2-b_2t_2-2\alpha_2+\mu_1t_1-\mu_2(t_1+t_2)+2\mu_3t_2}{\sigma\sqrt{t_2}},\\ \theta\frac{z_3-b_3t_3-2\alpha_3+2\alpha_2-\mu_1t_1+\mu_2(t_1+t_2)-\mu_3(t_2+t_3)}{\sigma\sqrt{t_3}};-\sqrt{t_1/t_2},-\sqrt{t_2/t_3}\end{array}\right]$$

$$-\exp\left(\frac{\lambda_7}{\sigma^2}\right)\Phi_3\left[\begin{array}{l}\theta\frac{z_1-b_1t_1-2a_1-2\mu_3t_1+2\mu_2t_1-\mu_1t_1}{\sigma\sqrt{t_1}},\\ \theta\frac{z_2-b_2t_2-2\alpha_2+2a_1+\mu_1t_1-\mu_2(t_1+t_2)+2\mu_3t_2}{\sigma\sqrt{t_2}},\\ \theta\frac{z_3-b_3t_3-2\alpha_3+2\alpha_2-2a_1-\mu_1t_1+\mu_2(t_1+t_2)-\mu_3(t_2+t_3)}{\sigma\sqrt{t_3}};\\ -\sqrt{t_1/t_2},-\sqrt{t_2/t_3}\end{array}\right]$$

where the function Φ_n *is a convolution of gaussian densities defined, for any* $n \in \mathbb{N}$, *by :*

$$\Phi_n\left[x_1,\ldots,x_n;\rho_1,\ldots,\rho_{n-1}\right] = \int_{D^n}\frac{\exp\left(-\frac{y_1^2}{2}-\sum_{i=1}^{n-1}\frac{\left(y_{i+1}-\rho_i y_i\right)^2}{2\left(1-\rho_i^2\right)}\right)}{\left(2\pi\right)^{n/2}\prod_{i=1}^{n-1}\sqrt{1-\rho_i^2}}dy_n\ldots dy_1 \tag{2.16}$$

$$D^n = \left]-\infty,x_1\right]\times\left]-\infty,x_2\right]\ldots\times\left]-\infty,x_n\right],\ x_i\in\mathbb{R},\ \rho_i\in\left]-1,1\right[\ ,\ i\in\left\{1,\ldots,n\right\}$$

The α_i *terms,* $i \in \{2,3\}$, *in (2.15) are given by :*

$$\alpha_2 = a_2\left(\mathbb{I}_{\{P_{[\ldots]}=P_{[AU2]}\}}+\mathbb{I}_{\{P_{[\ldots]}=P_{[GU2]}\}}+\mathbb{I}_{\{P_{[\ldots]}=P_{[AL2]}\}}+\mathbb{I}_{\{P_{[\ldots]}=P_{[GL2]}\}}\right)$$

$$+\left(a_2 - b_2 t_1\right)\left(\mathbb{I}_{\left\{P_{[\dots]}=P_{[AU1]}\right\}} + \mathbb{I}_{\left\{P_{[\dots]}=P_{[GU1]}\right\}} + \mathbb{I}_{\left\{P_{[\dots]}=P_{[AL1]}\right\}} + \mathbb{I}_{\left\{P_{[\dots]}=P_{[GL1]}\right\}}\right)$$

$$\alpha_3 = a_3\left(\mathbb{I}_{\left\{P_{[\dots]}=P_{[AU2]}\right\}} + \mathbb{I}_{\left\{P_{[\dots]}=P_{[GU2]}\right\}} + \mathbb{I}_{\left\{P_{[\dots]}=P_{[AL2]}\right\}} + \mathbb{I}_{\left\{P_{[\dots]}=P_{[GL2]}\right\}}\right)$$

$$+\left(a_3 - b_3 t_2\right)\left(\mathbb{I}_{\left\{P_{[\dots]}=P_{[AU1]}\right\}} + \mathbb{I}_{\left\{P_{[\dots]}=P_{[GU1]}\right\}} + \mathbb{I}_{\left\{P_{[\dots]}=P_{[AL1]}\right\}} + \mathbb{I}_{\left\{P_{[\dots]}=P_{[GL1]}\right\}}\right)$$

The λ_i *terms,* $i \in \{1,2,3,4,5,6,7\}$ *, in (2.15) are given by :*

$$\lambda_1 = 2\mu_1 a_1$$

$$\lambda_2 = 2\mu_2\alpha_2 - 2\mu_1\mu_2 t_1 + 2\mu_2^2 t_1$$

$$\lambda_3 = 2\mu_1 a_1 + 2\mu_2\alpha_2 - 4\mu_2 a_1 - 2\mu_1\mu_2 t_1 + 2\mu_2^2 t_1$$

$$\lambda_4 = 2\mu_3\alpha_3 + 2\mu_3^2 t_2 - 2\mu_1\mu_3 t_1 - 2\mu_2\mu_3\left(t_2 - t_1\right)$$

$$\lambda_5 = 2\mu_3\alpha_3 + 2\mu_1 a_1 - 4\mu_3 a_1 + 2\mu_3^2 t_2 - 2\mu_1\mu_3 t_1 - 2\mu_2\mu_3\left(t_2 - t_1\right)$$

$$\lambda_6 = 2\mu_3\alpha_3 + 2\mu_2\alpha_2 - 4\mu_3\alpha_2 + 2\left(\mu_3 - \mu_2\right)^2 t_1 + 2\mu_1\left(\mu_3 - \mu_2\right)t_1 + 2\mu_3^2\left(t_2 - t_1\right) - 2\mu_2\mu_3\left(t_2 - t_1\right)$$

$$\lambda_7 = 2\mu_1 a_1 + 2\mu_2\alpha_2 - 4\mu_3\alpha_2 + 2\mu_3\alpha_3 + 2\left(\mu_3 - \mu_2\right)^2 t_1 + 2\mu_3^2\left(t_2 - t_1\right)$$

$$-2\mu_2\mu_3\left(t_2 - t_1\right) + 2\left(\mu_3 - \mu_2\right)\left(2a_1 + \mu_1 t_1\right)$$

The z_i *terms,* $i \in \{1,2,3\}$ *, in (2.15) are given by :*

$$z_1 = \min\left(a_1 + b_1 t_1, k_1, a_2\right)\mathbb{I}_{\left\{P_{[\dots]}=P_{[AU1]}\right\}} + \min\left(a_1 + b_1 t_1, k_1, a_2 + b_2 t_1\right)\mathbb{I}_{\left\{P_{[\dots]}=P_{[AU2]}\right\}}$$

$$+\min\left(a_1 + b_1 t_1, \ln\left(k_1 / X_2(0)\right), a_2\right)\mathbb{I}_{\left\{P_{[\dots]}=P_{[GU1]}\right\}} + \min\left(a_1 + b_1 t_1, \ln\left(k_1 / X_2(0)\right), a_2 + b_2 t_1\right)\mathbb{I}_{\left\{P_{[\dots]}=P_{[GU2]}\right\}}$$

$$+\max\left(a_1 + b_1 t_1, k_1, a_2\right)\mathbb{I}_{\left\{P_{[\dots]}=P_{[AL1]}\right\}} + \max\left(a_1 + b_1 t_1, k_1, a_2 + b_2 t_1\right)\mathbb{I}_{\left\{P_{[\dots]}=P_{[AL2]}\right\}}$$

$$+\max\left(a_1 + b_1 t_1, \ln\left(k_1 / X_2(0)\right), a_2\right)\mathbb{I}_{\left\{P_{[\dots]}=P_{[GL1]}\right\}} + \max\left(a_1 + b_1 t_1, \ln\left(k_1 / X_2(0)\right), a_2 + b_2 t_1\right)\mathbb{I}_{\left\{P_{[\dots]}=P_{[GL2]}\right\}}$$

$$z_2 = \min\left(a_2 + b_2\left(t_2 - t_1\right), k_2, a_3\right)\mathbb{I}_{\left\{P_{[\dots]}=P_{[AU1]}\right\}} + \min\left(a_2 + b_2\left(t_2 - t_1\right), \ln\left(k_2 / X_2(0)\right), a_3\right)\mathbb{I}_{\left\{P_{[\dots]}=P_{[GU1]}\right\}}$$

$$+\min\left(a_2 + b_2 t_2, k_2, a_3 + b_3 t_2\right)\mathbb{I}_{\left\{P_{[\dots]}=P_{[AU2]}\right\}} + \min\left(a_2 + b_2 t_2, \ln\left(k_2 / X_2(0)\right), a_3 + b_3 t_2\right)\mathbb{I}_{\left\{P_{[\dots]}=P_{[GU2]}\right\}}$$

$$+\max\left(a_2 + b_2\left(t_2 - t_1\right), k_2, a_3\right)\mathbb{I}_{\left\{P_{[\dots]}=P_{[AL1]}\right\}} + \max\left(a_2 + b_2\left(t_2 - t_1\right), \ln\left(k_2 / X_2(0)\right), a_3\right)\mathbb{I}_{\left\{P_{[\dots]}=P_{[GL1]}\right\}}$$

$$+\max\left(a_2 + b_2 t_2, k_2, a_3 + b_3 t_2\right)\mathbb{I}_{\left\{P_{[\dots]}=P_{[AL2]}\right\}} + \max\left(a_2 + b_2 t_2, \ln\left(k_2 / X_2(0)\right), a_3 + b_3 t_2\right)\mathbb{I}_{\left\{P_{[\dots]}=P_{[GL2]}\right\}}$$

$$z_3 = \min\left(a_3 + b_3\left(t_3 - t_2\right), k_3\right)\mathbb{I}_{\{P_{[...]}=P_{[AU1]}\}} + \min\left(a_3 + b_3\left(t_3 - t_2\right), \ln\left(k_3 / X_2\left(0\right)\right)\right)\mathbb{I}_{\{P_{[...]}=P_{[GU1]}\}}$$

$$+\min\left(a_3 + b_3 t_3, k_3\right)\mathbb{I}_{\{P_{[...]}=P_{[AU2]}\}} + \min\left(a_3 + b_3 t_3, \ln\left(k_3 / X_2\left(0\right)\right)\right)\mathbb{I}_{\{P_{[...]}=P_{[GU2]}\}}$$

$$+\max\left(a_3 + b_3\left(t_3 - t_2\right), k_3\right)\mathbb{I}_{\{P_{[...]}=P_{[AL1]}\}} + \max\left(a_3 + b_3\left(t_3 - t_2\right), \ln\left(k_3 / X_2\left(0\right)\right)\right)\mathbb{I}_{\{P_{[...]}=P_{[GL1]}\}}$$

$$+\max\left(a_3 + b_3 t_3, k_3\right)\mathbb{I}_{\{P_{[...]}=P_{[AL2]}\}} + \max\left(a_3 + b_3 t_3, \ln\left(k_3 / X_2\left(0\right)\right)\right)\mathbb{I}_{\{P_{[...]}=P_{[GL2]}\}}$$

The μ_i terms, $i \in \{1,2,3\}$, in (2.15) are given by :

$$\mu_1 = \left(\mu - b_1\right)\left(\mathbb{I}_{\{P_{[...]}=P_{[AU1]}\}} + \mathbb{I}_{\{P_{[...]}=P_{[AL1]}\}} + \mathbb{I}_{\{P_{[...]}=P_{[AU2]}\}} + \mathbb{I}_{\{P_{[...]}=P_{[AL2]}\}}\right)$$

$$+\left(\mu - \frac{\sigma^2}{2} - b_1\right)\left(\mathbb{I}_{\{P_{[...]}=P_{[GU1]}\}} + \mathbb{I}_{\{P_{[...]}=P_{[GL1]}\}} + \mathbb{I}_{\{P_{[...]}=P_{[GU2]}\}} + \mathbb{I}_{\{P_{[...]}=P_{[GL2]}\}}\right)$$

$$\mu_2 = \left(\mu - b_2\right)\left(\mathbb{I}_{\{P_{[...]}=P_{[AU1]}\}} + \mathbb{I}_{\{P_{[...]}=P_{[AL1]}\}} + \mathbb{I}_{\{P_{[...]}=P_{[AU2]}\}} + \mathbb{I}_{\{P_{[...]}=P_{[AL2]}\}}\right)$$

$$+\left(\mu - \frac{\sigma^2}{2} - b_2\right)\left(\mathbb{I}_{\{P_{[...]}=P_{[GU1]}\}} + \mathbb{I}_{\{P_{[...]}=P_{[GL1]}\}} + \mathbb{I}_{\{P_{[...]}=P_{[GU2]}\}} + \mathbb{I}_{\{P_{[...]}=P_{[GL2]}\}}\right)$$

$$\mu_3 = \left(\mu - b_3\right)\left(\mathbb{I}_{\{P_{[...]}=P_{[AU1]}\}} + \mathbb{I}_{\{P_{[...]}=P_{[AL1]}\}} + \mathbb{I}_{\{P_{[...]}=P_{[AU2]}\}} + \mathbb{I}_{\{P_{[...]}=P_{[AL2]}\}}\right)$$

$$+\left(\mu - \frac{\sigma^2}{2} - b_3\right)\left(\mathbb{I}_{\{P_{[...]}=P_{[GU1]}\}} + \mathbb{I}_{\{P_{[...]}=P_{[GL1]}\}} + \mathbb{I}_{\{P_{[...]}=P_{[GU2]}\}} + \mathbb{I}_{\{P_{[...]}=P_{[GL2]}\}}\right)$$

θ is given by :

$$\theta = 1\left(\mathbb{I}_{\{P_{[...]}=P_{[AU1]}\}} + \mathbb{I}_{\{P_{[...]}=P_{[AU2]}\}} + \mathbb{I}_{\{P_{[...]}=P_{[GU1]}\}} + \mathbb{I}_{\{P_{[...]}=P_{[GU2]}\}}\right)$$

$$-1\left(\mathbb{I}_{\{P_{[...]}=P_{[AL1]}\}} + \mathbb{I}_{\{P_{[...]}=P_{[AL2]}\}} + \mathbb{I}_{\{P_{[...]}=P_{[GL1]}\}} + \mathbb{I}_{\{P_{[...]}=P_{[GL2]}\}}\right)$$

End of Formula 1.

From a numerical perspective, Formula 1 raises the question of the evaluation of the function Φ_3. A straightforward calculation yields the following integration rule :

$$\Phi_3\left[x_1, x_2, x_3; \rho_1, \rho_2\right] = \int_{y_2=-\infty}^{x_2} \frac{\exp\left(-y_2^2 / 2\right)}{\sqrt{2\pi}} N\left[\frac{x_1 - \rho_1 y_2}{\sqrt{1-\rho_1^2}}\right] N\left[\frac{x_3 - \rho_2 y_2}{\sqrt{1-\rho_2^2}}\right] dy_2 \qquad (2.17)$$

where the function $N[.]$ is the univariate standard normal cumulative distribution function.

Using (2.17), the numerical evaluation of the function Φ_3 is easy by means of a classical adaptive Gauss-Legendre quadrature. Alternatively, the following identities can be verified :

$$\Phi_3\left[x_1, x_2, x_3; \sqrt{t_1 / t_2}, \sqrt{t_2 / t_3}\right] = N_3\left[x_1, x_2, x_3; \sqrt{t_1 / t_2}, \sqrt{t_1 / t_3}, \sqrt{t_2 / t_3}\right] \qquad (2.18)$$

$$\Phi_3\left[x_1,x_2,x_3;-\sqrt{t_1/t_2},-\sqrt{t_2/t_3}\right]=N_3\left[x_1,x_2,x_3;-\sqrt{t_1/t_2},\sqrt{t_1/t_3},-\sqrt{t_2/t_3}\right] \quad (2.19)$$

$$\Phi_3\left[x_1,x_2,x_3;-\sqrt{t_1/t_2},\sqrt{t_2/t_3}\right]=N_3\left[x_1,x_2,x_3;-\sqrt{t_1/t_2},-\sqrt{t_1/t_3},\sqrt{t_2/t_3}\right] \quad (2.20)$$

$$\Phi_3\left[x_1,x_2,x_3;\sqrt{t_1/t_2},-\sqrt{t_2/t_3}\right]=N_3\left[x_1,x_2,x_3;\sqrt{t_1/t_2},-\sqrt{t_1/t_3},-\sqrt{t_2/t_3}\right] \quad (2.21)$$

where the function $N_3\left[.,.,.;.,.,.\right]$ is the trivariate standard normal cumulative distribution function, the numerical evaluation of which can be performed with double precision and computational time of approximately 0.01 second using the algorithm by Genz (2004).

A few numerical results are reported in Table 1, in which the $P_{[AU1]}$ survival probability is computed for increasing levels of the volatility coefficient σ and other parameters fixed as follows : $\mu = 0.01$, $t_1 = 0.25$, $t_2 = 0.5$, $t_3 = 1$, $k_1 = 0$, $k_2 = 0.02$, $k_3 = 0.03$, $a_1 = 0.22$, $b_1 = -0.12$, $a_2 = a_1 + b_1 t_1$, $b_2 = 0.16$, $a_3 = a_2 + b_2\left(t_2 - t_1\right)$, $b_3 = -0.24$. Notice that the absorbing boundary here is continuous at times t_1 and t_2, but non-continuous boundaries can be handled just as easily. Formula 1 is implemented using the algorithm by Genz (2004) for the computation of the trivariate standard normal cumulative distribution function. The results are compared with those obtained using the semi-analytical Monte Carlo algorithm devised by Wang and Pötzelberger (1997), denoted by WP simulation algorithm, that enables to draw only the endpoints of the time subintervals at each run, which is dramatically more efficient and accurate than a basic Monte Carlo simulation. Random numbers are drawn by the Mersenne Twister generator.

For all computed values, a 5-digit convergence can be observed between Formula 1 and the WP algorithm, on condition that a total of 100,000,000 stochastic simulations are performed. The latter method requires a computational time of 411 seconds on an i-7 4GHz personal computer. This is cut to 42 seconds when only 10,000,000 simulations are performed, which achieves 5-digit convergence in 2 cases out of 3 and 4-digit convergence in one case. The numerical computation of Formula 1 takes approximately 0.2 second. The efficiency of the implementation of the WP algorithm could probably be improved, for instance by resorting to low discrepancy sequences instead of a pseudo random number generator, but this is not the subject of this article.

Table 1. Numerical evaluation of the survival probability of an arithmetic Brownian under a one-sided piecewise affine, time-homogeneous, absorbing boundary, as a function of volatility

	Formula 1	WP simulation algorithm 10,000,000 runs	WP simulation algorithm 100,000,000 runs
Volatility = 20%	0.275332974	0.275318288	0.275387164
Volatility = 50%	0.257810712	0.257763484	0.257885116
Volatility = 80%	0.191728749	0.191718319	0.191716445

2.3 Proof of Formula 1

Only sequences of upper boundaries are tackled, since the results for sequences of lower boundaries ensue by symmetry of Brownian paths.

Let us deal with process X_1 first. Let us denote by p the sought probability when the boundary is defined by $g_1\left(t\right)$

in (2.3). The random variables $X_1(t_1)$, $X_1(t_2)$ and $X_1(t_3)$ are absolutely continuous random variables that admit known Gaussian density functions. At time t_1, X_1 must be located below $a_1 + b_1 t_1$, k_1 and a_2, in order not to be absorbed; at time t_2, it must stand underneath the points $a_2 + b_2(t_2 - t_1), k_2$ and a_3; at time t_3, it must end below $a_3 + b_3(t_3 - t_2)$ and k_3. Hence, by conditioning with respect to $X_1(t_1)$, $X_1(t_2)$ and $X_1(t_3)$, and by using the weak Markov property of $\{X_1(t), t \geq 0\}$, one can come up with the following integral formulation of the problem :

$$p = \int_{x_1=-\infty}^{z_1} \int_{x_2=-\infty}^{z_2} \int_{x_3=-\infty}^{z_3} \mathbb{P}\left(\left(X_1(t_1) \in dx_1\right) \cap \left(X_1(t) < a_1 + b_1 t, \forall 0 \leq t \leq t_1\right)\right) \tag{2.22}$$

$$\mathbb{P}\left(\left(X_1(t_2) \in dx_2\right) \cap \left(X_1(t) < a_2 + b_2(t - t_1), \forall t_1 \leq t \leq t_2\right) \middle| X_1(t_1) \in dx_1\right)$$

$$\mathbb{P}\left(\left(X_1(t_3) \in dx_3\right) \cap \left(X_1(t) < a_3 + b_3(t - t_2), \forall t_2 \leq t \leq t_3\right) \middle| X_1(t_2) \in dx_2\right) dx_3 dx_2 dx_1$$

$$= \int_{x_1=-\infty}^{z_1 - b_1 t_1} \int_{x_2=-\infty}^{z_2 - b_2 t_2} \int_{x_3=-\infty}^{z_3 - b_3 t_3} f_1(x_1) f_2(x_1, x_2) f_3(x_2, x_3) dx_3 dx_2 dx_1 \tag{2.23}$$

where the functions $f_1(x_1)$, $f_2(x_1, x_2)$ and $f_3(x_2, x_3)$ are defined by :

$$f_1(x_1) = \mathbb{P}\left(Y(t_1) \in dx_1, \sup_{0 \leq t \leq t_1} Y(t) < a_1\right) \tag{2.24}$$

$$f_2(x_1, x_2) = \mathbb{P}\left(Y(t_2) \in dx_2, \sup_{t_1 \leq t \leq t_2} Y(t) < a_2 - b_2 t_1 \middle| Y(t_1) \in dx_1\right) \tag{2.25}$$

$$f_3(x_2, x_3) = \mathbb{P}\left(Y(t_3) \in dx_3, \sup_{t_2 \leq t \leq t_3} Y(t) < a_3 - b_3 t_2 \middle| Y(t_2) \in dx_2\right) \tag{2.26}$$

and the process $\{Y(t), t \geq 0\}$ is defined by :

$$dY(t) = \begin{cases} \mu_1 dt + \sigma dB(t), \ \forall 0 \leq t < t_1 \\ \mu_2 dt + \sigma dB(t), \ \forall t_1 \leq t \leq t_2 \\ \mu_3 dt + \sigma dB(t), \ \forall t_2 \leq t \leq t_3 \end{cases} \tag{2.27}$$

$$\mu_i = \mu - b_i, i \in \{1, 2, 3\}$$

The function $f_1(x_1)$ is obtained by differentiating the classical formula for the joint distribution of the maximum of Brownian motion with drift and its endpoint over the closed time interval $[0, t_1]$ (see, e.g., Karatzas and Shreve, 1991).

To obtain the functions $f_2(x_1, x_2)$ and $f_3(x_2, x_3)$, the following lemma is introduced.

Lemma 1 *Let* $\{Y(t), t \geq 0\}$ *be an arithmetic Brownian motion with constant drift* $\mu \in \mathbb{R}$ *and volatility* $\sigma \in \mathbb{R}_+$ *under a given probability measure* $\mathbb{P}$. *Let* t_i *and* t_j *be two non-random times such that* $t_j > t_i > t_0 = 0$.

Then, if x_i, x_j *and* h *are real constants with* $x_i < h$ *and* $x_j < h$ *, we have, at time* t_0 *:*

$$\mathbb{P}\left(Y(t_i) \leq x_i, Y(t_j) \leq x_j, \sup_{t_i \leq t \leq t_j} Y(t) \leq h\right) \tag{2.28}$$

$$= N_2\left[\frac{x_i - \mu t_i}{\sigma\sqrt{t_i}}, \frac{x_j - \mu t_j}{\sigma\sqrt{t_j}}; \sqrt{\frac{t_i}{t_j}}\right] - \exp\left(\frac{2\mu h}{\sigma^2}\right) N_2\left[\frac{x_i + \mu t_i}{\sigma\sqrt{t_i}}, \frac{x_j - 2h - \mu t_j}{\sigma\sqrt{t_j}}; -\sqrt{\frac{t_i}{t_j}}\right]$$

where the function $N_2[x_1, x_2; \rho]$ *is the bivariate standard normal cumulative distribution function with upper bounds* x_1 *and* x_2 *and correlation coefficient* ρ

Proof of lemma 1

$$\mathbb{P}\left(Y(t_i) \leq x_i, Y(t_j) \leq x_j, \sup_{t_i \leq t \leq t_j} Y(t) \leq h\right)$$

$$= \int_{-\infty}^{x_i}\int_{-\infty}^{x_j} \mathbb{P}\left(Y(t_i) \in dy, Y(t_j) \in dz\right)\mathbb{P}\left(\sup_{t_i \leq t \leq t_j} Y(t) \leq h \middle| Y(t_i) \in dy, Y(t_j) \in dz\right) dy dz \tag{2.29}$$

The pair $(Y(t_i), Y(t_j))$ is bivariate normal with correlation coefficient equal to $\sqrt{t_i / t_j}$. The conditional cumulative distribution function of $\sup_{t_i \leq t \leq t_j} Y(t)$ is given by Wang and Pötzelberger (1997) and can be written as follows :

$$\mathbb{P}\left(\sup_{t_i \leq t \leq t_j} Y(t) \leq h \middle| Y(t_i) \in dy, Y(t_j) \in dz\right) = 1 - \exp\left(\frac{2(h-y)(z-h)}{\sigma^2(t_j - t_i)}\right) \tag{2.30}$$

One can then solve the integration problem in (2.29) to obtain (2.28).

□

Differentiating the right-hand side of (2.29) and dividing by the density function of $Y(t_i)$, one can obtain :

$$\phi(x_i, x_j, h, \mu, \sigma, t_i, t_j) = \mathbb{P}\left(Y(t_j) \in dx_j, \sup_{t_i \leq t \leq t_j} Y(t) \leq h \middle| Y(t_i) \in dx_i\right)$$

$$= \exp\left(-\frac{1}{2}\left(\frac{x_j - x_i - \mu(t_j - t_i)}{\sigma\sqrt{(t_j - t_i)}}\right)^2\right) / \left(\sigma\sqrt{2\pi(t_j - t_i)}\right) \tag{2.31}$$

$$-\exp\left(\frac{2\mu(h - x_i)}{\sigma^2}\right)\left(\exp\left(-\frac{1}{2}\left(\frac{x_j - 2h + x_i - \mu(t_j - t_i)}{\sigma\sqrt{(t_j - t_i)}}\right)^2\right) / \left(\sigma\sqrt{2\pi(t_j - t_i)}\right)\right)$$

Plugging :

$$f_2(x_1, x_2) = \phi(x_1, x_2, a_2 - b_2 t_1, \mu_2, \sigma, t_1, t_2) \tag{2.32}$$

$$f_3(x_2, x_3) = \phi(x_2, x_3, a_3 - b_3 t_2, \mu_3, \sigma, t_2, t_3) \tag{2.33}$$

into (2.23), the rest of the proof, whose details are omitted, then consists in performing the necessary calculations to solve the triple integral in (2.22) and obtain the linear combination of eight trivariate cumulative distribution functions given by Formula 1. Elementary modifications provide the survival probability when the boundary is defined by the function $g_2(t)$ in (2.4). A basic application of Ito's lemma to $\ln(X_2(t) / X_2(0))$ shows that the survival probability of the process X_2 is given by the formula for the survival probability of the process X_1 with the two following adjustments : the drift coefficients become $\mu_i = \mu - b_i - \sigma^2 / 2$, $i \in \{1,2,3\}$ and k_i becomes $\ln(k_i / X_2(0))$.

□

2.4 Generalization to Higher Dimension

Similar exact formulae can be derived for $n > 3$ but they become more and more cumbersome. In general, for any $n \in \mathbb{N}$, they will involve a number 2^n of the $n-$ variate cumulative distribution functions of Gaussian type given by (2.16). For an arithmetic Brownian motion subject to the absorbing boundary $g_1(t)$, the integration problem to solve is the following :

$$\int_{D^n} \prod_{i=0}^{n-1} \phi(x_i, x_{i+1}, a_{i+1} - b_{i+1} t_i, \mu_{i+1}, \sigma, t_i, t_{i+1})\, dx_n dx_{n-1} ... dx_1 \tag{2.35}$$

where $x_0 = 0$ and

$$D^n = \left]-\infty, \min(a_1 + b_1 t_1, k_1, a_2)\right] \times \left]-\infty, \min(a_2 + b_2 t_2, k_2, a_3)\right] \times ... \times \left]-\infty, \min[a_{n-1} + b_{n-1} t_{n-1}, k_n]\right]$$

The main issue is numerical rather than analytical : evaluating the Gaussian integral given by (2.16) in high dimension is not easy. Rewriting it in terms of the standard normal cumulative distribution function of order n, as was done in (2.18) – (2.21) for $n = 3$, does not solve the numerical issue, as there does not exist an algorithm capable of evaluating the

n – variate standard normal cumulative distribution function with arbitrary precision in "reasonable" time as soon as $n = 4$. For more background on this topic, the reader may refer to Genz and Bretz (2009).

However, for $n = 4$, it can be verified that the following integration rule holds :

$$\Phi_4\left[x_1, x_2, x_3, x_4; \rho_1, \rho_2, \rho_3\right] \tag{2.36}$$

$$= \int_{y_2=-\infty}^{x_2} \int_{y_3=-\infty}^{\frac{x_3-\rho_2 y_2}{\sqrt{1-\rho_2^2}}} \frac{1}{2\pi} \exp\left(-\frac{\left(y_2^2+y_3^2\right)}{2}\right) N\left[\frac{x_1-\rho_1 y_2}{\sqrt{1-\rho_1^2}}\right] N\left[\frac{x_4-\rho_3\sqrt{1-\rho_2^2}\,y_3-\rho_3\rho_2 y_2}{\sqrt{1-\rho_3^2}}\right] dy_2 dy_3$$

More generally, the actual numerical dimension of the function Φ_n can always be reduced by a factor of 2 by using :

$$\Phi_n\left[x_1, x_2, \ldots, x_{n-1}, x_n; \rho_1, \ldots, \rho_{n-2}, \rho_{n-1}\right]$$

$$= \int_{y_2=-\infty}^{x_2} \int_{y_3=-\infty}^{x_3} \ldots \int_{y_{n-1}=-\infty}^{x_{n-1}} N\left[\frac{x_1-\rho_1 y_2}{\sqrt{1-\rho_1^2}}\right] N\left[\frac{x_n-\rho_{n-1} y_{n-1}}{\sqrt{1-\rho_{n-1}^2}}\right] \tag{2.37}$$

$$\frac{\exp\left(-\frac{y_2^2}{2} - \frac{1}{2\left(1-\rho_2^2\right)}\left(y_3-\rho_2 y_2\right)^2 \ldots - \frac{1}{2\left(1-\rho_{n-2}^2\right)}\left(y_{n-1}-\rho_{n-2} y_{n-2}\right)^2\right)}{\prod_{i=2}^{n-2}\sqrt{\left(1-\rho_i^2\right)}\left(2\pi\right)^{\frac{n-2}{2}}} dy_2 dy_3 \ldots dy_{n-1}$$

Given the smoothness of the integrand in (2.37), it should be possible to attain a combination of accuracy and efficiency that would be satisfactory for all practical purposes in "moderate" dimension, roughly speaking, by applying adaptive Gauss-Legendre quadrature combined with a Kronrod rule (Kronrod, 1964; Calvetti et al., 2000) to reduce the number of required iterations. These are standard numerical techniques and it is easy to find available code or built-in functions in the usual scientific computing software. The dimension n at which the use of a closed form formula analogous to Formula 1 ceases to be "competitive" with regard to a conditional Monte Carlo scheme should be numerically investigated. It must be emphasized that, even in "high" dimension, where Monte Carlo simulation becomes the method of last resort, exact formulae valid in lower dimension remain useful in two ways : they provide benchmarks with respect to which the accuracy of the numerical algorithms can be checked, and they can be used as control variates that substantially reduce the variance of the Monte Carlo estimates.

3. Survival Probability of an Arithmetic or a Geometric Brownian Motion under and above a Sequence of Two-sided affine or Exponential Boundaries over a Finite Time Interval

3.1 Definitions

Let us consider a finite time interval $\left[t_0, t_2\right]$ divided in two subintervals $\left[t_0, t_1\right]$ and $\left[t_1, t_2\right]$, $t_2 \geq t_1 \geq t_0 = 0$. The absorbing boundary now consists of two parallel upper and lower curves in each time interval, these curves being line segments when dealing with process X_1, or exponential curves when dealing with process X_2. More specifically, let $P_{[\ldots]}\left(\mu, \sigma, a_1, a_2, a_3, a_4, b_1, b_2, k_1, k_2, t_1, t_2\right)$ be defined as one of the following four cumulative distribution functions, where k_1 and k_2 are real constants :

$$\mathrm{P}_{[\mathrm{AUL1}]}\left(\mu,\sigma,a_1,a_2,a_3,a_4,b_1,b_2,k_1,k_2,t_1,t_2\right),\ a_1\in\mathbb{R}_+, a_2\in\mathbb{R}_-, \left(a_3,a_4,b_1,b_2,k_1,k_2\right)\in\mathbb{R}^8,$$
$$a_3>a_4, a_3>a_2+b_1t_1,\ t_2\geq t_1\geq t_0=0 \tag{3.1}$$

$$=\mathbb{P}\begin{pmatrix}\left(X_1(t)<a_1+b_1t,\forall 0\leq t\leq t_1\right)\cap\left(X_1(t)>a_2+b_1t,\forall 0\leq t\leq t_1\right)\cap X_1(t_1)<k_1\\ \cap\left(X_1(t)<a_3+b_2(t-t_1),\forall t_1\leq t\leq t_2\right)\cap\left(X_1(t)>a_4+b_2(t-t_1),\forall t_1\leq t\leq t_2\right)\cap X_1(t_2)<k_2\end{pmatrix}$$

$$\mathrm{P}_{[\mathrm{AUL2}]}\left(\mu,\sigma,a_1,a_2,a_3,a_4,b_1,b_2,k_1,k_2,t_1,t_2\right),\ a_1\in\mathbb{R}_+, a_2\in\mathbb{R}_-, \left(a_3,a_4,b_1,b_2,k_1,k_2\right)\in\mathbb{R}^8,$$
$$a_3>a_4, a_3+b_2t_1>a_2+b_1t_1,\ t_2\geq t_1\geq t_0=0 \tag{3.2}$$

$$=\mathbb{P}\begin{pmatrix}\left(X_1(t)<a_1+b_1t,\forall 0\leq t\leq t_1\right)\cap\left(X_1(t)>a_2+b_1t,\forall 0\leq t\leq t_1\right)\cap X_1(t_1)<k_1\\ \cap\left(X_1(t)<a_3+b_2t,\forall t_1\leq t\leq t_2\right)\cap\left(X_1(t)>a_4+b_2t,\forall t_1\leq t\leq t_2\right)\cap X_1(t_2)<k_2\end{pmatrix}$$

$$\mathrm{P}_{[\mathrm{GUL1}]}\left(\mu,\sigma,a_1,a_2,a_3,a_4,b_1,b_2,k_1,k_2,t_1,t_2\right),\ a_1\in\mathbb{R}_+, a_2\in\mathbb{R}_-, \left(a_3,a_4,b_1,b_2,k_1,k_2\right)\in\mathbb{R}^8,$$
$$a_3>a_4, a_3>a_2+b_1t_1,\ t_2\geq t_1\geq t_0=0 \tag{3.3}$$

$$=\mathbb{P}\begin{pmatrix}\left(X_2(t)<X_2(0)\exp\left(a_1+b_1t\right),\forall 0\leq t\leq t_1\right)\cap\left(X_2(t)>X_2(0)\exp\left(a_2+b_1t\right),\forall 0\leq t\leq t_1\right)\\ \cap X_2(t_1)<k_1\cap\left(X_2(t)<X_2(0)\exp\left(a_3+b_2(t-t_1)\right),\forall t_1\leq t\leq t_2\right)\\ \cap\left(X_2(t)>X_2(0)\exp\left(a_4+b_2(t-t_1)\right),\forall t_1\leq t\leq t_2\right)\cap X_2(t_2)<k_2\end{pmatrix}$$

$$\mathrm{P}_{[\mathrm{GUL2}]}\left(\mu,\sigma,a_1,a_2,a_3,a_4,b_1,b_2,k_1,k_2,t_1,t_2\right),\ a_1\in\mathbb{R}_+, a_2\in\mathbb{R}_-, \left(a_3,a_4,b_1,b_2,k_1,k_2\right)\in\mathbb{R}^8,$$
$$a_3>a_4, a_3+b_2t_1>a_2+b_1t_1,\ t_2\geq t_1\geq t_0=0 \tag{3.4}$$

$$=\mathbb{P}\begin{pmatrix}\left(X_2(t)<X_2(0)\exp\left(a_1+b_1t\right),\forall 0\leq t\leq t_1\right)\cap\left(X_2(t)>X_2(0)\exp\left(a_2+b_1t\right),\forall 0\leq t\leq t_1\right)\\ \cap X_2(t_1)<k_1\cap\left(X_2(t)<X_2(0)\exp\left(a_3+b_2t\right),\forall t_1\leq t\leq t_2\right)\\ \cap\left(X_2(t)>X_2(0)\exp\left(a_4+b_2t\right),\forall t_1\leq t\leq t_2\right)\cap X_2(t_2)<k_2\end{pmatrix}$$

3.2 Statement of Formula 2

Formula 2 *Let* $\mathrm{P}_{[\ldots]}\left(\mu,\sigma,a_1,a_2,a_3,a_4,b_1,b_2,k_1,k_2,t_1,t_2\right)$ *be defined as in Subsection 3.1. Then,*

$$\mathrm{P}_{[\ldots]}\left(\mu,\sigma,a_1,a_2,a_3,a_4,b_1,b_2,k_1,k_2,t_1,t_2\right)$$

$$=\sum_{m=-\infty}^{\infty}\sum_{n=-\infty}^{\infty}\exp\left(\frac{2\mu_1}{\sigma^2}m\theta+\frac{2\mu_2}{\sigma^2}n\phi\right) \tag{3.5}$$

$$\left\{\begin{array}{l} N_2\left[\frac{\beta_1-2m\theta+\lambda_1}{\sigma\sqrt{t_1}},\frac{\beta_3-2m\theta-2n\phi+\lambda_2}{\sigma\sqrt{t_2}};\sqrt{\frac{t_1}{t_2}}\right]\\ -N_2\left[\frac{\beta_2-2m\theta+\lambda_1}{\sigma\sqrt{t_1}},\frac{\beta_3-2m\theta-2n\phi+\lambda_2}{\sigma\sqrt{t_2}};\sqrt{\frac{t_1}{t_2}}\right]\\ -N_2\left[\frac{\beta_1-2m\theta+\lambda_1}{\sigma\sqrt{t_1}},\frac{\beta_4-2m\theta-2n\phi+\lambda_2}{\sigma\sqrt{t_2}};\sqrt{\frac{t_1}{t_2}}\right]\\ +N_2\left[\frac{\beta_2-2m\theta+\lambda_1}{\sigma\sqrt{t_1}},\frac{\beta_4-2m\theta-2n\phi+\lambda_2}{\sigma\sqrt{t_2}};\sqrt{\frac{t_1}{t_2}}\right]\end{array}\right\}$$

$$-\sum_{m=-\infty}^{\infty}\sum_{n=-\infty}^{\infty}\exp\left(\frac{2\mu_1}{\sigma^2}m\theta+\frac{2\mu_2}{\sigma^2}\left(\beta_4-n\phi-2m\theta\right)+\frac{2}{\sigma^2}\left(\mu_2^2-\mu_1\mu_2\right)t_1\right)$$

$$\left\{\begin{array}{l} N_2\left[\frac{\beta_1-2m\theta+\lambda_3}{\sigma\sqrt{t_1}},\frac{\beta_3-2\beta_4+2m\theta+2n\phi+\lambda_4}{\sigma\sqrt{t_2}};-\sqrt{\frac{t_1}{t_2}}\right]\\ -N_2\left[\frac{\beta_2-2m\theta+\lambda_3}{\sigma\sqrt{t_1}},\frac{\beta_3-2\beta_4+2m\theta+2n\phi+\lambda_4}{\sigma\sqrt{t_2}};-\sqrt{\frac{t_1}{t_2}}\right]\\ -N_2\left[\frac{\beta_1-2m\theta+\lambda_3}{\sigma\sqrt{t_1}},\frac{-\beta_4+2m\theta+2n\phi+\lambda_4}{\sigma\sqrt{t_2}};-\sqrt{\frac{t_1}{t_2}}\right]\\ +N_2\left[\frac{\beta_2-2m\theta+\lambda_3}{\sigma\sqrt{t_1}},\frac{-\beta_4+2m\theta+2n\phi+\lambda_4}{\sigma\sqrt{t_2}};-\sqrt{\frac{t_1}{t_2}}\right]\end{array}\right\}$$

$$-\sum_{m=-\infty}^{\infty}\sum_{n=-\infty}^{\infty}\exp\left(\frac{2\mu_1}{\sigma^2}\left[a_2-m\theta\right]+\frac{2\mu_2}{\sigma^2}\left[n\phi\right]\right)$$

$$\left\{\begin{array}{l} N_2\left[\frac{\beta_1-2a_2+2m\theta+\lambda_1}{\sigma\sqrt{t_1}},\frac{\beta_3-2a_2+2m\theta-2n\phi+\lambda_2}{\sigma\sqrt{t_2}};\sqrt{\frac{t_1}{t_2}}\right]\\ -N_2\left[\frac{\beta_2-2a_2+2m\theta+\lambda_1}{\sigma\sqrt{t_1}},\frac{\beta_3-2a_2+2m\theta-2n\phi+\lambda_2}{\sigma\sqrt{t_2}};\sqrt{\frac{t_1}{t_2}}\right]\\ -N_2\left[\frac{\beta_1-2a_2+2m\theta+\lambda_1}{\sigma\sqrt{t_1}},\frac{\beta_4-2a_2+2m\theta-2n\phi+\lambda_2}{\sigma\sqrt{t_2}};\sqrt{\frac{t_1}{t_2}}\right]\\ +N_2\left[\frac{\beta_2-2a_2+2m\theta+\lambda_1}{\sigma\sqrt{t_1}},\frac{\beta_4-2a_2+2m\theta-2n\phi+\lambda_2}{\sigma\sqrt{t_2}};\sqrt{\frac{t_1}{t_2}}\right]\end{array}\right\}$$

$$+\sum_{m=-\infty}^{\infty}\sum_{n=-\infty}^{\infty}\exp\left(\frac{2\mu_1}{\sigma^2}\left(a_2-m\theta\right)+\frac{2\mu_2}{\sigma^2}\left(\beta_4-n\phi+2m\theta-2a_2\right)+\frac{2}{\sigma^2}\left(\mu_2^2-\mu_1\mu_2\right)t_1\right)$$

$$\left\{\begin{array}{l} N_2\left[\dfrac{\beta_1 - 2a_2 + 2m\theta + \lambda_3}{\sigma\sqrt{t_1}}, \dfrac{\beta_3 - 2\beta_4 + 2a_2 + 2n\phi - 2m\theta + \lambda_4}{\sigma\sqrt{t_2}}; -\sqrt{\dfrac{t_1}{t_2}}\right] \\ -N_2\left[\dfrac{\beta_2 - 2a_2 + 2m\theta + \lambda_3}{\sigma\sqrt{t_1}}, \dfrac{\beta_3 - 2\beta_4 + 2a_2 + 2n\phi - 2m\theta + \lambda_4}{\sigma\sqrt{t_2}}; -\sqrt{\dfrac{t_1}{t_2}}\right] \\ -N_2\left[\dfrac{\beta_1 - 2a_2 + 2m\theta + \lambda_3}{\sigma\sqrt{t_1}}, \dfrac{-\beta_4 + 2a_2 + 2n\phi - 2m\theta + \lambda_4}{\sigma\sqrt{t_2}}; -\sqrt{\dfrac{t_1}{t_2}}\right] \\ +N_2\left[\dfrac{\beta_2 - 2a_2 + 2m\theta + \lambda_3}{\sigma\sqrt{t_1}}, \dfrac{-\beta_4 + 2a_2 + 2n\phi - 2m\theta + \lambda_4}{\sigma\sqrt{t_2}}; -\sqrt{\dfrac{t_1}{t_2}}\right] \end{array}\right\}$$

where the following notations hold :

$$\mu_1 = (\mu - b_1)\left(\mathbb{I}_{\{P_{[\ldots]}=P_{[AUL1]}\}} + \mathbb{I}_{\{P_{[\ldots]}=P_{[AUL2]}\}}\right) + \left(\mu - \frac{\sigma^2}{2} - b_1\right)\left(\mathbb{I}_{\{P_{[\ldots]}=P_{[GUL1]}\}} + \mathbb{I}_{\{P_{[\ldots]}=P_{[GUL2]}\}}\right)$$

$$\mu_2 = (\mu - b_2)\left(\mathbb{I}_{\{P_{[\ldots]}=P_{[AUL1]}\}} + \mathbb{I}_{\{P_{[\ldots]}=P_{[AUL2]}\}}\right) + \left(\mu - \frac{\sigma^2}{2} - b_2\right)\left(\mathbb{I}_{\{P_{[\ldots]}=P_{[GUL1]}\}} + \mathbb{I}_{\{P_{[\ldots]}=P_{[GUL2]}\}}\right)$$

$$\theta = a_1 - a_2$$

$$\phi = a_3 - a_4$$

$$\beta_1 = \left(\min\left(a_1 + b_1t_1, k_1, a_3\right) - b_1t_1\right)\mathbb{I}_{\{P_{[\ldots]}=P_{[AUL1]}\}} + \left(\min\left(a_1 + b_1t_1, k_1, a_3 + b_2t_1\right) - b_1t_1\right)\mathbb{I}_{\{P_{[\ldots]}=P_{[AUL2]}\}}$$

$$+\left(\min\left(a_1 + b_1t_1, \ln\left(k_1 / X_2(0)\right), a_3\right) - b_1t_1\right)\mathbb{I}_{\{P_{[\ldots]}=P_{[GUL1]}\}}$$

$$+\left(\min\left(a_1 + b_1t_1, \ln\left(k_1 / X_2(0)\right), a_3 + b_2t_1\right) - b_1t_1\right)\mathbb{I}_{\{P_{[\ldots]}=P_{[GUL2]}\}}$$

$$\beta_2 = \left(\max\left(a_2 + b_1t_1, a_4\right) - b_1t_1\right)\left(\mathbb{I}_{\{P_{[\ldots]}=P_{[AUL1]}\}} + \mathbb{I}_{\{P_{[\ldots]}=P_{[GUL1]}\}}\right)$$

$$+\left(\max\left(a_2 + b_1t_1, a_4 + b_2t_1\right) - b_1t_1\right)\left(\mathbb{I}_{\{P_{[\ldots]}=P_{[AUL2]}\}} + \mathbb{I}_{\{P_{[\ldots]}=P_{[GUL2]}\}}\right)$$

$$\beta_3 = \left(\min\left(a_3 + b_2\left(t_2 - t_1\right), k_2\right) - b_2t_2\right)\mathbb{I}_{\{P_{[\ldots]}=P_{[AUL1]}\}} + \left(\min\left(a_3 + b_2t_2, k_2\right) - b_2t_2\right)\mathbb{I}_{\{P_{[\ldots]}=P_{[AUL2]}\}}$$

$$+\left(\min\left(a_3 + b_2\left(t_2 - t_1\right), \ln\left(k_2 / X_2(0)\right)\right) - b_2t_2\right)\mathbb{I}_{\{P_{[\ldots]}=P_{[GUL1]}\}}$$

$$+\left(\min\left(a_3 + b_2t_2, \ln\left(k_2 / X_2(0)\right)\right) - b_2t_2\right)\mathbb{I}_{\{P_{[\ldots]}=P_{[GUL2]}\}}$$

$$\beta_4 = \left(a_4 - b_2t_1\right)\left(\mathbb{I}_{\{P_{[\ldots]}=P_{[AUL1]}\}} + \mathbb{I}_{\{P_{[\ldots]}=P_{[GUL1]}\}}\right) + a_4\left(\mathbb{I}_{\{P_{[\ldots]}=P_{[AUL2]}\}} + \mathbb{I}_{\{P_{[\ldots]}=P_{[GUL2]}\}}\right)$$

$$\lambda_1 = -\mu_1t_1,\ \lambda_2 = -\mu_1t_1 - \mu_2\left(t_2 - t_1\right),\ \lambda_3 = -\mu_1t_1 + 2\mu_2t_1,\ \lambda_4 = \mu_1t_1 - 2\mu_2t_1 - \mu_2\left(t_2 - t_1\right)$$

End of Formula 2.

A few numerical values are reported in Table 2 for various levels of volatility and other parameters fixed as follows :

$\mu = 0.01$, $t_1 = 0.25$, $t_2 = 0.5$, $t_3 = 1$, $k_1 = 0$, $k_2 = 0.02$, $a_1 = 0.36$, $b_1 = 0.15$, $a_2 = -0.42$,

$b_2 = 0.15$, $a_3 = a_1 + b_1 t_1$, $b_3 = -0.12$, $a_4 = a_2 + b_2 t_1$, $b_4 = -0.12$. A comparison is made with results obtained using the algorithm by Pötzelberger and Wang (2001), denoted by PW, specifically designed for two-sided boundaries. The infinite double series in Formula 2 is truncated to summation operators ranging from $m = -4$ to $m = 4$ and from $n = -4$ to $n = 4$, since adding more terms does not modify the obtained numerical results at least up to the 8th digit. Computational time is approximately 0.3 second. In general, the infinite double series can be truncated in a simple manner by setting a convergence threshold such that no further terms are added once the difference between two successive finite sums becomes smaller than that prespecified level.

Table 2. Numerical evaluation of the survival probability of an arithmetic Brownian under a two-sided piecewise affine, time-homogeneous, absorbing boundary, as a function of volatility

	Formula 2	PW simulation algorithm 10,000,000 runs	PW simulation algorithm 100,000,000 runs
Volatility = 20%	0.377958716	0.377923023	0.377971427
Volatility = 50%	0.12468422	0.124786191	0.124625324
Volatility = 80%	0.02140513	0.021276911	0.021382631

3.3 Proof of Formula 2

Let us consider the calculation of $P_{[AUL1]}$. Since the upper and the lower sides of the boundary grow at the same rate in each time interval, i.e. at the rate b_1 both from below and from above in $[t_0, t_1]$ and at the rate b_2 both from below and from above in $[t_1, t_2]$, the same technique can be applied as in the beginning of the proof of Formula 1, i.e. the initial boundary crossing problem is turned into one where the boundary and the drift of the process become piecewise constant. Hence, denoting by p the sought probability, the problem can be formulated as follows :

$$p = \int_{\beta_2}^{\beta_1}\int_{\beta_4}^{\beta_3} f_1(x_1) f_2(x_1, x_2)\, dx_2 dx_1 \tag{3.6}$$

where the functions $f_1(x_1)$ and $f_2(x_1, x_2)$ are defined by :

$$f_1(x_1) = \mathbb{P}\left(\sup_{0 \le t \le t_1} Y(t) < a_1, \inf_{0 \le t \le t_1} Y(t) > a_2 \,\middle|\, Y(t_1) \in dx_1\right) \tag{3.7}$$

$$f_2(x_1, x_2) = \mathbb{P}\left(\sup_{t_1 \le t \le t_2} Y(t) < a_3 - b_2 t_1, \inf_{t_1 \le t \le t_2} Y(t) > a_4 - b_2 t_1, Y(t_2) \le x_2 \,\middle|\, Y(t_1) \in dx_1\right) \tag{3.8}$$

and the process $\{Y(t), t \ge 0\}$ is defined by :

$$dY(t) = \begin{cases} \mu_1 dt + \sigma dB(t), \forall 0 \le t < t_1 \\ \mu_2 dt + \sigma dB(t), \forall t_1 \le t \le t_2 \end{cases} \tag{3.9}$$

$$\mu_i = \mu - b_i \,,\ i \in \{1,2\}$$

The function $f_1(x_1)$ results from the differentiation of the classical formula for the joint distribution of the maximum, the minimum and the endpoint of a Brownian motion (see, e.g., Cox & Miller, 1965). To obtain $f_2(x_1, x_2)$, the following lemma is introduced.

Lemma 2 *Let* $Y(t)$ *be an arithmetic Brownian motion with constant drift* $\mu \in \mathbb{R}$ *and volatility* $\sigma \in \mathbb{R}_+$ *under a given probability measure* $\mathbb{P}$. *Let* q *be the conditional probability defined, at time* $t_0 = 0$, *by :*

$$q = \mathbb{P}\left(\sup_{t_i \le t \le t_j} Y(t) \le b,\ \inf_{t_i \le t \le t_j} Y(t) \ge a, Y(t_j) \le x_j \middle| Y(t_i) \in dx_i \right) \tag{3.10}$$

where x_i, x_j, a *and* b *are real constants such that :* $b > a$, $b \ge x_i > a$, $b \ge x_j > a$, *and* t_i *and* t_j *are two non-random times such that :* $t_j > t_j \ge 0$. *Then,*

$$q = \sum_{n=-\infty}^{\infty} \exp\left(\frac{2n\mu(b-a)}{\sigma^2}\right)\left[N\left(\frac{x_j - x_i - \mu(t_j - t_i) - 2n(b-a)}{\sigma\sqrt{t_j - t_i}}\right) - N\left(\frac{a - x_i - \mu(t_j - t_i) - 2n(b-a)}{\sigma\sqrt{t_j - t_i}}\right)\right] \tag{3.11}$$

$$- \sum_{n=-\infty}^{\infty} \exp\left(\frac{2\mu(a - x_i - n(b-a))}{\sigma^2}\right)\left[N\left(\frac{x_j - 2a + x_i - \mu(t_j - t_i) + 2n(b-a)}{\sigma\sqrt{t_j - t_i}}\right) - N\left(\frac{-a + x_i - \mu(t_j - t_i) + 2n(b-a)}{\sigma\sqrt{t_j - t_i}}\right)\right]$$

Proof of lemma 2

$$\mathbb{P}\left(Y(t_i) \le x_i,\ \sup_{t_i \le t \le t_j} Y(t) \le b,\ \inf_{t_i \le t \le t_j} Y(t) \ge a, Y(t_j) \le x_j \right) \tag{3.12}$$

$$= \int_{y=a}^{x_i} \int_{z=a}^{x_j} \mathbb{P}\left(Y(t_i) \in dy, Y(t_j) \in dz\right) \mathbb{P}\left(\sup_{t_i \le t \le t_j} Y(t) \le b,\ \inf_{t_i \le t \le t_j} Y(t) \ge a \middle| Y(t_i) \in dy, Y(t_j) \in dz \right) dz dy$$

The following result can be found in Guillaume (2010) :

$$\mathbb{P}\left(\sup_{t_i \le t \le t_j} Y(t) \le b,\ \inf_{t_i \le t \le t_j} Y(t) \ge a \middle| Y(t_i) \in dy, Y(t_j) \in dz \right) \tag{3.13}$$

$$= \sum_{n=-\infty}^{\infty} \exp\left(\frac{2n(b-a)(z - y - n(b-a))}{\sigma^2 (t_j - t_i)}\right) - \exp\left(\frac{2(b - y - n(b-a))(z - b + n(b-a))}{\sigma^2 (t_j - t_i)}\right)$$

Plugging (3.13) into (3.12) yields :

$$\mathbb{P}\left(Y(t_i) \le x_i, \sup_{t_i \le t \le t_j} Y(t) \le b, \inf_{t_i \le t \le t_j} Y(t) \ge a, Y(t_j) \le x_j\right)$$

$$= \sum_{n=-\infty}^{\infty} \exp\left(\frac{2n\mu(b-a)}{\sigma^2}\right)\left\{N_2\left[\frac{x_i - \mu t_i}{\sigma\sqrt{t_i}}, \frac{x_j - 2n(b-a) - \mu t_j}{\sigma\sqrt{t_j}}; \sqrt{\frac{t_i}{t_j}}\right]\right. \quad (3.14)$$

$$-N_2\left[\frac{x_i - \mu t_i}{\sigma\sqrt{t_i}}, \frac{a - 2n(b-a) - \mu t_j}{\sigma\sqrt{t_j}}; \sqrt{\frac{t_i}{t_j}}\right] - N_2\left[\frac{a - \mu t_i}{\sigma\sqrt{t_i}}, \frac{x_j - 2n(b-a) - \mu t_j}{\sigma\sqrt{t_j}}; \sqrt{\frac{t_i}{t_j}}\right]$$

$$\left.+N_2\left[\frac{a - \mu t_i}{\sigma\sqrt{t_i}}, \frac{a - 2n(b-a) - \mu t_j}{\sigma\sqrt{t_j}}; \sqrt{\frac{t_i}{t_j}}\right]\right\}$$

$$-\sum_{n=-\infty}^{\infty} \exp\left(\frac{2\mu(a - n(b-a))}{\sigma^2}\right)\left\{N_2\left[\frac{x_i + \mu t_i}{\sigma\sqrt{t_i}}, \frac{x_j - 2a + 2n(b-a) - \mu t_j}{\sigma\sqrt{t_j}}; -\sqrt{\frac{t_i}{t_j}}\right]\right.$$

$$-N_2\left[\frac{x_j + \mu t_i}{\sigma\sqrt{t_i}}, \frac{-a + 2n(b-a) - \mu t_j}{\sigma\sqrt{t_j}}; -\sqrt{\frac{t_i}{t_j}}\right] - N_2\left[\frac{a + \mu t_i}{\sigma\sqrt{t_i}}, \frac{x_j - 2a + 2n(b-a) - \mu t_j}{\sigma\sqrt{t_j}}; -\sqrt{\frac{t_i}{t_j}}\right]$$

$$\left.+N_2\left[\frac{a + \mu t_i}{\sigma\sqrt{t_i}}, \frac{-a + 2n(b-a) - \mu t_j}{\sigma\sqrt{t_j}}; -\sqrt{\frac{t_i}{t_j}}\right]\right\}$$

The interchange between summation and integral is a straightforward application of Tonelli's theorem to non-negative measurable functions, where the measures are the counting measure on $\mathbb{Z}$ and the Lebesgue measure on $\mathbb{R}$. Lemma 2 ensues by differentiating (3.14) and dividing by the density function of $Y(t_i)$.

□

Applying Lemma 2, the function $f_2(x_1, x_2)$ can be plugged in (3.6). Then, performing the necessary calculations, Formula 2 can be obtained.

□

4. Conclusion

In this paper, new formulae were obtained for the probability of absorption of generalised Brownian motion through sequences of affine or exponential one-sided or two-sided boundaries. It was shown that the method could be applied to higher numbers of successive one-sided boundaries. However, such an extension may not be commendable in the case of two-sided boundaries, as the resulting analytical formulae will involve a quickly increasing number of summation operators, thus slowing down the process of numerical convergence.

References

Anderson, T. W. (1960). A Modification of the Sequential Probability Ratio Test to Reduce the Sample Size. *Ann. Math. Statist.*, *31*, 165-197. http://dx.doi.org/10.1214/aoms/1177705996

Calvetti, D., Golub, G.H., Gragg, W.B., & Reichel, L. (2000). Computation of Gauss-Kronrod Quadrature Rules. *Math. Comput.*, *69*, 1035-1052. http://dx.doi.org/10.1090/S0025-5718-00-01174-1

Choi, C., & Nam, D. (2003). Some Boundary Crossing Results for Linear Diffusion Processes. *Stat. Probab. Letters*, *62*, 281-291. http://dx.doi.org/10.1016/S0167-7152(03)00015-4

Cox, D.R. & Miller, H.D. (1965). *The Theory of Stochastic Processes*. London, Methuen Daniels, H.E. (1996). Approximating the First Crossing Time Density for a Curved Boundary, *Bernoulli*, *2*, 133-143.

http://dx.doi.org/10.2307/3318547

Doob, J. L. (1949). Heuristic Approach to the Kolmogorov-Smirnov Theorem, *Ann. Math. Statist.*, *20*, 393-403

Genz, A. (2004). Numerical Computation of Rectangular Bivariate and Trivariate Normal and *t* Probabilities. *Stat. Comput.*, *14*, 151-160. http://dx.doi.org/10.1214/aoms/1177729991

Genz, A., & Bretz, F. (2009). *Computation of Multivariate Normal and t Probabilities*, Berlin Heidelberg: Springer-Verlag

Groeneboom, P. (1989). Brownian Motion with a Parabolic Drift and Airy Functions, *Prob. Theory Rel. Fields*, *81*, 79-109. http://dx.doi.org/10.1007/BF00343738

Guillaume, T. (2010). Step double barrier options. *J. derivatives, 18* (1), 59-79. http://dx.doi.org/10.3905/jod.2010.18.1.059

Jeanblanc, M., Yor, M., & Chesney, M. (2009). *Mathematical Methods for Financial Markets*, London: Springer-Verlag

Karatzas, I., & Shreve, S. (1991). *Brownian Motion and Stochastic Calculus*. New York : Springer-Verlag Kronrod, A.S. (1964), (Russian), *Doklady Akad. Nauk SSSR, 154*, 283-286.

Levy, P. (1948). *Processus Stochastiques et Mouvement Brownien*, Paris : Gauthier-Villars Novikov, A. Frishling, V. , & Kordzakhia, N. (1999). Approximations of Boundary Crossing Probabilities for a Brownian Motion, *J.Appl.Prob.*, *34*, 1019-1030. http://dx.doi.org/10.1239/jap/1032374752

Pötzelberger, K., & Wang, L. (2001). Boundary Crossing Probability for Brownian Motion, *J.Appl.Prob.*, *38*, 152-164. http://dx.doi.org/10.1239/jap/996986650

Salminen, P. (1988). On the First Hitting Time and Last Exit Time for a Brownian Motion to/from a Moving Boundary, *Adv.Appl.Prob.*, *20*, 411-426. http://dx.doi.org/10.2307/1427397

Scheike, T.H. (1992). A boundary crossing result for Brownian motion. *J.Appl.Prob.*, *29*, 448-453. http://dx.doi.org/10.2307/3214581

Siegmund, D. (1986). Boundary Crossing Probabilities and Statistical Applications, *Ann. Statist.*, *14*, 361-404. http://dx.doi.org/10.1214/aos/1176349928

Wang, L., & Pötzelberger, K. (1997). Boundary Crossing Probability for Brownian Motion and General Boundaries, *J.Appl.Prob.*, *34*, 54-65. http://dx.doi.org/10.2307/3215174

Wang, L., & Pötzelberger, K. (2007). Crossing Probabilities for Diffusion Processes with Piecewise Continuous Boundaries, *Methodol.Comput. Appl.Probab.*, *9*, 21-40. http://dx.doi.org/10.1007/s11009-006-9002-6

5

Linear Hybrid Deterministic Dynamic Modeling for Time-to-Event Processes: State and Parameter Estimations

E. A. Appiah[1] & G. S. Ladde[1]

[1] Department of Mathematics and Statistics, University of South Florida, Tampa, FL, USA.

Correspondence: G.S. Ladde, Department of Mathematics and Statistics, University of South Florida, 4202 East Fowler Avenue, CMC 342, Tampa, FL 33620-5700, USA. E-mail: gladde@usf.edu

Abstract

In this work, we initiate an innovative alternative modeling approach for time-to-event dynamic processes. The proposed approach is composed of the following basic components: (1) development of continuous-time state of dynamic process, (2) introduction of discrete-time dynamic intervention process, (3) formulation of continuous and discrete-time interconnected dynamic system, (4) utilizing Euler-type discretized schemes, and (5) introduction of conceptual and computational state and parameter estimation procedures. The presented approach is motivated by state and parameter estimation of time-to-event processes in biological, chemical, engineering, epidemiological, medical, military, multiple-markets and social dynamic processes under the influence of discrete-time intervention processes. The role and scope of our approach is exhibited by presenting several well-known hazard/risk rate and survival function estimates as special cases. Moreover, conceptual algorithms are illustrated by time-series data sets under the influence of intervention processes.

Keywords: Kaplan-Meier estimator, hazard/risk rate function, piecewise exponential estimator, time-to-event closed process, totally discrete-time hybrid system

1. Introduction

In the survival and reliability data analysis, the main interest is focused on a nonnegative random variable, say T which describes a time-to-event process characterizing an occurrence of time until a certain event. Historically well-known time-to-event processes are deaths in population dynamic and component failures in mechanical systems (Kalbfleisch & Prentice, 2011). The human mobility, electronic communications, technological changes, advancements in engineering, medical, and social sciences have diversified the role and scope of time-to-event processes in cultural, epidemiological, financial, military and social sciences (Ladde, 2015; Chandra & Ladde, 2014; Ladde & Ladde, 2012; Wanduku & Ladde, 2011; Anis, 2009).

The study of survival analysis rests on the concept of time-to-event. The mathematical statistics development of time-to-event analysis is based on the probabilistic approach and the concept of hazard rate. Moreover, the time-to-event is described by the closed form expressions of survival function that is determined by the concept of hazard rate (Kalbfleisch & Prentice, 2011; Lawless, 2011; Miller, 2011). We note that in general, hazard rate is unknown. This leads to a problem of determining hazard rate function. This is based on a feasible approach of collecting data set for the time-to-event processes in biological, chemical, engineering, epidemiological, medical, multiple-markets and social sciences. The hazard/risk rate and survival function estimation problems in the survival and reliability analysis are centered around the idea of "right censored data" (Miller, 2011). In fact, the common conventional understanding for resolving ties between censored and uncensored observations is adopted by shifting the censored observations slightly to the left of uncensored observations (Whittemore & Keller, 1983). In short, the items/individuals/objects in a given sample are decomposed into two mutually exclusive groups, namely, (a) deaths/failure /removal/non-operational/inactive, and (b) censored/losses/ withdrawals.

In the survival and reliability data analysis, parametric and nonparametric methods are applied to estimate the hazard/risk rate and survival functions (Kalbfleisch & Prentice, 2011; Lawless, 2011). A parametric approach is based on the assumption that the underlying survival distribution belongs to some specific family of distributions (e.g. normal, Weibull, exponential). On the other hand, a nonparametric approach is centered around the best-fitting member of a class of survival distribution functions (Kaplan & Meier, 1958). Moreover, Kaplan-Meier(KME) (Kaplan & Meier, 1958) and Nelson-Aalen (Aalen, 1978; Nelson, 1969) type nonparametric approach do not assume neither distribution class, nor closed-form distributions. In fact, it just depends on a data. The Kaplan-Meier and Nelson-Aalen type nonparametric estimation approaches are systematically analyzed by our totally discrete-time hybrid dynamic modeling process.

In the existing literature (Kalbfleisch & Prentice, 2011; Lawless, 2011), the closed-form expression for a survival function is based on the usage of probabilistic analysis approach. The closed-form representation of the survival function coupled with mathematical statistics method (parametric approach) is used to estimate both survival and hazard/risk rate functions. In fact, the parametric approach/model has advantages of simplicity, the availability of likelihood based inference procedures and the ease of use for a description, comparison, prediction, or decision (Lawless, 2011). In this work, we initiate an innovative alternative approach for modeling time-to-event dynamic processes. This approach leads to the development for estimating survival and hazard/risk rate functions. The presented approach is motivated by a simple observation regarding the probabilistic definition of the survival function (Kalbfleisch & Prentice, 2002). Moreover, this approach does not require a knowledge of either a closed-form solution distribution or a class of distributions.

Historically, exponential distributions have been widely used in analyzing survival/reliability data (Lawless, 2011; Davis, 1952). This was partly due to the mathematical simplicity and the availability of simple statistical methods. An application of the exponential model with covariates to medical survival data was initiated in Feigl and Zelen (1965). The assumption of a constant hazard/risk rate function is very restrictive. In fact, it is often violated. This is due to the fact that in some real life applications, sudden changes in the hazard rate at unknown times can be encountered due to a major maintenance in a mechanical system or a new treatment procedure in medical sciences (Anis, 2009). For example, usually a machine component functions with a constant hazard/risk rate function λ_1, until it suffers a shock. After this shock, the component may continue to operate but with a different constant hazard/risk rate function λ_2. In the medical field, there is usually a high initial risk after a major operation which settles down to a lower constant long-term risk rate (Anis, 2009). This type of change could occur in multiple times. In view of this, one is often interested in detecting the locations of such changes and estimating the sizes of the detected changes. Recently, several authors (Han, Schell & Kim 2014; He & Su, 2013; Fang & Su, 2011, Goodman, Li, & Tiwari, 2011) have proposed estimators based on change point hazard models. A Bayesian approach for estimating the piecewise exponential distribution (Gamerman, 1994) and estimating the grid of time-points (Demarqui, Loschi, & Colosimo, 2008) for the piecewise exponential model are also available in the literature. In order to incorporate these types of sudden changes (intervention process) in the hazard rate function, we modify the developed continuous state dynamic model to an interconnected hybrid dynamic model that is composed of both continuous time state and discrete time state (intervention process) dynamic processes.

Employing the total time on test (TTT) for undefined censored data beyond the last observation, the idea of Piecewise Exponential Estimator (PEXE) of a survival function was introduced by (Kitchin, Langberg, & Proschan, 1980) and applied for estimating life distribution from incomplete data. The PEXE has been modified to address the issues regarding the presence of ties in the data by Whittemore and Keller (1983).

The comparison of the PEXE with the KME (Kim & Proschan, 1991) exhibits the advantage of the PEXE over the KME. For example, the PEXE is a continuous survival function. Moreover, it exhibits the complete information that is coming from the censored data. Using a total time test and the PEXE based approach, the estimators of the hazard/risk rate and cumulative distribution functions on the left closed pairwise consecutive failure time intervals are determined in Kulasekera and White (1996). The PEXE is further extended by Malla and Mukerjee (2010) with an exponential tail extension in the framework of the Kaplan and Meier (1958) nonparametric estimator approach. Under the presented dynamic framework, we develop the PEXE and new PEXE of Malla and Mukerjee (2010) types in a systematic and unified way. In short, the presented novel approach incorporates all the existing features such as: incomplete data, issues regarding the ties, exponential tail extensions in the framework of Kaplan and Meier (1958), and so on in a coherent manner.

The organization of the presented work is as follows. In Section 2, recognizing the classical probabilistic analysis model of time-to-event as a dynamic process, we initiate a linear hybrid deterministic dynamic model for time-to-event processes. Moreover, a fundamental mathematical result that provides a basis for interconnected continuous-discrete-time and totally discrete-time dynamic processes, is developed. Utilizing the dynamic model and the main result developed in Section 2, basic conceptual analytic algorithms and its special cases for interconnected continuous-discrete-time and totally discrete-time linear hybrid dynamic models for time-to-event processes are presented in Section 3. In Section 4, we outline conceptual computational schemes. In Section 5, we present a very general conceptual and computational algorithm for estimating a hazard/risk rate function for multiple censoring times between consecutive failure times. These general results include the presented results in Section 4 as special cases. In Section 6, conceptual computational and simulation algorithms are developed. The developed computational schemes are applied to estimate hazard/risk rate and survival functions in a systematic and unified way. Moreover, several well-known results are exhibited as special cases. A few conclusions are drawn in Section 7 to exhibit the role and scope of linear hybrid deterministic modeling for time-to-event processes. Moreover, further extensions and generalizations to both deterministic and stochastic nonlinear and non-stationary hybrid modeling for time-to-event processes are currently underway. In addition, currently, a complex time-to-event dynamic analysis is also undertaken by the authors. These results will appear elsewhere. Finally, proofs of theorems and corollaries in Sections 2, 3, 4 and 5 are outlined in supplementary Section 8.

2. Linear Hybrid Dynamic Modeling of Time-to-event Process

In this section, based on the probabilistic definition of the survival function, we develop a model for time-to-event dynamic processes. From the probabilistic definition of the survival function (Kalbfleisch & Prentice, 2011; Lawless, 2011; Miller, 2011) and differential calculus (Apostol, 1967), we recognize that

$$\lambda(t)\Delta t \approx \frac{S(t) - S(t+\Delta t)}{S(t)}, \tag{1}$$

where S and λ are survival and hazard/risk rate functions, respectively. Moreover, from (1) and differential calculus (Apostol, 1976), we have

$$dS = -\lambda(t)S\,dt, \quad S(t_0) = S_0, \quad t \in [t_0, \infty), \tag{2}$$

where dS is a differential of a survival function S. In fact, (2) is a differential equation, and it is an initial value problem (IVP) (Ladde & Ladde, 2012). Based on continuous-time dynamic modeling (Ladde & Ladde, 2012), (2) represents a continuous-time linear dynamic model of time-to-event processes. In fact, we consider time-to-event processes to be probabilistic dynamic processes. The state of the process is represented by survival/infective/operational/radical and its complementary state, failure/removal/death/non-operational/normal, and it is measured by a probability distribution function. Employing Newtonian modeling approach, the instantaneous rate of change of survival state is directly proportional to the magnitude of the survival. The negative sign in (2) signifies that the state of survival is decaying/diminishing/decreasing. λ is a positive constant of proportionality. In general, it is a function of time. This is because of the fact that in general, the time-to-event processes are non-stationary. The solution of (2) on the interval $[t_0, \infty)$ is given by

$$S(t) = S_0 \exp\left[-\Lambda(t)\right], \tag{3}$$

where

$$\Lambda(t) = \int_0^t \lambda(u)du, \tag{4}$$

and it is the cumulative hazard/risk rate function.

Remark 2.1. If $\lambda(t) = \lambda$ for $t \geq 0$, $t_0 = 0$, $S(0) = 1$, then (3) reduces to the following well-known exponential distribution function:

$$S(t) = \exp[-\lambda t], \quad t \in [0, \infty), \tag{5}$$

and a complementary state of the survival state of time-to-event process is represented by

$$F(t) = 1 - S(t) = 1 - \exp[-\lambda t], \quad t \in [0, \infty),$$

and it is referred as a failure distribution function. Furthermore, we note that survival state dynamic model (2) signifies that the time-to-event process is closed (Rosen, 1970), that is, $S(t) + F(t) = 1$. It is analogous to epidemiological dynamic modeling process without removal (Ladde & Ladde, 2012; Wanduku & Ladde, 2011).

The presented motivational observation coupled with the introduction of the idea of continuous-time state dynamic process (2) operating under the discrete-time intervention processes further leads to a development of a linear hybrid dynamic model (Ladde & Ladde, 2012) for time-to-event processes. It is known (Ladde & Ladde, 2012) that many real world time-to-event dynamic processes are subject to intervention processes (internal or external). Therefore, it is natural that time-to-event dynamic processes undergo state adjustment processes. This causes a modification of the presented state dynamic processes that are described by simple state dynamic model (2). We note that the dynamic state adjustment processes are caused by periodic changes in science, technology, medicine, culture, socio-economic, environmental conditions and general behavior.

In the following, we introduce a type of hazard/risk rate function. Moreover, using dynamic approach, we present a development of PEXE (Kitchin et al., 1980; Kim & Proschan, 1991) in a systematic and unified way.

Definition 2.1. Let $\tau_0 < \tau_1 < \tau_2 < \ldots < \tau_k < \tau_{k+1}$ be a given partition of a time interval $[\tau_0, \mathscr{T}]$, with $\tau_0 = 0$ and $\tau_{k+1} = \infty$. Let $\lambda_1, \lambda_2, \ldots, \lambda_{k+1}$ be model parameters. A hazard/risk rate function for a nonnegative random variable T that characterizes time-to-event processes, is of the following form:

$$\lambda(t) = \sum_{i=1}^{k+1} = \lambda_j I_{[\tau_{j-1},\tau_j)}(t)\,, \quad t \in \mathbf{R}_+ = [0,\infty)\,, \tag{6}$$

where λ_j are positive real numbers for $j \in I(1, k+1)$, $(I(1,l) = \{1, 2, \ldots, l\})$; $I_{[\tau_{j-1},\tau_j)}$ is the characteristic function with respect to $[\tau_{j-1}, \tau_j)$. Moreover, T is said to have a piecewise constant hazard function.

Definition 2.2. $\prod_{i|\tau_j \le t}$ denotes the symbol for a product of objects for all positive integers $i \in I(1,\infty)$ that satisfy the conditions $\tau_i \le \tau_j$ and $\tau_j \le t < \tau_{j+1}$ for some $j \in I(1,n)$ and for $\tau_i, \tau_{j-1}, \tau_{j+1}, t \in [\tau_0, \mathscr{T}]$.

From Definition 2.1, we recognize that the sudden changes in the hazard/risk rate function are encountered due to various types of intervention processes (internal or external) (Ladde & Ladde, 2012). This causes to interrupt the current continuous-time state dynamic process (2). Following the linear hybrid dynamic model (Ladde & Ladde, 2012), a modified version of time-to-event dynamic model (2) is represented by:

$$\begin{cases} dS = -\lambda(t) S\, dt\,, & S(\tau_{j-1}) = S_{j-1}\,, \quad t \in [\tau_{j-1}, \tau_j)\,, \\ S_j = S(\tau_j^-, \tau_{j-1}, S_{j-1})\,, & S(\tau_0) = S_0\,, \quad j \in I(1, k+1)\,, \end{cases} \tag{7}$$

where $S(\tau_j^-) = S(\tau_j^-|\lambda, \tau_{j-1}, S_{j-1})$ describes a very simpler form of intervention process generated at an intervention time τ_j; τ_j^- stands for $t \in [\tau_{j-1}, \tau_j)$, that is less than τ_j and very close to τ_j. We note that system (7) is interconnected hybrid dynamic system composed of both continuous and discrete time state dynamic systems. Imitating the procedure described in Ladde and Ladde (2012), the solution process of the IVP (7) is as follows:

$$S(t, \tau_{j-1}, S_{j-1}|\lambda) = S_{j-1} \exp\left[-\int_{\tau_{j-1}}^{t} \lambda(u) du\right]\,, \text{ for all } t \in [\tau_{j-1}, \tau_j)\,. \tag{8}$$

Furthermore, the solution process of the overall time-to-event dynamic process (7) on $[\tau_0, \mathscr{T})$ is

$$S(t, \tau_{j-1}, S_0|\lambda) = S_0 \prod_{m=1}^{j-1} \exp\left[-\int_{\tau_{m-1}}^{\tau_m} \lambda(u) du\right] \exp\left[-\int_{\tau_{j-1}}^{t} \lambda(u) du\right], \; t \in [\tau_0, \mathscr{T})\,, j \in I(1, k+1)\,. \tag{9}$$

Remark 2.2. From (7) and (8), we note that the solution process (8) is indeed PEXE (Kitchin et al., 1980; Kim & Proschan, 1991).

In the following, we present a very simple fundamental auxiliary result that would be used, subsequently. Moreover, it exhibits an analytic unified bridge and basis for (7) and its complete discrete-time version.

Theorem 2.1. *Let $\{\tau_j\}_0^n$ be a partition of $[0, \mathscr{T}]$ and let β be a monotonic nondecreasing function defined by*

$$\beta(t) = \begin{cases} 0, & t \in [\tau_{j-1}, \tau_j)\,, \\ 1, & t = \tau_j\,, \end{cases} \tag{10}$$

for each $j \in I(1,n)$. Let x be a state dynamic process in biological, engineering, epidemiological, human, medical, military, physical and social sciences under the influence of time-to-event processes. Let x be described by:

$$\begin{cases} dx = [-\alpha(t)x + \gamma(t)]\, d\beta(t), & t \in [\tau_{j-1}, \tau_j)\,, \\ x_j = (1 - \alpha_j) x(\tau_j^-, \tau_{j-1}, x_{j-1}) + \gamma_j\,, & x(\tau_0) = x_0\,, \end{cases} \tag{11}$$

where α and γ are real-valued continuous functions defined on $[0,\infty)$; $\alpha_j = \alpha(\tau_j)$ and $\gamma_j = \gamma(\tau_j)$. Then

$$x(t) = \prod_{k|\tau_j \le t} (1 - \alpha_k) x_0 + \sum_{i=1}^{j-1} \Phi(t, \tau_i)\gamma_i + \gamma_j\,, \quad \text{for} \quad t \ge \tau_0\,, \tag{12}$$

where j is the largest integer so that $\tau_j \le t < \tau_{j+1}$, $\tau_k \le \tau_j$ and

$$\Phi(t, \tau_i) = \prod_{\tau_i \le \tau_j \le t} (1 - \alpha_i)\,, \quad \Phi(\tau_i, \tau_i) = 1 \quad \text{for} \quad i \in I(0, n)\,.$$

Proof. The proof of Theorem 2.1 is given in the supplementary Section 8.

Remark 2.3. From (10), the hybrid dynamic system (11), is equivalent to the hybrid dynamic system

$$\begin{cases} \mathrm{d}x = 0\,\mathrm{d}t\,, \quad x(\tau_{j-1}) = x_{j-1}\,, \quad t \in [\tau_{j-1}, \tau_i)\,, \\ x_j = (1-\alpha_j)x(\tau_j^-, \tau_{j-1}, x_{j-1}) + \gamma_j\,, \quad x(\tau_0) = x_0\,, \end{cases} \tag{13}$$

for $j \in I(1,n)$. The solution process of (13) is represented in (12).

In the following, we present a couple of special cases of Theorem 2.1. These special cases illustrate a systematic way for exhibiting the existing results in Kaplan and Meier (1958), Nelson (1969), Aalen (1978) and Malla and Mukerjee (2010) in the framework of presented innovative dynamic approach.

Corollary 2.1. *If functions α and γ in Theorem 2.1 are replaced by functions λ and $\gamma = 0$, then (12) reduces to*

$$x(t) = \prod_{j|\tau_j \le t} (1-\lambda_j)x_0\,, \quad t \ge \tau_0\,. \tag{14}$$

Corollary 2.2. *If $\alpha = 0$ and $x_0 = 0$ in Theorem 2.1, then the conclusion of Theorem 2.1 reduces to*

$$x(t) = \sum_{i|\tau_{j-1} \le t} \gamma_i\,, \quad t \ge \tau_0 \quad \textit{and} \quad t \in [\tau_{j-1}, \tau_j)\,. \tag{15}$$

In the following, we present a definition of cumulative jump process (Malla & Mukerjee, 2010) in the framework of hybrid dynamic model.

Example 2.1. Let $T_1, T_2, \ldots, T_n$ be discrete failure times for the discrete-time event process, and $0 = a_0 < a_1 \le a_2 \le \ldots \le a_m$ be jumps of a survival function in magnitude. Then the dynamic for the cumulative jump process is as described in Corollary 2.2, and its solution process is exhibited in (15).
In this example, applying Corollary 2.2 in the context of $\gamma_0 = 0$, $\gamma_i = a_i$, the cumulative jump process is represented by

$$x(t) = \begin{cases} A_{j-1} = \sum\limits_{i=1}^{j-1} a_i\,, & \text{for} \quad t \in [\tau_{j-1}, \tau_j)\,, \\ A_j = \sum\limits_{i=1}^{j} a_i\,, & t = \tau_j\,. \end{cases} \tag{16}$$

From (16), we recognize that the cumulative jump defined in Malla and Mukerjee (2010) is indeed recast as the discrete time intervention process described by the hybrid dynamic system illustrated in Corollary 2.2 at the discrete time τ_j for $j \in I(1,m)$ with $\gamma_0 = a_0 = 0$ and $\gamma_i = a_i$.

Example 2.2. Under the conditions of Example 2.1, the magnitude of the survival function at the failure times is represented by

$$S(t) = \begin{cases} 1 - A_{j-1}\,, & \text{for} \quad t \in [\tau_{j-1}, \tau_j)\,, \\ 1 - A_j\,, & t = \tau_j, \quad j \in I(1,m)\,, \end{cases} \tag{17}$$

where $\gamma_0 = 1$ and $x(\tau_j) = A_j$. The $S(t)$ in (17) is the magnitude of the survival function determined by the cumulative jump (Malla & Mukerjee, 2010) process described in Example 2.1.

Remark 2.4. We remark that the continuous-time dynamic model can be exhibited by the cumulative hazard/risk rate function. In fact, from (2), we have

$$\mathrm{d}\ln S = -\lambda(t)\mathrm{d}t\,, \quad \ln S(\tau_0) = S_0\,. \tag{18}$$

Based on the solution processes of (2) and (7), the solution process of (18) can be represented as:

$$-\ln\left[\frac{S(t)}{S(\tau_0)}\right] = \Lambda(t, \tau_0, S_0|\lambda) = \int_{\tau_0}^{t} \lambda(u)\mathrm{d}u\,. \tag{19}$$

and

$$-\ln\left[\frac{S(t)}{S(\tau_0)}\right] = \Lambda(t, \tau_0|\lambda) = \sum_{m=1}^{j-1} \int_{\tau_{m-1}}^{\tau_m} \lambda(u)\mathrm{d}u + \int_{\tau_{j-1}}^{t} \lambda(u)\mathrm{d}u\,,\ t \in [\tau_{j-1}, \tau_j)\,. \tag{20}$$

respectively. Furthermore, we set $x = \ln S$, $S_0 = 1$ and $\gamma(t) = -\lambda(t)$ where S and λ are defined in (18). From Corollary 2.2, we have

$$\ln S(t) = -\Lambda(t)\,, \tag{21}$$

where $\Lambda(t) = \sum_{i|\tau_i \le t} \lambda_i$ is a cumulative hazard function.

Remark 2.5. We remark that if x is replaced by survival function, S in Corollary 2.1, and x and γ are replaced by S and λ in Corollary 2.2, then (14) and (15) are replaced by:

$$S(t) = \prod_{j|\tau_j \le t} (1 - \lambda_j) S_0\,, \quad t \ge \tau_0 \tag{22}$$

and

$$S(t) = \sum_{i|\tau_i \le t} \lambda_i\,, \quad t \ge \tau_0\,, \tag{23}$$

respectively. Moreover, (22) is the solution process of the discrete-time dynamic system described by Corollary 2.1. Furthermore, dynamic system outlined in Corollary 2.1 provides an innovative alternative approach for finding the discrete-time survival function (Kaplan & Meier, 1958) in a systematic manner.

We utilize the above presented concepts and results in subsequent sections in a systematic and unified way.

3. Fundamental Results for Continuous and Discrete-Time to Event Dynamic Processes

In this section, we utilize hybrid dynamic model (7) and fundamental analytic Theorem 2.1 for time-to-event process to develop a general fundamental result. The developed result provides basic analytic and computational tools for estimating survival state and parameters. The presented approach also provides a systematic and unified way of estimating the parameters and survival functions.

Let $x(t)$ be the total number of units/individuals operating/alive (or survivals) at time t, for $t \in [\tau_0, \mathcal{T}]$. It is described by (11). Let λ and S be hazard/risk rate and survival functions of the units/patients/infectives/species/individuals, respectively. Employing a dynamic model for number of units/species/ individuals coupled with survival state dynamic model (2) or (7), we present an interconnected hybrid dynamic model below.

Following the argument used in developing dynamic models (Ladde & Ladde, 2012), we introduce the following interconnected system of differential equations:

$$\begin{cases} dS = -\lambda(t) S\, dt\,, & t \in [\tau_{j-1}, \tau_j)\,, \\ S_j = (1 - \beta_j) S(\tau_j^-, \tau_{j-1}, S_{j-1})\,, & S(\tau_0) = 1\,, \\ dx = (-\alpha(t)x + \gamma(t))d\beta(t)\,, & x(\tau_0) = x_0\,, \quad t \in [\tau_{j-1}, \tau_j)\,, \\ x_j = (1 - \alpha_j) x(\tau_j^-, \tau_{j-1}, x_{j-1}) + \gamma_j\,, \end{cases} \tag{24}$$

Remark 3.1. We outline a few important observations that exhibit the role and scope of dynamic approach to illustrate the existing results (Han et al., 2014; Kim & Proschan, 1991; Thaler, 1984; Kitchin et al., 1980; Kaplan & Meier, 1958) as special cases.

(i) Dynamic system (24) in the context of (13) (Remark 2.3) is reduced to

$$\begin{cases} dS = -\lambda(t) S\, dt\,, & t \in [\tau_{j-1}, \tau_j)\,, \\ S_j = (1 - \beta_j) S(\tau_j^-, \tau_{j-1}, S_{j-1})\,, & S(\tau_0) = 1\,, \\ dx = 0\, dt\,, \quad x(\tau_0) = x_0\,, & t \in [\tau_{j-1}, \tau_j)\,, \\ x_j = (1 - \alpha_j) x(\tau_j^-, \tau_{j-1}, x_{j-1}) + \gamma_j\,. \end{cases} \tag{25}$$

(ii) From Corollary 2.1 in the context of Remark 2.5, in particular (22), system (24) becomes:

$$\begin{cases} dS = 0\, dt\,, & t \in [\tau_{j-1}, \tau_j)\,, \\ S_j = (1 - \lambda_j) S_{j-1}\,, \\ dx = 0\, dt\,, & x(\tau_0) = x_0\,, \\ x_j = (1 - \alpha_j) x_{j-1} + \gamma_j\,. \end{cases} \tag{26}$$

We note that (26) is a special version of (24). In addition, we refer to system (26) as a totally discrete-time hybrid dynamic system.

Now, we are ready to present a basic result regarding continuous and discrete time interconnected dynamic of survival species or objects or thoughts operating under the time-to-event intervention processes. Prior to the formulation of the fundamental result, we introduce a concept of number of survivals.

Definition 3.1. Let z be a function defied by $z(t) = x(t)S(t)$, where S and x are solution process of (24) for $t \in [\tau_0, \mathscr{T}]$. Moreover, for each $t \in [\tau_0, \mathscr{T}]$, $z(t)$ stands for the number of survivals at t under an influence of time-to-event process.

Theorem 3.1. *Let (x, S) be a solution process of (24). Then the interconnected hybrid dynamic population model for time-to-event process (24) and corresponding intervention iterative process are described by:*

$$\begin{cases} dz = -\lambda(t)z dt, & z(\tau_{j-1}) = z_{j-1}, \quad \textit{for} \quad t \in [\tau_{j-1}, \tau_j), \quad j \in I(1,k), \\ z(\tau_j) = (1-\alpha_j)(1-\beta_j)z(\tau_j^-) + \gamma_j(1-\beta_j), \end{cases} \tag{27}$$

and

$$z(\tau_j) = (1-\lambda(\tau_j)\Delta\tau_j)(1-\alpha_j)(1-\beta_j)z(\tau_{j-1}) + \gamma_j(1-\beta_j). \tag{28}$$

respectively, where z is defined in Definition 3.1 and $\Delta\tau_j = \tau_j - \tau_{j-1}$ for $j \in I(1,k)$.

Proof. For the detailed proof of Theorem 3.1, the readers are encouraged to read the supplementary Section 8.

In the following, we present a few special/trivial cases that exhibit existing results in the framework of hybrid dynamic of time-to-event interconnected system.

Corollary 3.1. *Let us consider a very special/trivial case of Theorem 3.1 as follows:*

$$\begin{cases} dS = -\lambda(t)S\,dt, & t \geq \tau_0, \\ dx = 0\,dt, & t \geq \tau_0, \\ x(\tau_j) = x(\tau_j^-, \tau_{j-1}, x_{j-1}), & x(\tau_0) = x_0, \quad j \in I(1,k). \end{cases} \tag{29}$$

Applying Theorem 3.1 and using (27) and (28), (29) reduces to

$$\begin{cases} dz = -\lambda(t)z dt, & z(\tau_{j-1}) = z_{j-1}, \quad t \in [\tau_{j-1}, \tau_j), \\ z(\tau_j) = z(\tau_j^-, \tau_{j-1}, z_{j-1}) = z(\tau_{j-1}), & j \in I(1,k), \end{cases} \tag{30}$$

and

$$z(\tau_j) = \left(1-\lambda(\tau_j)\Delta\tau_j\right)z(\tau_{j-1}). \tag{31}$$

Corollary 3.2. *Let us consider a special case of (24) as follows:*

$$\begin{cases} dS = -\lambda(t)S\,dt, & S(\tau_{j-1}) = S_{j-1}, \quad t \in [\tau_{j-1}, \tau_j), \\ S(\tau_j) = S(\tau_j^-, \tau_{j-1}, S_{j-1}), \end{cases} \tag{32}$$

where a_j is defined in Example 2.1. Then applying Euler-type discretization scheme (Atkinson, 2008) on $[\tau_{j-1}, \tau_j^-]$, yields

$$S(\tau_j^-) - S(\tau_{j-1}) = -\lambda(\tau_{j-1})\Delta\tau_j S(\tau_{j-1}). \tag{33}$$

Moreover, from (32) and (33), we have

$$S(\tau_j) - S(\tau_{j-1}) = -\lambda(\tau_j)\Delta\tau_j S(\tau_{j-1}). \tag{34}$$

Corollary 3.3. *Under the assumptions of Theorem 3.1 in the context of Remark 3.1(ii), (26) becomes:*

$$\begin{cases} dz = 0\,dt, & z(\tau_{j-1}) = z_{j-1}, \quad t \in [\tau_{j-1}, \tau_j), \\ z(\tau_j) = (1-\lambda_j)(1-\alpha_j)z_{j-1} + \gamma_j, \end{cases} \tag{35}$$

and

$$z(\tau_j) = (1-\lambda_j)(1-\alpha_j)z(\tau_{j-1}) + \gamma_j. \tag{36}$$

This corollary is indeed a totally discrete-time version of hybrid dynamic system operating under discrete-time intervention process.

Using Definition 3.1 and the discrete-time iterative process (28), we introduce a couple of definitions.

Definition 3.2. Let τ_{j-1} and τ_j be a pair of consecutive observation times belonging to $[0, \mathscr{T}]$. $z(\tau_{j-1})$ stands for the number of survivals at the time τ_{j-1} for each $j \in I(1,k)$. Moreover, $z(\tau_{j-1})$ is the number of survivals under observation over the sub-interval of time $[\tau_{j-1}, \tau_j)$. $z(\tau_{j-1})\Delta\tau_j$ is the amount of time spent under observation/testing/evaluation by $z(\tau_{j-1})$ survivals over the length $\Delta\tau_j$ of time interval $[\tau_{j-1}, \tau_j)$.

Definition 3.3. For $j \in I(1,k)$, $z(\tau_{j-1}) - z(\tau_j)$ stands for the change in number of survivals over the interval of time $[\tau_{j-1}, \tau_j]$ of length $\Delta\tau_j$.

Remark 3.2. The discrete-time processes (28), (31), (34) and (36) are referred as our numerical schemes with respect to interconnected hybrid dynamic models for a survival population dynamic processes. Moreover, from (28), we will introduce three more special numerical schemes, namely, time-to-event: (i) failure/death/removal/infective, (ii) censored/withdrawn, and (iii) admission/joining/susceptible/relapsed processes. We further note that the presented numerical schemes allow "ties" with deaths/failure or censored/quiting process. In addition, the population under the presented observation/-supervision process includes the patient/objects population as a special case.

(i) For each $j \in I(1,k)$, let us assume that either τ_{j-1} and τ_j are consecutive failure/death/removal/infective times of individual/machine/species, or τ_{j-1} and τ_j are censored and failure times, respectively. For $\alpha_j = \gamma_j = \beta_j = 0$, the numerical scheme (28) for failure/death/removal/infective/etc process data set is described by

$$z(\tau_j) = (1 - \lambda(\tau_j)\Delta\tau_j)z(\tau_{j-1}), \tag{37}$$

and hence

$$z(\tau_j) - z(\tau_{j-1}) = -\lambda(\tau_j)z(\tau_{j-1})\Delta\tau_j, \tag{38}$$

where τ_{j-1} is either the failure or censored time.

Moreover, $\alpha_j = \gamma_j = \beta_j = 0$ in (28) coupled with (94) is equivalent to the Kaplan and Meier (1958) assumption, namely,

$$x(\tau_j^-) - x(\tau_j) = \text{the number of deaths at } \tau_j .$$

That is

$$z(\tau_{j-1}) - z(\tau_j^-) = 0 \quad \text{and} \quad z(\tau_j) = z(\tau_j^+) .$$

This implies that $z(t)$ is left discontinuous and right continuous at τ_j.

(ii) Let us assume that either τ_{j-1} and τ_j are consecutive censored times, or τ_{j-1} and τ_j are failure and censored times, respectively. For $\alpha_j = \beta_j = 0$, and γ_j^c stands for the number of censored objects/infectives/etc at a time τ_j. The numerical scheme (28) for censored/listed/identified process data set is described by

$$z(\tau_j) = \left(1 - \lambda(\tau_j)\Delta\tau_j\right)z(\tau_{j-1}) - \gamma_j^c, \tag{39}$$

where τ_{j-1} is either a failure or censored time.

Thus

$$z(\tau_j) - z(\tau_{j-1}) = -\lambda(\tau_j)z(\tau_{j-1})\Delta\tau_j - \gamma_j^c \tag{40}$$

Again, we note that $\alpha_j = \beta_j = 0, \gamma_j^c$, in the context of (94) is equivalent to the Kaplan and Meier (1958) assumption, namely,

$$z(\tau_j) = z(\tau_j^-) \quad \text{and} \quad z(\tau_j) - z(\tau_j^+) = \gamma_j^c .$$

This implies that $z(t)$ is left continuous and right discontinuous at τ_j.

(iii) Let us assume that τ_{j-1} is either failure or censored time, and τ_j is a joining/admitting/relapsing time. For $\alpha_j = 0$ and γ_j^a denoting the number of objects/infectives that joined the observation process at time τ_j. The numerical scheme (28) for admission/joining/sustainable/recruiting/relapsing process is

$$z(\tau_j) = \left(1 - \lambda(\tau_j)\Delta\tau_j\right)z(\tau_{j-1}) + \gamma_j^a . \tag{41}$$

The scheme determined by $\alpha_j = 0$ in (28) with (94) and the addition γ_j^a in (41) is equivalent to $z(\tau_j) - z(\tau_j^-) = \gamma_j^a$ and $z(\tau_j) = z(\tau_j^+)$.

(iv) Remarks (i), (ii) and (iii) remain valid for the iterative processes (28), (31) and (36).

(I) For $\alpha_j = 0 = \beta_j = \gamma_j$ in (28), (34) reduces to (38); for $\alpha_j = 0 = \beta_j = \gamma_j$, (36) reduces to $z(\tau_j) = (1-\lambda_j)z(\tau_{j-1})$.

(II) For $\alpha_j = 0 = \beta_j$ and $\gamma_j = -\gamma_j^c$ in (28), (28) reduces to (40); for $\alpha_j = 0 = \lambda_j$ and $\gamma_j = -\gamma_j^c$, (36) becomes

$$z(\tau_j) - z(\tau_{j-1}) = (1-\lambda_j)z(\tau_{j-1}) - \gamma_j^c\,. \tag{42}$$

(III) For $\alpha_j = 0 = \beta_j$ and $\gamma_j = \gamma_j^a$ in (28), and $\alpha_j = 0 = \lambda_j$ and $\gamma_j = \gamma_j^a$ in (36), (28) reduces to (41), and (36) reduces to

$$z(\tau_j) - z(\tau_{j-1}) = (1-\lambda_j)z(\tau_{j-1}) + \gamma_j^a. \tag{43}$$

4. Estimations of Risk Rate and Survival Functions

Now, we are ready to find an estimate for the hazard/risk rate and survival functions for interconnected continuous and discrete-time survival state dynamic processes. For the sake of completeness and clarity, we first introduce a couple of definitions.

Definition 4.1. For $j \in I(1,k)$, let τ_{j-1} and τ_j be consecutive change times under continuous-time state survival dynamic process. The parameter estimate at τ_j is defined by the quotient of change of objects over the consecutive time change interval $[\tau_{j-1}, \tau_j)$ and the total time spent by the objects under observation over the time interval of length $\Delta\tau_j$.

Definition 4.2. For $j \in I(1,k)$, let τ_{j-1} and τ_j be consecutive change times for discrete-time state survival dynamic process. The parameter estimate at τ_j is defined by the quotient of the change in the number of survival state over the consecutive time change interval $[\tau_{j-1}, \tau_j)$ and the number of objects at the immediate past time, that is, either the change time or the censored time.

Remark 4.1. We observe that the Definitions 4.1 and 4.2 are consistent with each other. This statement can be justified in the context of discrete-time iterative scheme (95) and the continuous and discrete-time hybrid-type descriptions of survival state dynamic model (25) and totally discrete-time hybrid dynamic system (26).

Now, we are ready to present a main result regarding parameter and survival state estimation problems. This result includes several existing results as special cases. In the following, we simply state a conceptual computational algorithm. The detailed proof is given in the supplementary section.

Theorem 4.1. *Let us assume that the conditions of Theorem 3.1 in the context of Remarks 3.1 and 3.2(i),(ii) are satisfied.*

(a) For $j \in I(1,k)$, if τ_{j-1} and τ_j are consecutive risk/failure/removal/death/non-operational times in $[\tau_0, \mathscr{T}]$ then an estimate for the hazard/risk rate function at τ_j is determined by:

$$\hat{\lambda}(\tau_j) = \frac{z(\tau_{j-1}) - z(\tau_j)}{z(\tau_{j-1})\Delta\tau_j}\,, \tag{44}$$

and an estimate for the hazard/risk rate function is

$$\hat{\lambda}(t) = \hat{\lambda}(\tau_j)\,, \quad \textit{for} \quad t \in [\tau_{j-1}, \tau_j) \quad \textit{and} \quad j \in I(1,k)\,. \tag{45}$$

(b) For $j \in I(1,k)$, if $\tau_{j-1} < \tau_j^c < \tau_j$, and τ_j^c is censored time between a pair of consecutive failure times τ_{j-1} and τ_j in $[\tau_0, \mathscr{T})$, then,

(i) a change in the number of items/subjects/thoughts that are under observation over the subinterval $[\tau_{j-1}, \tau_j)$ of the time interval of study $[\tau_0, \mathscr{T}]$ is

$$z(\tau_{j-1}) - z(\tau_j) - \gamma_j^c\,. \tag{46}$$

(ii) a total amount of time spent under the observation/testing/evaluation of $z(\tau_{j-1}) - z(\tau_j) - \gamma_j^c$ items/patients/infectives/radicals/subjects over the time interval $[\tau_{j-1}, \tau_j)$ is

$$z(\tau_{j-1})\Delta\tau_j^c + z(\tau_j^c)\Delta\tau_{jc}\,, \quad \Delta\tau_{jc} = \tau_j - \tau_j^c. \tag{47}$$

(iii) an estimate for the hazard/risk rate function at τ_j *is defined as:*

$$\hat{\lambda}(\tau_j) = \frac{z(\tau_{j-1}) - z(\tau_j) - \gamma_j^c}{z(\tau_{j-1})\Delta\tau_j^c + z(\tau_j^c)\Delta\tau_{jc}}, \tag{48}$$

and an estimate for the hazard/risk rate function is

$$\hat{\lambda}(t) = \hat{\lambda}(\tau_j), \quad \textit{for} \quad t \in [\tau_{j-1}, \tau_j) \quad \textit{and} \quad j \in I(1,k). \tag{49}$$

(iv) Moreover, an estimate for the survival function in (24) *is*

$$\hat{S}(t) = S_0 \exp\left[\sum_{m=1}^{j-1} \hat{\lambda}_m(\tau_m - \tau_{m-1}) + \hat{\lambda}_j\left(t - \tau_{j-1}\right)\right], \; t \in [\tau_{j-1}, \tau_j). \tag{50}$$

Remark 4.2. We note that if $\tau_j^c = \tau_j$ in Theorem 4.1(b), then we have "ties" between censored and failure times. In this case, $\Delta\tau_j^c = \Delta\tau_j$ and $\Delta\tau_{jc} = 0$. From this, (47) and (48) reduce to

$$z(\tau_{j-1})\Delta\tau_j, \tag{51}$$

and

$$\hat{\lambda}(\tau_j) = \frac{z(\tau_{j-1}) - z(\tau_j) - \gamma_j^c}{z(\tau_{j-1})\Delta\tau_j} \quad \text{for} \quad j \in I(1,k). \tag{52}$$

This observation justifies Remark 3.2 regarding the mixed "ties."

In the following, we exhibit the role and scope of Theorem 4.1. This is achieved by presenting the well-known hazard/risk rate and survival functions as special cases.

Corollary 4.1. *Let us assume that conditions of Corollary 3.3 in the context of Remark 3.2(iv)(I) and (II) are satisfied.*

(a) For $j \in I(1,k)$, *if* τ_{j-1} *and* τ_j *are consecutive risk/failure times in* $[\tau_0, \mathscr{T}]$, *then employing Remark 3.2(iv)(I) and Definitions 3.2, 3.3 and 4.2, an estimate for the risk/hazard rate function at* τ_j *is determined by:*

$$\hat{\lambda}(\tau_j) = \frac{z(\tau_{j-1}) - z(\tau_j)}{z(\tau_{j-1})}, \tag{53}$$

and

$$\lambda(t) = \hat{\lambda}(\tau_j), \quad t \in [\tau_{j-1}, \tau_j). \tag{54}$$

Substituting (53) into (22), an estimate for the survival function is obtained as:

$$S(t) = \prod_{i|\tau_{j-1}\le t} \left(1 - \hat{\lambda}_i\right) = \prod_{i|\tau_{j-1}\le t} \left(1 - \frac{z(\tau_{i-1}) - z(\tau_i)}{z(\tau_{i-1})}\right)$$
$$= \prod_{i|\tau_{j-1}\le t} \left(1 - \frac{d_i}{z(\tau_{i-1})}\right), \quad t \ge \tau_0, \tag{55}$$

where $d_i = z(t_{i-1}) - z(\tau_i)$ *is the number of deaths over the consecutive risk/failure time interval* $[\tau_{i-1}, \tau_i)$, $\tau_i \le \tau_{j-1} \le t < \tau_j$ *for some* $j \in I(1,k)$.

(b) For $j \in I(1,k)$, *if* $\tau_{j-1} < \tau_j^c < \tau_j$, *and* τ_j^c *is censored time between a pair of consecutive risk/failure times* τ_{j-1} *and* τ_j *in* $[\tau_0, \mathscr{T})$, *then, employing Remark 3.2(iv)(II) and Definitions 3.2, 3.3 and 4.2, an estimate for the risk/hazard rate function at* τ_j *is determined by:*

$$\hat{\lambda}(\tau_j) = \frac{z(\tau_{j-1}) - z(\tau_j) - \gamma_j^c}{z(\tau_j^c)}, \tag{56}$$

and

$$\lambda(t) = \hat{\lambda}(\tau_j), \quad t \in [\tau_{j-1}, \tau_j). \tag{57}$$

Substituting (56) into (22), an estimate for the survival function when τ_j^c is a censored time between consecutive failure times, τ_{j-1} and τ_j is given by:

$$S(t) = \prod_{i|\tau_{j-1} \le t} \left(1 - \hat{\lambda}_i\right) = \prod_{i|\tau_{j-1} \le t} \left(1 - \frac{z(\tau_{i-1}) - z(\tau_i) - \gamma_i^c}{z(\tau_i^c)}\right)$$
$$= \prod_{i|\tau_{j-1} \le t} \left(1 - \frac{d_i}{z(\tau_i^c)}\right), \quad t \ge \tau_0 , \tag{58}$$

where i runs over the positive integers for which $\tau_i \le \tau_{j-1}$, $\tau_{j-1} \le t < t$ for some $j \in I(1,k)$; τ_{i-1}, τ_i are consecutive failure times for $i \in I(1, j)$, and $d_i = z(t_{i-1}) - z(\tau_i) - \gamma_i^c$ is the number of deaths over the consecutive failure time interval $[\tau_{j-1}, \tau_j)$.

Remark 4.3. (a) We remark that (55) and (58) are indeed the Kaplan and Meier (1958)-type survival estimate functions.

(b) In the literature (Kalbfleisch & Prentice, 2011; Lawless, 2011), the numbers in the denominator of (55) and (58) are referred to as the number of individuals at rist at τ_{j-1} and τ_j^c respectively. Denoting this by n_j, we can write both (55) and (58) as

$$S(t) = \prod_{i|\tau_{j-1} \le t} \left(\frac{n_i - d_i}{n_i}\right) . \tag{59}$$

This is the well-known formula cited in the literature (Kalbfleisch & Prentice, 2011; Lawless, 2011).

(c) From Remark 2.4, we obtain

$$\hat{\Lambda}(t) = \sum_{\tau_j \le t} \hat{\lambda}_j = \sum_{\tau_j \le t} \frac{d_j}{n_j}, \quad t \ge \tau_0 , \tag{60}$$

where

$$n_j = \begin{cases} z(\tau_{j-1}) & \text{if there are no censors in} \quad [\tau_{j-1}, \tau_j) , \\ z(\tau_j^c) & \text{if} \quad \tau_j^c \quad \text{is a censored time in} \quad [\tau_{j-1}, \tau_j) . \end{cases} \tag{61}$$

This is the estimator introduced by Nelson (1969) and Aalen (1978). These special cases exhibit the role and scope of the presented innovative alternative dynamic approach.

In the following, we state a corollary that further illustrates the role and scope of our dynamic approach. Further details regarding the proof is outlined in the supplementary section.

Corollary 4.2. *Let us assume that the conditions of Corollary 3.2 and Example 2.1 in the context of Remark 3.2(iii) are satisfied. For $j \in I(1, n)$, if τ_{j-1} and τ_j are consecutive risk/failure times in $[\tau_0, \mathscr{T}]$, then employing Definitions 3.2, 3.3 and 4.2, an estimate for the risk/hazard rate function at τ_j is determined by:*

$$\hat{\lambda}(\tau_j) = \frac{a_j}{(1 - A_{j-1})\Delta\tau_j} , \tag{62}$$

and

$$\hat{\lambda}(t) = \hat{\lambda}(\tau_j) , \quad t \in [\tau_{j-1}, \tau_j) , \tag{63}$$

where a_j and A_{j-1} are defined in Example 2.1.

Moreover, an estimate for the survival function is represented by

$$\hat{S}(t) = S_{j-1} \exp\left[-\hat{\lambda}_j(t - \tau_{j-1})\right] , \quad \textit{for} \quad t \in [\tau_{j-1}, \tau_j) . \tag{64}$$

Remark 4.4. The PEXE of Kitchin et al. (1980), as well as Kim and Proschan (1991) is undefined beyond the last observed failure time. To rectify that, Malla and Mukerjee (2010) provided the following exponential tail hazard/risk rate estimate:

$$\hat{\lambda}_{\text{tail}} = \frac{\exp(-\hat{\Lambda}_m)}{\sum_{i=1}^{m} (I_j - J_j)} \tag{65}$$

where

$$I_j = \int_{\tau_{j-1}}^{\tau_j} \hat{S}^{KM}(t)\mathrm{d}t = (1 - A_{j-1})(\tau_j - \tau_{j-1})$$

and

$$J_j = \int_{\tau_{j-1}}^{\tau_j} \hat{S}^{MN}(t) = \exp(-\hat{\Lambda}_{j-1})\frac{(1 - A_{j-1})(\tau_j - \tau_{j-1})}{a_j}\left[1 - \exp\left(-\frac{a_j}{1 - A_{j-1}}\right)\right].$$

Thus, under the following assumptions: (i) no ties among the failure times, (ii) the last observation is uncensored, a new PEXE of Malla and Mukerjee (2010) is given by

$$S(t) = \begin{cases} \exp(-\Lambda_{j-1})\exp\left(\frac{-a_j(t-\tau_{j-1})}{(1-A_{j-1})(\tau_j-\tau_{j-1})}\right), & \tau_{j-1} \le t < \tau_j, \quad j \in I(1,m) \\ \exp(-\hat{\Lambda}_m)\exp(-\hat{\lambda}_{\text{tail}}(t - \tau_m)), & \tau_m \le t < \infty\,. \end{cases} \tag{66}$$

We further note that the presented dynamic approach does not require the failure function to be invertible.

5. Multiple Censored Times between Consecutive Failure Times

In this section, we further apply the conceptual dynamic results developed in Sections 2 and 3 to multiple censored times between consecutive failure times. We present a result that provides a very general algorithm for estimating a hazard rate function for multiple censoring times between consecutive failure times τ_{j-1} and τ_j with $\tau_{j-1}, \tau_j \in [\tau_0, \mathscr{T})$. We further note that the presented results in this section extend the results of Section 4 in a systematic and unified manner.

Theorem 5.1. *Let the hypotheses of Theorem 3.1 in the context of Remarks 3.1, 3.2(i) and 3.2(ii) be satisfied. For each $j \in I(1,m)$, let τ_{j-1} and τ_j be consecutive failure times. Let $\{\tau_{j-1l}\}_{l=1}^{k_j}$ be a finite sequence of censored time observations over a time interval $[\tau_{j-1}, \tau_j]$. Let γ_j^l be the number of objects censored at time τ_{j-1l}, for $l \in I(1,k_j)$ and $\{\gamma_j^l\}_{l=1}^{k_j}$ be a corresponding sequence of observed number of objects/species/patients/etc. Then*

1. *$z(\tau_{j-1}) - z(\tau_j) - \sum_{l=1}^{k_j} \gamma_j^l$ is a change in the number of items/subjects that is under the observation over the sub-interval $[\tau_{j-1}, \tau_j]$ of the time interval of study $[\tau_0, \mathscr{T}]$*
2. *$\sum_{l=1}^{k_j+1} z(\tau_{j-1l-1})\Delta(\tau_{j-1l})$ is a total amount of time spent under the observation/testing/evaluation/monitoring of $z(\tau_{j-1l-1})$ items/patients/ infectives/subjects on the interval $[\tau_{j-1l-1}, \tau_{j-1l})$ for $l \in I(1,k_j))$ and $j \in I(1,n)$.*
3. *an estimate for the hazard rate function at τ_j is determined by*

$$\hat{\lambda}(\tau_j) = \frac{z(\tau_{j-1}) - z(\tau_j) - \sum_{l=1}^{k_j} \gamma_j^l}{\sum_{l=1}^{k_j+1} z(\tau_{j-1l-1})\Delta(\tau_{j-1l})}, \tag{67}$$

and an estimate for the hazard rate function is

$$\hat{\lambda}(t) = \hat{\lambda}(\tau_j), \quad for \quad t \in [\tau_{j-1}, \tau_j) \quad and \quad j \in I(1,n) \tag{68}$$

Proof. The detailed proof of Theorem 5.1 is given in the supplementary section 8.

Corollary 5.1. *Under the conditions of Theorem 5.1 and assumptions of Corollary 3.3 in the context of Remark 3.2(iv), an estimate for the hazard rate function at τ_j is determined by*

$$\hat{\lambda}(\tau_j) = \frac{z(\tau_{j-1}) - z(\tau_j) - \sum_{l=1}^{k_j} \gamma_j^l}{z(\tau_{j-1k_j})}, \tag{69}$$

and an estimate for the hazard rate function is $\hat{\lambda}(t) = \hat{\lambda}(\tau_j)$, for $t \in [\tau_{j-1}, \tau_j)$ and $j \in I(1,n)$. An estimate for the survival function is thus given by

$$\hat{S}(t) = \prod_{i|\tau_{j-1}<t} (1 - \hat{\lambda}(\tau_i)),\ t \ge \tau_0,\ \tau_i \le \tau_{j-1} \le t < \tau_j \text{ for some } j \in I(1,n). \tag{70}$$

Corollary 5.2. *Under the conditions of Theorem 5.1 and estimate for the cumulative hazard/risk rate and survival functions are represented by:*

$$\hat{\Lambda}(t,\tau_0) = \sum_{m=1}^{j-1} \hat{\lambda}_m(\tau_m - \tau_{m-1}) + \hat{\lambda}_j\left(t - \tau_{j-1}\right), \ t \in [\tau_{j-1}, \tau_j)$$

and

$$\hat{S}(t,\tau_0) = S_0 \exp\left[\sum_{m=1}^{j-1} \hat{\lambda}_m(\tau_m - \tau_{m-1}) + \hat{\lambda}_j\left(t - \tau_{j-1}\right)\right], \ t \in [\tau_{j-1}, \tau_j)$$

for $t \geq \tau_0$, $\tau_{j-1} \leq t < \tau_j$ *for some* $j \in I(1,n)$.

Remark 5.1. (a) We remark that the innovative dynamic approach for the development of computational parameter estimation algorithm (67) is an alternative approach for the algorithm proposed by Kim and Proschan (1991).

(b) The estimates (67) in the context of (20) yields the estimate obtained by Kulasekera and White (1996) as special cases.

(c) For continuous-time interconnected hybrid state survival dynamic process, if $k_j = 0$, for some $j \in I(1,n)$, then $l = 0$ and $\gamma_j^0 = 0$ and (67) reduces to (44). On the other hand, if $k_j = 1$ for some $j \in I(1,n)$, then $l = 0$ and $\gamma_j^1 = \gamma_j^c$ and (67) implies (48).

(d) For discrete-time interconnected hybrid state survival dynamic process, if $k_j = 0$, for some $j \in I(1,n)$, then $l = 0$ and $\gamma_j^0 = 0$ and (69) reduces to (53). On the other hand, if $k_j = 1$, for some $j \in I(1,n)$, then $l = 0$ and $\gamma_j^1 = \gamma_j^c$ and (69) implies (56).

The presented innovative approach of parameter and state estimation includes the Thaler (1984)-type hazard rate estimation problem as a particular case. To justify this statement, we first introduce a concept of hazard/risk rate function for responder and non-responder states. In addition, we state a corollary of Theorem 5.1 without its proof. The proof is outlined in the supplementary section.

Definition 5.1. For $i \in I(0,1)$, Let $\lambda_0(t)$ and $\lambda_1(t)$ represent the hazard/risk rate functions in the non-responder and responder states, respectively, at time t (Thaler, 1984).

Corollary 5.3. *Let us assume that the conditions of Corollary 3.1 in the context of Remark 3.2(i) are satisfied. For* $j \in I(1,n_0)$, *let* τ_{j-1} *and* τ_j *be consecutive risk/failure times in state 0. For* $j' \in (1,n_1)$, *let* $\tau_{j'-1}$ *and* $\tau_{j'}$ *be consecutive failure times in state 1. Let* $z_0(\tau_j)$ *be the number of survivals at* τ_j *in state* 0. *Let* $z_1(\tau_{j'})$ *be the number of survivals at* $\tau_{j'}$ *in state* 1. *Then an estimate for the hazard/risk rate function at* τ_j *is determined by:*

$$\hat{\lambda}_0(\tau_j) = \frac{\sum_{m=1}^{j} [z_0(\tau_{m-1}) - z_0(\tau_m)]}{\sum_{m=1}^{j} z_0(\tau_{m-1})\Delta\tau_m} = \frac{\sum_{m=1}^{j} d_{0j}}{\sum_{m=1}^{j} z_0(\tau_{m-1})\Delta\tau_m}, \tag{71}$$

where d_{0j} *is the number of deaths/failures at the jth distinct failure time in state i, and an estimate for the hazard rate function is*

$$\hat{\lambda}_0(t) = \hat{\lambda}_0(\tau_j), \quad for \quad t \in [\tau_{j-1}, \tau_j) \quad and \quad j \in I(1,n_0)\,. \tag{72}$$

An estimate for the hazard/risk rate function at $\tau_{j'}$ *is determined by:*

$$\hat{\lambda}_1(\tau_{j'}) = \frac{\sum_{m=1}^{j'} [z_1(\tau_{m-1}) - z_1(\tau_m)]}{\sum_{m=1}^{j'} z_1(\tau_{m-1})\Delta\tau_m} = \frac{\sum_{m=1}^{j} d_{1j'}}{\sum_{m=1}^{j} z_1(\tau_{m-1})\Delta\tau_m}, \tag{73}$$

where $d_{1j'}$ *is the number of deaths/failures at the j'th distinct failure time in state* 1, *and an estimate for the hazard rate function is*

$$\hat{\lambda}_1(t) = \hat{\lambda}_1(\tau_{j'}), \quad for \quad t \in [\tau_{j'-1}, \tau_{j'}) \quad and \quad j' \in I(1,n_1)\,. \tag{74}$$

The hazard/risk ratio rate function estimate is given by:

The corresponding estimate of the log *hazard/risk rate ratio function for patients currently in a response compared to a nonresponse state is given by:*

$$\hat{\rho}(t) = \ln\left[\frac{\hat{\lambda}_0(\tau_j)}{\hat{\lambda}_1(\tau_{j'})}\right] \text{ for }, \tau_{j-1} < t \leq \tau_j \text{ and } \tau_{j'-1} \leq t < \tau_{j'} . \tag{75}$$

Remark 5.2. We remark that (71), (73) and (75) are identical to the result obtained in Thaler (1984). Moreover, the estimates in (71), (73) and (75) were obtained in the framework of an innovative dynamic approach.

In the following, we state a general theorem that provides a theoretical estimate for the hazard/risk rate function between two successive change point times, τ_{j-1} and τ_j.

Theorem 5.2. *Let the hypothesis of Theorem 5.1 be satisfied. Let $\{T_i^j\}_{i=1}^n$ be a sequence of times(failure/ censor/arrival) that fall between the change point times τ_{j-1} and τ_j for $j = I(1,k)$. Then an estimate for the hazard rate function at τ_j is determined by*

$$\hat{\lambda}(\tau_j) = \hat{\lambda}(\tau_j) = \frac{z(\tau_{j-1}) - z(\tau_j) - \sum_{m=1}^{l} \eta_m^j}{\sum_{m=1}^{l+1} z(T_m^j)\Delta(T_m^j)} , \quad j \in I(1,k+1) . \tag{76}$$

where

$$\eta_m^j = \begin{cases} 0 & \text{if } T_m^j \text{ is failure time} \\ \gamma_m^{jc} & \text{if } T_m^j \text{ is censored time} \\ -\gamma_m^{ja} & \text{if } T_m^j \text{ is arrival time} \end{cases} ; \tag{77}$$

γ_m^{jc} is the number of objects/items/individuals censored at time T_m^j; γ_m^{ja} is the number of objects/items/individuals joining/arriving at time T_m^j, and an estimate for the hazard rate function is $\lambda(t) = \hat{\tau}_j$ for $t \in [\tau_{j-1}, \tau_j)$.

Proof. The proof of Theorem 5.2 is outlined in the supplementary section.

6. Computational Algorithms

In this section, we outline very general conceptual computational, data organizational and simulation schemes. The computational and simulation algorithms are based on fundamental theoretical result (Theorem 5.1) developed in Section 5.

6.1 Conceptual Computational Parameter and State Estimation Scheme

The theoretical computational algorithm for interconnected continuous-time hybrid dynamic process (24), is as follows:

$$z(\tau_{j-1}) - z(\tau_j) - \sum_{l=1}^{k_j} \gamma_j^l = \hat{\lambda}(\tau_j) \sum_{l=1}^{k_j+1} z(\tau_{j-1l-1})\Delta(\tau_{j-1l}), \tag{78}$$

and the conceptual computational algorithm for totally discrete-time hybrid dynamic process (26) is

$$z(\tau_{j-1}) - z(\tau_j) - \sum_{l=1}^{k_j} \gamma_j^l = \hat{\lambda}(\tau_j) z(\tau_{j-1k_j}) . \tag{79}$$

Here $\mathscr{P}_0^{\mathscr{T}} : \tau_0 < \tau_1 < \ldots < \tau_{j-1} < \tau_j < \ldots < \tau_n$ is a partition of failure times over the time interval $[0, \mathscr{T})$. Let $\mathscr{P}_j$ be a partition corresponding to a given finite sequence of censored times over the failure time interval $[\tau_{j-1}, \tau_j)$, and let it be represented by

$$\mathscr{P}_j : \tau_{j-1} = \tau_{j-10} < \tau_{j-11} < \ldots < \tau_{j-1l-1} < \tau_{j-1l} < \ldots < \tau_{j-1k_{j-1}} < \tau_{j-1k_j} . \tag{80}$$

For $j \in I(1,n)$, λ is the hazard rate function; $z(t)$ stands for the number of survivals at time t; γ_j^l denotes the number of objects censored at the time τ_{j-1l}, $j \in I(1,m)$ and $l \in I(0,k_j)$, $k_j \in I(0,\infty)$.

For the continuous-time hybrid dynamic process (24), an estimate of the survival function is represented by

$$\hat{S}(t,\tau_0) = S_0 \exp\left[\sum_{m=1}^{j-1} \hat{\lambda}_m(\tau_m - \tau_{m-1}) + \hat{\lambda}_j\left(t - \tau_{j-1}\right)\right] , \ t \in [\tau_{j-1}, \tau_j) \text{ for } t \geq \tau_0 . \tag{81}$$

For the totally discrete-time hybrid dynamic process (26), an estimate of the survival function is represented by

$$\hat{S}(t) = \prod_{i|\tau_{j-1}<t} (1 - \hat{\lambda}(\tau_i)),\ t \geq \tau_0. \tag{82}$$

First, we construct a detailed flowchart for the general conceptual computational algorithm developed in Section 5.

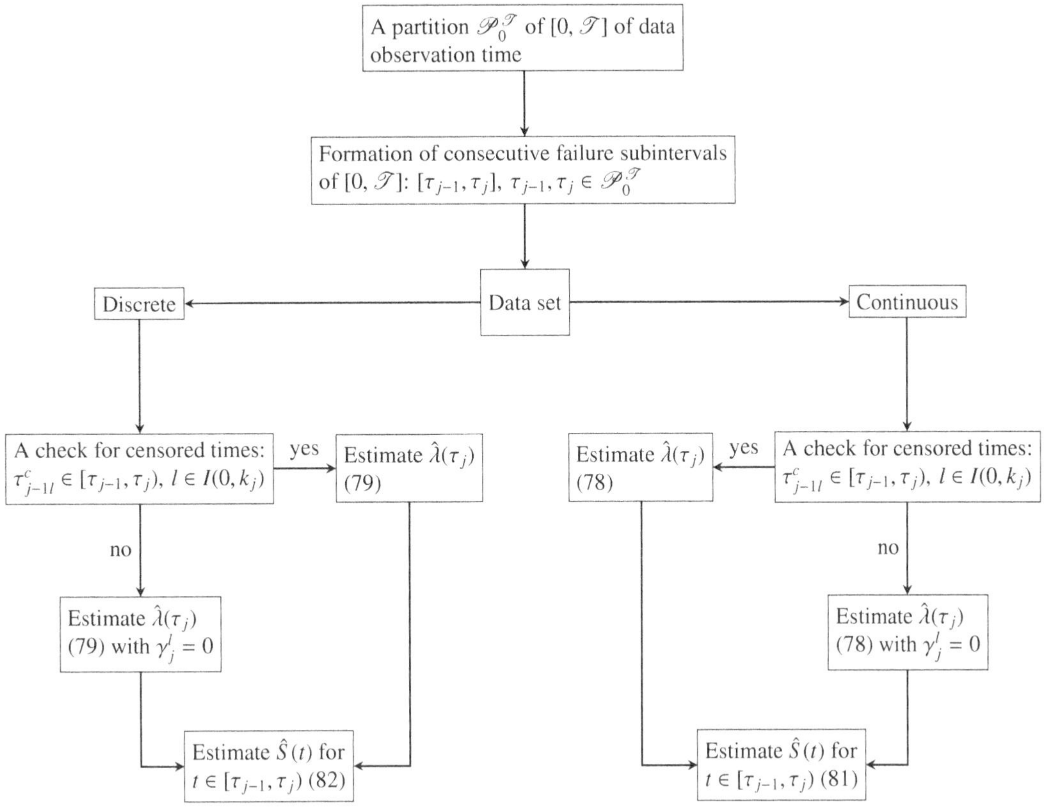

Flowchart 1. Conceptual Computational Algorithm

We observe that the conceptual computational algorithm (Flowchart 1) is composed of two sub-conceptual computational algorithms, namely, continuous-time and discrete-time hybrid dynamic processes.

6.2 Conceptual and Computational Simulation Algorithms

A pseudocode for a simulation scheme for both interconnected continuous-time and totally discrete-time hybrid dynamic processes are outlined below:

for $j = 1$ to N **do**
Compute k_j, $z(\tau_{j-1}), z(\tau_j)$
  if $k_j = 0$ **then**
Compute $z(\tau_{j-1})\Delta\tau_j$
  else
Compute $\sum_{l=1}^{k_j} \gamma^l_j$, $\sum_{l=1}^{k_j+1} z(\tau_{j-1l-1})\Delta(\tau_{j-1l})$
  end if
Compute $\hat{\lambda}(\tau_j)$, $\hat{S}(t)$
end for

Simulation Scheme 1a. Pseudocode for interconnected continuous-time hybrid dynamic process

for $j = 1$ to N **do**
Compute k_j, $z(\tau_{j-1}), z(\tau_j)$
  if $k_j = 0$ **then**
Compute $z(\tau_{j-1})$
  else
Compute $\sum_{l=1}^{k_j} \gamma^l_j$, $z(\tau_{j-1k_j})$
  end if
Compute $\hat{\lambda}(\tau_j)$, $\hat{S}(t)$
end for

Simulation Scheme 1b. Pseudocode for totally discrete-time hybrid dynamic process

Moreover, a flowchart for the simulation algorithm for parameter and state estimation problems for interconnected continuous-time (24) and discrete-time (26) hybrid dynamic processes are provided in Flowchart 2.

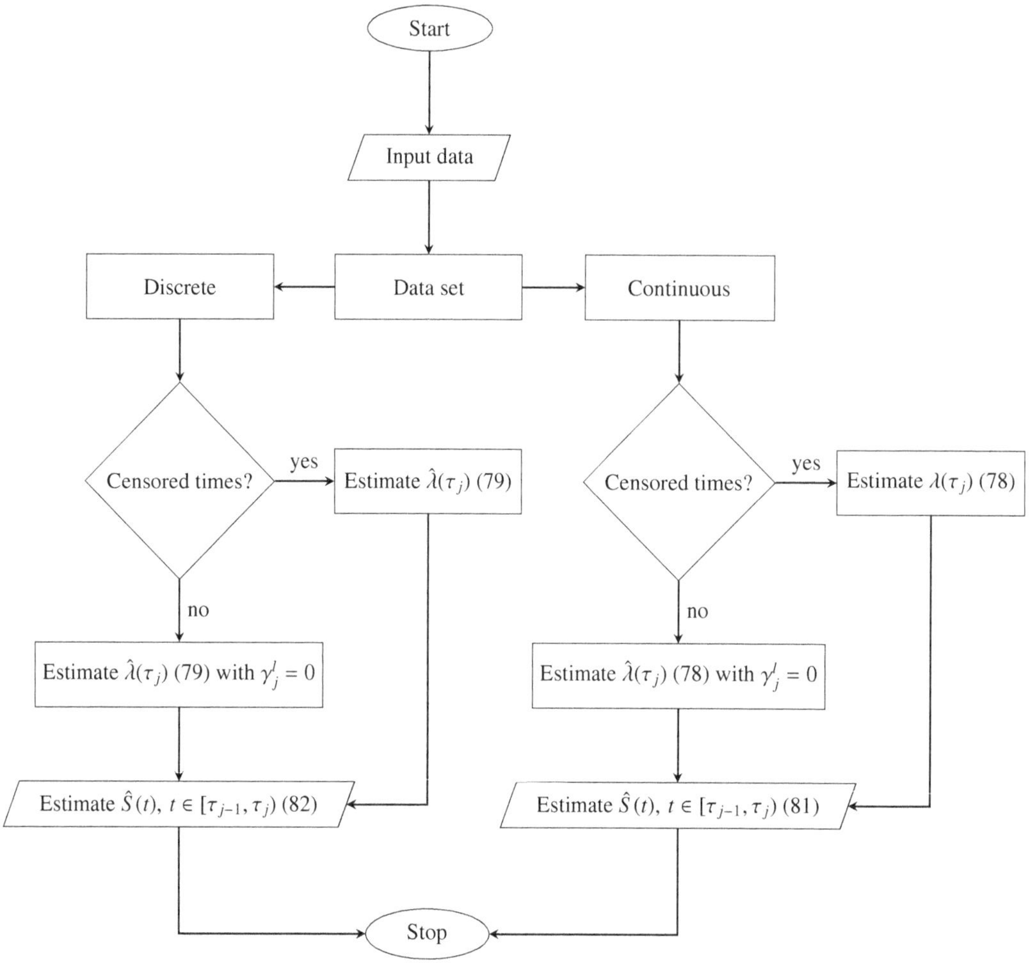

Flowchart 2. Simulation Algorithm for interconnected hybrid dynamic processes

We note that flowchart for simulation algorithm (Flowchart 2) is composed of two sub-simulation algorithms, namely, continuous-time and totally discrete-time hybrid dynamic processes.

In the following, using the conceptual computational algorithm, we exemplify our theoretical procedure by estimating hazard rate and survival functions of two data sets in a systematic and unified way. The first data set can be found in Kaplan and Meier (1958).

Illustration 6.1. Suppose that out of a sample of 8 items the following are observed:

Table 1. Dataset used by Kaplan and Meier (1958)

Order of Observation	Time of Cessation of Observation	Cause of Cessation	Time Notation
1	0.8	Failure	τ_1
2	1.0	Censored	τ_{11}
3	2.7	Censored	τ_{12}
4	3.1	Failure	τ_2
5	5.4	Failure	τ_3
6	7.0	Censored	τ_{31}
7	9.2	Failure	τ_4
8	12.1	Censored	

We note that the data set in Table 1 is for the totally discrete-time hybrid time-to-event dynamic process (26). In view of this, we apply the totally discrete-time parameter and state estimation schemes (79) and (82). In short, we utilize the discrete-time conceptual computational sub-algorithm (Simulation Scheme 1b) "pseudocode" and simulation sub-algorithm (Flowchart 2).

For $t \in [\tau_0, \tau_1)$, there are no censored times between $[\tau_0, \tau_1)$. Therefore, $k_j = 0$, and from Remark 5.1(d) and hence using (79) we have

$$\hat{\lambda}(\tau_1) = \hat{\lambda}_1 = \frac{z(\tau_0) - z(\tau_1)}{z(\tau_0)} = \frac{1}{8}.$$

Utilizing (82), the corresponding survival function is given by

$$\hat{S}(t) = \begin{cases} 1, & \text{for} \quad t \in [\tau_0, \tau_1), \\ 1 - \lambda_1 = \frac{7}{8}, & \text{for} \quad t = \tau_1. \end{cases}$$

For $t \in [\tau_1, \tau_2)$, we note that there are two censored times between τ_1 and τ_2. So, $k_j = k_2 = 2$. Hence

$$\sum_{l=1}^{2} \gamma_2^l = \gamma_2^1 + \gamma_2^2 = 1 + 1 = 2.$$

Also, $z(\tau_{j-1k_j}) = z(\tau_{12}) = 5$. Thus, from Remark 5.1(d) and hence applying (79), we have

$$\hat{\lambda}(\tau_2) = \hat{\lambda}_2 = \frac{z(\tau_1) - z(\tau_2) - \sum\limits_{l=1}^{2} \gamma_2^l}{z(\tau_{12})} = \frac{1}{5}.$$

Utilizing (82), the corresponding survival function is thus given by

$$\hat{S}(t) = \begin{cases} \frac{7}{8}, & \text{for} \quad t \in [\tau_1, \tau_2), \\ \prod\limits_{k|\tau_j \le t} (1 - \hat{\lambda}_j) = \prod\limits_{j=1}^{2} (1 - \hat{\lambda}_j) = \frac{7}{10}, & \text{for} \quad t = \tau_2. \end{cases}$$

There is no censoring time between the interval $[\tau_2, \tau_3) = [3.1, 5.4)$. Therefore, $k_j = 0$, and from Remark 5.1(d) and hence using (79) we obtain

$$\hat{\lambda}(\tau_3) = \frac{z(\tau_2) - z(\tau_3)}{z(\tau_2)} = \frac{1}{4}.$$

Once again, utilizing (82), the corresponding survival function is thus given by

$$\hat{S}(t) = \begin{cases} \frac{7}{10}, & \text{for} \quad t \in [\tau_2, \tau_3), \\ \prod\limits_{j=1}^{3} (1 - \hat{\lambda}_j) = \frac{21}{40}, & \text{for} \quad t = \tau_3. \end{cases}$$

Continuing in this manner, we record the estimates for hazard rate and survival functions in the following table with the last column exhibiting the survival function estimate as obtained by Kaplan and Meier (1958).

Table 2. Kaplan and Meier Survival estimates for data set given in Kaplan and Meier (1958).

Failure Times τ_j	Survivals $z(\tau_j)$	Hazard Rate Function $\hat{\lambda}(\tau_j)$	Survival Function $\hat{S}(\tau_j)$
0.8	7	1/8	7/8
3.1	4	1/5	7/10
5.4	3	1/4	21/40
9.2	1	1/2	21/80
(12.1)	0	1/2	21/80

Using the dataset in Kim and Proschan (1991) and theoretical computational algorithm, Theorem 5.1, we illustrate the estimation of hazard rate and survival functions, systematically.

Illustration 6.2. Suppose that seven items (new) are put on test at time 0. Each item is observed until it fails or until it is withdrawn, whichever occurs first. The resulting set of observation (Kim & Proschan, 1991) is shown in Table 3 in order of occurrence.

Table 3. Data from Kim and Proschan (1991)

Order of Observation	Time of Cessation of Observation	Cause of Cessation	Time Notation	Finite sequence of censored Time	Size of sequence	Number of Censored
0	0					
1	2.0	Failure	$\tau_1 = \tau_{01} = \tau_{10}$			
2	3.5	Censored	τ_{11}	$\{\tau_{j-1l}\}_{l=1}^{2}$	$k_2 = 2$	$\{\gamma_2^l\}_{l=1}^{2}$
3	4.5	Censored	τ_{12}			
4	6.2	Failure	$\tau_2 = \tau_{13} = \tau_{20}$			
5	8.0	Censored	τ_{21}	$\{\tau_{j-1l}\}_{l=1}^{1}$	$k_3 = 1$	$\{\gamma_3^l\}_{l=1}^{2}$
6	8.8	Failure	$\tau_3 = \tau_{22}$			
7	11.3	Failure	τ_4			

The data set in Table 3 is for the interconnected continuous-time hybrid dynamic time-to-event dynamic process (24). In view of this, we apply the continuous-time parameter and state estimation schemes (78) and (81). In short, we utilize the continuous-time conceptual computational sub-algorithm (Simulation Scheme 1a) "pseudocode" and simulation sub-algorithm (Flowchart 2).

For $[0, \tau_1)$, since there are no censored times in between $[0, \tau_1)$, $k_j = k_1 = 0$. Thus from Remark 5.1(c) and using (78) we have

$$\hat{\lambda}(\tau_1) = \frac{z(\tau_0) - z(\tau_1)}{z(\tau_0)(\tau_{01} - \tau_0)} = \frac{1}{14}.$$

Thus $\hat{\lambda}(t) = \frac{1}{14} \approx 0.0714$ for $t \in [\tau_0, \tau_1) = [0, 2.0)$.

For the estimate on $[\tau_1, \tau_2) = [2.0, 6.2)$, we note that there are two censoring times between $[\tau_1, \tau_2)$, hence $k_j = k_2 = 2$ and

$$\sum_{l=1}^{2} \gamma_2^l = \gamma_2^1 + \gamma_2^2 = 1 + 1 = 2.$$

Thus from Remark 5.1(c) and thus applying (78), we have

$$\hat{\lambda}(\tau_2) = \frac{z(\tau_1) - z(\tau_2) - \sum_{l=1}^{k_2} \gamma_2^l}{\sum_{l=1}^{k_2+1} z(\tau_{1l-1})\Delta\tau_{1l}} = \frac{z(\tau_1) - z(\tau_2) - \sum_{l=1}^{2} \gamma_2^l}{\sum_{l=1}^{3} z(\tau_{1l-1})\Delta\tau_{1l}} = \frac{1}{20.8}.$$

Thus, $\hat{\lambda}(t) = \frac{1}{20.8}$, for $t \in [2.0, 6.2)$.

On the interval $[\tau_2, \tau_3) = [6.2, 8.8)$, we have only one censoring time in between the two failure times. So, $k_j = k_3 = 1$. Thus from Remark 5.1(c) and hence, using (67), we obtain

$$\hat{\lambda}(\tau_3) = \frac{z(\tau_2) - z(\tau_3) - \sum_{l=1}^{1} \gamma_3^l}{\sum_{l=1}^{2} z(\tau_{2l-1})\Delta\tau_{2l}} = \frac{3 - 1 - 1}{z(\tau_{20})\Delta\tau_{21} + z(\tau_{21})\Delta\tau_{22}} = \frac{1}{7}.$$

Hence, $\hat{\lambda}(t) = \frac{1}{7}$, for $t \in [6.2, 8.0)$.

There is no censoring in the interval $[\tau_3, \tau_4)$. Thus,

$$\hat{\lambda}(\tau_4) = \frac{z(\tau_3) - z(\tau_4)}{z(\tau_3)\Delta\tau_4} = \frac{1}{2.5},$$

which implies that $\hat{\lambda}(t) = \frac{1}{2.5} = 0.4$, for $t \in [8.0, 11.3)$.

Following this estimation procedure we have

$$\hat{\lambda}(t) = \begin{cases} 0.0714 & 0 \le t < \tau_1 = 2 \\ 0.0481 & \tau_1 \le t < \tau_2 = 6.8 \\ 0.1429 & \tau_2 \le t < \tau_3 = 8.8 \\ 0.4 & \tau_3 \le t < \tau_4 = 11.3\,. \end{cases} \tag{83}$$

To obtain the estimate of survival function, we use (81) or we apply the solution process described in Section 2 regarding (7) and obtain exponential pieces on successive intervals between failure times that are joined to form a continuous function. Thus,

$$\hat{S}(t) = \begin{cases} \exp(-0.0714t)\,, & 0 \le t < 2 \\ \exp[-0.1429 - 0.0481(t-2)]\,, & 2 \le t < 6.2 \\ \exp[0.3448 - 0.1429(t-6.2)]\,, & 6.2 \le t < 8.8 \\ \exp[0.4591 - 0.4(t-8.8)]\,, & 8.8 \le t < 11.3 \\ \text{no estimator}, & t \ge 11.3 \end{cases} \tag{84}$$

Remark 6.1. These are the same results obtained by using the method proposed Kim and Proschan (1991).

7. Conclusions

Most of the research work in the area of survival and reliability analysis is centered around the probabilistic analysis approach. In general, a closed-form solution is not feasible. In addition, a hazard rate function is nonlinear in covariate state processes and non-stationary. The presented linear hybrid deterministic dynamic modeling is more suitable for a complex time-to-event processes. This innovative approach does not require a closed-form solution distribution. The influence of both continuous and discrete-time states can be easily incorporated as an interconnected hybrid dynamic model for time-to-event processes. In fact, it allows to have a time-varying covariate state influence on the dynamic of a complex survival/reliability of systems. The influence of human mobility, electronic communications, rapid technological changes, advancements in biological, engineering, medical, military, physical and social sciences is motivated to initiate, formulate and to develop an innovative interconnected alternative modeling approach for time-to-event processes in biological, chemical, engineering, epidemiological, medical, multiple-markets and social dynamic processes through discrete-time intervention processes. The presented innovative modeling approach further enhanced our motivation to develop state and parameter estimation procedures. Moreover, the parameter and state estimation approach is dynamic. The dynamic nature rather than the existing algebraic approach plays a very significant role in state and parameter estimation problems in systematic and unifying way. The discrete-time dynamic is exhibited by the two flowcharts and Simulation algorithms 1(a) and 1(b). Furthermore, the significance of the conceptual computational algorithms are also exhibited by illustrations. At the initial level of our objective, we began with a very simple observation of the probabilistic definition of the survival function. This has led to the development of this approach. The role and scope of the presented dynamic approach is exhibited through several existing results (Han et al., 2014; Malla & Mukerjee, 2010; Kim & Proschan, 1991; Thaler, 1984; Aalen, 1978; Nelson, 1969; Kaplan & Meier, 1958) as corollaries, illustrations and remarks. In fact, the full force of the role and scope of hybrid deterministic modeling for time-to-event processes is currently being explored (Appiah E. A. *Time-To-Event Dynamic Processes: Modeling, Methods and Estimations*-Ph.D Dissertation, 2017) for both deterministic and stochastic nonlinear and non-stationary hybrid modeling for time-to-event processes. Furthermore, a complex time-to-event dynamic study is also currently undertaken by Ladde and his team. These developed results will be reported elsewhere.

8. Supplements: Proofs of Theorems

In this supplementary section, proofs of a few theorems and corollaries stated in sections 2, 3, 4 and 5 are presented.

***Proof of Theorem 2.1*:** The theorem is proved by the principle of mathematical induction (PMI) (Ladde & Ladde, 2012). From (11), for $j = 1$, we have

$$\mathrm{d}x = [-\alpha(t)\,x + \gamma(t)]\mathrm{d}\beta(t),\ \ x(t_0) = x_0,\ \ t \in [\tau_0, \tau_1)\,.$$

From (10) and the definition of Riemann-Stieltjes integral (Apostol, 1974), we have

$$x(t) - x(\tau_0) = \int_{\tau_0}^{t} [-\alpha(s)\,x(s) + \gamma(s)]\mathrm{d}\beta(s) = 0,\ \text{for}\ t \in [\tau_0, \tau_1)\,. \tag{85}$$

We define

$$x(t) = x(t, \tau_0, x_0) = x_0(t, \tau_0, x_0), \quad x_0(\tau_0) = x_0, \text{ for } t \in [\tau_0, \tau_1) . \tag{86}$$

From (10), (11), (85), and $x_0(t, \tau_0, x_0) = x_0(\tau_1^-, \tau_0, x_0)$ for $t \in [\tau_0, \tau^-]$, we have

$$x_0(\tau_1) - x_0(\tau_0) = 0 + \int_{\tau_1^-}^{t} [-\alpha(s)\, x(s) + \gamma(s)]\, d\beta(s), \text{ for } t \in [\tau_0, \tau_1] .$$

From this, the continuity of α and γ, the definitions of Riemann-Stieltjes integral (Apostol, 1974) and the initial value problem (Ladde & Ladde, 2012), we have

$$\begin{aligned} x_0(\tau_1, \tau_0, x_0) &= x_0(\tau_0) + \beta(\tau_1)[-\alpha(t_1^*)x(t_1^*) + \gamma(t_1^*)] - \beta(t_1^*)[-\alpha(t_1^*)x(t_1^*) + \gamma(t_1^*)] \\ &= x_0(\tau_0) - \alpha_1 x_0(\tau_1^-, \tau_0, x_0) + \gamma_1 , \end{aligned} \tag{87}$$

for $t_1^* \in [\tau_1^-, \tau_1]$. From (87) and setting $x_0(\tau_1, \tau_0, x_0) = x(\tau_1) = x_1$ and again $x(\tau_1^-, \tau_0, x_0) = x_0$, we obtain

$$\begin{aligned} x_1 &= x(\tau_1^-, \tau_0, x_0) - \alpha_1 x(\tau_1^-, \tau_0, x_0) + \gamma_1 \\ &= (1 - \alpha_1)x_0 + \gamma_1 . \end{aligned} \tag{88}$$

Continuing the above argument, we can establish the induction hypothesis (Ladde & Ladde, 2012) as:

$$x_j = \Phi(\tau_j, \tau_0)x_0 + \sum_{i=1}^{j} \Phi(\tau_j, \tau_i)\gamma_i , \quad \text{for} \quad x(\tau_j) = x_j ,$$

where

$$\Phi(\tau_j, \tau_i) = \prod_{k=i}^{j}(1 - \alpha_k) , \Phi(\tau_i, \tau_i) = 1 \quad \text{for} \quad i \in I(0, n) .$$

Now, we consider

$$dx = [-\alpha(t)\, x + \gamma(t)]\, d\beta(t), \quad x(\tau_j) = x_j,\ t \in [\tau_j, \tau_{j+1}) .$$

From the definitions of x_j and Φ, and using the above argument, one can establish the following:

$$x_j(t) = x(t, \tau_j, x_j) = \prod_{k=1}^{j}(1 - \alpha_k)x_0 + \sum_{i=1}^{j-1} \Phi(\tau_j, \tau_i)\gamma_i + \gamma_j \quad \text{for } t \in [\tau_j, \tau_{j+1}) . \tag{89}$$

Hence

$$\begin{cases} x(\tau_{j+1}^-, \tau_j, x_j) = \prod_{k=1}^{j}(1 - \alpha_k)x_0 + \sum_{i=1}^{j} \Phi(\tau_j, \tau_i)\gamma_i , \\ x_{j+1}(\tau_{j+1}, \tau_j, x_j) = (1 - \alpha_{j+1})x_j + \gamma_{j+1} . \end{cases} \tag{90}$$

Therefore, from (89) and (90), we have

$$\begin{aligned} x_{j+1} &= (1 - \alpha_{j+1})x_j + \gamma_{j+1} \\ &= \prod_{k=1}^{j+1}(1 - \alpha_k)x_0 + \sum_{i=1}^{j+1} \Phi(\tau_{j+1}, \tau_i)\gamma_i . \end{aligned}$$

By the application of PMI and the definition of the IVP regarding hybrid dynamic system (Ladde & Ladde, 2012), we have

$$x(t) = \prod_{k|\tau_j \le t} (1 - \alpha_k)x_0 + \sum_{i=1}^{j-1} \Phi(t, \tau_i)\gamma_i + \gamma_j ,$$

for $t \ge \tau_0$ and $t \in [\tau_{j-1}, \tau_{j+1})$. This establishes the proof of the theorem.

***Proof of Theorem 3.1*:** For $t \in [\tau_{j-1}, \tau_j),\ j \ge 1$, from Definition 3.1, Remark 3.1 and the nature of S, we have

$$dz(t) = -\lambda(t)z(t)dt . \tag{91}$$

This establishes the continuous-time dynamic equation in (27). The proof of the discrete-time dynamic part in (27) and iterative process in (28) are outlined below.

Multiplying the discrete-time iterative process in (24) by $S(\tau_j^-)$ and noting the fact that $S(\tau_j) = S(\tau_j^-)$, we obtain

$$x(\tau_j)S(\tau_j) = (1-\alpha_j)(1-\beta_j)x(\tau_j^-)S(\tau_j^-) + \gamma_j(1-\beta_j)S(\tau_j^-)\,. \tag{92}$$

Moreover, using the definition of z, (92) reduces to

$$z(\tau_j) = (1-\alpha_j)(1-\beta_j)z(\tau_j^-) + \gamma_j(1-\beta_j)\,. \tag{93}$$

This establishes (27).

Applying the Euler-type numerical scheme (Atkinson, 2008) to (91) over an interval $[\tau_{j-1}, \tau_j^-]$, we obtain

$$z(\tau_j^-) - z(\tau_{j-1}) = -\lambda(\tau_{j-1})z(\tau_{j-1})\Delta\tau_j\,. \tag{94}$$

From (93) and (94) , we have

$$z(\tau_j) = (1-\lambda(\tau_j)\Delta\tau_j)(1-\alpha_j)(1-\beta_j)z(\tau_{j-1}) + \gamma_j(1-\beta_j)\,. \tag{95}$$

(95) exhibits the discrete time dynamic for survival process corresponding to the continuous-time dynamic process described in (27) and the discrete-time intervention process. Moreover, (95) exhibits the validity of (28). This establishes proof of Theorem 3.1.

***Proof of Theorem 4.1*:**

(a) Using the discrete-time iterative scheme (28), Remark 3.2(i)(38) and Definitions 3.2, 3.3 and 4.1, we have

$$\lambda(t) = \hat{\lambda}(\tau_j) = \frac{z(\tau_{j-1}) - z(\tau_j)}{z(\tau_{j-1})\Delta\tau_j}$$

for $t \in [\tau_{j-1}, \tau_j)$ and $j \in I(1,k)$. This establishes (a).

(b) Let τ_j^c be a censoring time between two consecutive risk/failure times, τ_{j-1} and τ_j. We consider a partition of $[\tau_{j-1}, \tau_j]$: $\tau_{j-1} < \tau_j^c < \tau_j$.

Employing iterative processes in (40) and (38) on respective subintervals $[\tau_{j-1}, \tau_j^c]$ and $[\tau_j^c, \tau_j]$, we have

$$\begin{aligned} z(\tau_j) - z(\tau_{j-1}) &= z(\tau_j^c) - z(\tau_{j-1}) + z(\tau_j) - z(\tau_j^c) \\ &= -\lambda(\tau_{j-1})\Delta\tau_j^c - \gamma_j^c - \lambda(\tau_j)z(\tau_j^c)\Delta\tau_{jc} \\ &= -\lambda(\tau_j)\left[z(\tau_{j-1})\Delta\tau_j^c + z(\tau_j^c)\Delta\tau_{jc}\right] - \gamma_j^c\,. \end{aligned} \tag{96}$$

From (96), we obtain:

$$z(\tau_{j-1}) - z(\tau_j) - \gamma_j^c = \lambda(\tau_j)\left[z(\tau_{j-1})\Delta\tau_j^c + z(\tau_j^c)\Delta\tau_{jc}\right]\,. \tag{97}$$

From (97) and knowing that $\lambda(\tau_j)$ is the hazard/risk rate of change per unit time per unit object/subject, we conclude that $z(\tau_{j-1}) - z(\tau_j) - \gamma_j^c$ is the number of failure/non-operating objects and $z(\tau_{j-1})\Delta\tau_j^c + z(\tau_j^c)\Delta\tau_{jc}$ denotes the total amount of time spent by $z(\tau_{j-1}) - z(\tau_j) - \gamma_j^c$ over the the interval $[\tau_{j-1}, \tau_j)$. This establishes (i) and (ii).

To complete the proofs of (iii) and (iv), we utilize Definition 4.1 and (97), and obtain

$$\hat{\lambda}(\tau_j) = \frac{z(\tau_{j-1}) - z(\tau_j) - \gamma_j^c}{z(\tau_{j-1})\Delta\tau_j^c + z(\tau_j^c)\Delta\tau_{jc}} \quad \text{for} \quad j \in I(1,k)\,.$$

and hence

$$\lambda(t) = \hat{\lambda}(\tau_j), \quad t \in [\tau_{j-1}, \tau_j), \quad j \in I(1,k)\,.$$

This establishes proof of the theorem.

Proof of Corollary 4.2: Under the conditions of Example 2.1 and using the relationship between S, the cumulative jumps in Example 2.2, Corollary 3.2(in particular (34)), an estimate for the risk/hazard rate function at τ_j is obtained as:

$$\hat{\lambda}(\tau_j) = \frac{a_j}{(1 - A_{j-1})\Delta\tau_j}, \tag{98}$$

and an estimate for the risk/hazard rate function is

$$\hat{\lambda}(t) = \hat{\lambda}(\tau_j), \quad \text{for} \quad t \in [\tau_{j-1}, \tau_j) \quad \text{and} \quad j \in I(1, m) \tag{99}$$

From (32), using (8) and (99), an estimate for the survival function is given by:

$$\hat{S}(t) = \exp(-\Lambda_{j-1})\exp\left(\frac{-a_j(t - \tau_{j-1})}{(1 - A_{j-1})(\tau_j - \tau_{j-1})}\right), \quad \tau_{j-1} \le t < \tau_j, \tag{100}$$

where

$$\Lambda_j = \sum_{i=1}^{j} \frac{a_i}{1 - A_{i-1}}, \; 1 \le j \le m, \; \Lambda_0 := 0,$$

and Λ_j is the cumulative hazard function. This establishes the proof of the corollary.

Proof of Theorem 5.1: For each $j \in I(1, n)$ and $\tau_{j-1}, \tau_j \in \mathscr{P}_0^{\mathscr{T}}$, objects/subjects are censored k_j times over a partition of $[\tau_{j-1}, \tau_j]$ of consecutive failure times. Let $\mathscr{P}_j$ be a partition corresponding to a given finite sequence of censored times over the failure time interval $[\tau_{j-1}, \tau_j)$, and let it be represented by

$$\mathscr{P}_j : \tau_{j-1} = \tau_{j-10} < \tau_{j-11} < \ldots < \tau_{j-1l-1} < \tau_{j-1l} < \ldots < \tau_{j-1k_{j-1}} < \tau_{j-1k_j}. \tag{101}$$

where $\mathscr{P}_j$ is a partition of $[\tau_{j-1}, \tau_j]$.

For each $j \in I(1, n)$, using the iterative schemes (38) and (40) we have

$$\begin{aligned} z(\tau_j) - z(\tau_{j-1}) &= \sum_{l=1}^{k_j} \left[z(\tau_{j-1l}) - z(\tau_{j-1l-1})\right] + [z(\tau_j) - z(\tau_{j-1k_j})] \\ &= -\lambda(\tau_j)\left[\sum_{l=1}^{k_j+1} z(\tau_{j-1l-1})\Delta\tau_{j-1l}\right] - \sum_{l=1}^{k_j} \gamma_j^l, \end{aligned} \tag{102}$$

and hence

$$z(\tau_{j-1}) - z(\tau_j) - \sum_{l=1}^{k_j} \gamma_j^l = \lambda(\tau_j) \sum_{l=1}^{k_j+1} z(\tau_{j-1l-1})\Delta(\tau_{j-1l}). \tag{103}$$

Thus, $z(\tau_{j-1}) - z(\tau_j) - \sum_{l=1}^{k_j} \gamma_j^l$ is a change in the number of items/subjects that are under observation over the subinterval $[\tau_{j-1}, \tau_j]$, and $\sum_{l=1}^{k_j+1} z(\tau_{j-1l-1})\Delta(\tau_{j-1l})$ is a total amount of time spent under the observation/testing/evaluation/monitoring of $z(\tau_{j-1l})$ items/patients/infectives/subjects on the interval $[\tau_{j-1l-1}, \tau_{j-1l})$ for $l \in I(1, k_j))$ and $j \in I(1, n)$. These statements establish conclusions 1 and 2 of Theorem 5.1.

Finally, from Definition 4.1, we obtain an estimate for a hazard rate function at $\tau_j \in [\tau_0, \mathscr{T})$ as:

$$\hat{\lambda}(\tau_j) = \frac{z(\tau_{j-1}) - z(\tau_j) - \sum_{l=1}^{k_j} \gamma_j^l}{\sum_{l=1}^{k_{j+1}} z(\tau_{j-1l-1})\Delta(\tau_{j-1l})}.$$

This establishes (67).

Moreover,

$$\hat{\lambda}(t) = \hat{\lambda}(\tau_j), \quad \text{for} \quad t \in [\tau_{j-1}, \tau_j) \quad \text{and} \quad j \in I(1, n). \tag{104}$$

This completes the proof of the theorem.

***Proof of Theorem 5.2*:**

Let $0 = \tau_0 < \tau_1 < \tau_2 < \ldots < \tau_{j-1} < \tau_j < \ldots < \tau_k$ be the partition of $[\tau_0, \mathscr{T})$ corresponding to change point times. For $j = 1, 2, \ldots, k$, we consider a partition of $[\tau_{j-1}, \tau_j]$ as follows:

$$\mathscr{P}_j^\tau : \tau_{j-1} = T_0^j < T_1^j < T_2^j < T_3^j < \ldots < T_{l-1}^j < T_l^j < \ldots < T_{n-1}^j < T_n^j < T_{n+1}^j = \tau_j \,. \tag{105}$$

Imitating the proof of Theorem 5.1, we have

$$\begin{aligned} z(\tau_j) - z(\tau_{j-1}) &= \sum_{m=1}^{l} \left[z(T_m^j) - z(T_{m-1}^j) \right] + [z(\tau_j) - z(T_l^j)] \\ &= \sum_{m=1}^{l} \left[-\lambda(T_{m-1}^j) z(T_{m-1}^j) \Delta T_m^j - \eta_m^j \right] + [-\lambda(T_l^j) z(T_l^j) \Delta \tau_j] \\ &\quad - \lambda(\tau_j) \left[\sum_{m=1}^{l} z(T_{m-1}^j) \Delta T_m^j \right] - \sum_{m=1}^{l} \eta_m^j - \lambda(\tau_j) z(t_l^j) \Delta \tau_j \\ &= -\lambda(\tau_j) \left[\sum_{m=1}^{l+1} z(T_{m-1}^j) \Delta T_m^j \right] - \sum_{m=1}^{l} \eta_m^j \,, \end{aligned} \tag{106}$$

and hence

$$z(\tau_{j-1}) - z(\tau_j) - \sum_{m=1}^{l} \eta_m^j = \lambda(\tau_j) \sum_{m=1}^{l+1} z(T_{m-1}^j) \Delta T_m^j \tag{107}$$

Thus, $z(\tau_{j-1}) - z(\tau_j) - \sum\limits_{m=1}^{l} \eta_m^j$ is a change in the number of items/subjects that is under the observation over the subinterval $[\tau_{j-1}, \tau_j]$ of the time interval of study $[\tau_0, \mathscr{T}]$ and $\sum\limits_{m=1}^{l+1} z(T_m^j) \Delta T_m^j$ is a total amount of time spent under the observation/testing/evaluation of $z(T_m^j)$ items/patients/infectives/subjects on the interval $[T_{m-1}^j, T_m^j)$ for $m \in I(1, l))$ and $j \in I(1, k)$. These statements establish conclusions 1 and 2 of Theorem 5.1.

Finally, from Definition 4.1, we obtain an estimate for a hazard rate function at $\tau_j \in [\tau_0, \mathscr{T})$ as:

$$\hat{\lambda}(\tau_j) = \frac{z(\tau_{j-1}) - z(\tau_j) - \sum\limits_{m=1}^{l} \eta_m^j}{\sum\limits_{m=1}^{l+1} z(T_{m-1}^j) \Delta T_m^j} \,,$$

Moreover,

$$\hat{\lambda}(t) = \hat{\lambda}(\tau_j) \,, \quad \text{for} \quad t \in [\tau_{j-1}, \tau_j) \quad \text{and} \quad j \in I(1, k) \,. \tag{108}$$

This establishes proof of the theorem.

Proof of Corollary 5.3: Let $\tau_0 < \tau_1 < \ldots < \tau_{m-1} < \tau_m < \ldots < \tau_{j-1} < \tau_j < \ldots < \tau_n = \mathscr{T}$ be a partition of $[\tau_0, \mathscr{T}]$. Using (31), for fixed $i = 0$ and $j \in I(1, n_0)$, we have

$$z_0(\tau_m) - z_0(\tau_{m-1}) = -\lambda_0(\tau_m) z_0(\tau_{m-1}) \Delta \tau_m \,. \tag{109}$$

Summing (109) from $m = 1$ to j, we obtain

$$\begin{aligned} \sum_{m=1}^{j} [z_0(\tau_m) - z_0(\tau_{m-1})] &= \sum_{m=1}^{j} -\lambda_0(\tau_m) z_0(\tau_{m-1}) \Delta_m \\ &= -\lambda_0(\tau_j) \sum_{m=1}^{j} z_0(\tau_{m-1}) \Delta \tau_m \,. \end{aligned} \tag{110}$$

Rearranging (110) establishes (71). The proof of (73) is similar to the proof of (71). (75) is obtained by taking the natural log of the ratio of (71) and (73) . This establishes the proof of the corollary.

Acknowledgments

This research is supported by the Mathematical Sciences Division, U.S. Army Research Office, Grant No: W911NF-15-1-0182.

References

Aalen, O. (1978). Nonparametric inference for a family of counting processes. *The Annals of Statistics*, 701-726. http://dx.doi.org/10.1214/aos/1176344247

Anis, M. Z. (2009). Inference on a sharp jump in hazard rate: a review. *Economic Quality control, 24*(2), 213-229. http://dx.doi.org/10.1515/EQC.2009.213

Apostol, T. M. (1967). *Calculus, vol 1: one-variable calculus, with an introduction to linear algebra.* New York, NY: John Wiley & Sons.

Apostol, T. M. (1974). *Mathematical Analysis.* Reading, MA: Addison Wesley.

Atkinson, K. E. (2008). *An introduction to numerical anlaysis.* New York, NY: John Wiley & Sons.

Chandra, J., & Ladde, G. S. (2014). Multi-cultural dynamics on social networks under external random perturbations. *International Journal of Communications, Network and System Sciences, 7*(06), 181-195. http://dx.doi.org/10.4236/ijcns.2014.76020

Davis, D. J. (1952). An analysis of some failure data. *Journal of the American Statistical Association, 47*(258), 113-150. http://dx.doi.org/10.1080/01621459.1952.10501160

Demarqui, F. N., Loschi, R. H., & Colosimo, E. A. (2008). Estimating the grid of time-points for the piecewise exponential model. *Lifetime Data Analysis, 14*(3), 333-356. http://dx.doi.org/10.1007/s10985-008-9086-0

Fang, L., & Su, Z. (2011). A hybrid approach to predicting events in clinical trials with time-to-event outcomes. *Contemporary Clinical Trials, 32*(5), 755-759. http://dx.doi.org/10.1016/j.cct.2011.05.013

Feigl, P., & Zelen, M. (1965). Estimation of exponential survival probabilities with concomitant information. *Biometrics*, 826-838. http://dx.doi.org/10.2307/2528247

Gamerman, D. (1994). Bayes estimation of the piece-wise exponential distribution. *IEEE Transactions on Reliability, 43*(1), 128-131. http://dx.doi.org/10.1109/24.285126

Goodman, M. S., Li, Y., & Tiwari, R. C. (2011). Detecting Multiple change points in piecewise constant hazard functions. *Journal of Applied Statistics, 33*(11), 2523-2532. http://dx.doi.org/10.1080/02664763.2011.559209

Han, G., Schell, M. J., & Kim, J. (2014). Improved survival modeling in cancer research using a reduced piecewise exponential approach. *Statistics in Medicine, 33*(1), 59-73. http://dx.doi.org/10.1002/sim.5915

He, P., Kong, G., & Su, Z. (2013). Estimating the survival functions for right-censored and interval-censored data with piecewise constant hazard functions. *contemporary Clinical Trials, 35*(2), 122-127. http://dx.doi.org/10.1016/j.cct.2013.04.009

Kalbfleish, J. D., & Prentice, R. L. (2011). *The statistical analysis of failure time data* (Vol. 360). New Jersey: John Wiley & Sons.

Kaplan, E. L., & Meier, P. (1958). Nonparametric estimation from incomplete observations. *Journal of the American Statistical Association, 53*(282), 457-481. http://dx.doi.org/10.1080/01621459.1958.10501452

Kim, J. S., & Proschan, F. (1991). Piecewise exponential estimator of the survivor function. *IEEE Transactions on Reliability, 40*(2), 134-139. http://dx.doi.org/10.1109/24.87112

Kitchin, J., Langberg, N. A., & Proschan, F. (1980). *A new method for estimating life distributions from incomplete data* (No. FSU - Statistics-M548). Florida State Uni Tallahassee, Dept of Statistics.

Kulasekera, K., & White, W. H. (1996). Estimation of the survival function from censored data: a method based on the total time on test. *Commnications in Statistics-Simulation and Computation, 25*(1), 189-200. http://dx.doi.org/10.1080/03610919608813306

Ladde, A. G., & Ladde, G. S. (2012). *An introduction to differential equations. Deterministic Modeling, Methods and Anlaysis*(Vol. 1). Singapore: World Scientific

Ladde, G. S. (2015). *Network dynamic processes under stochastic perturbations.* Technical report, U.S. Army Research Office, Mathematical Sciences Division, Research Triangle Park, NC.

Lawless, J. F. (2011). *Statistical models and methods for lifetime data* (Vol. 362). New Jersey: John Wiley & Sons.

Malla, G., & Mukerjee, H. (2010). A new piecewise exponential estimator of a survival function. *Statistics and Probability Letters, 80*(23), 1911-1917. http://dx.doi.org/10.1016/j.spl.2010.08.019

Miller R, G. (2011). *Survival Analysis* (Vol 66). New York, NY: John Wiley & Sons.

Nelson, W. (1969). Hazard plotting for incomplete failure data. *Journal of Quality Control, 1*(1), 27-52.

Rosen, R. (1970). *Dynamical system theory biology: stability theory and its applications.* New York, NY: John Wiley & Sons.

Thaler, H. (1984). Nonparametric estimation of the hazard ratio. *Journal of the American Statistical Association, 79*(386), 290-293 http://dx.doi.org/10.1080/01621459.1984.10478043

Wanduku, D., & Ladde, G. S. (2011). A two-scale network dynamic model for human mobility process. *Mathematical Biosciences, 229*(1), 1-15. http://dx.doi.org/10.1016/j.mbs.2010.11.003

Whittemore, A. S., & Keller, J. B. (1983). *Survival estimation with censored data.* Department of Statistics, Stanford University, Stanford, CA.

6

On Sequential Learning for Parameter Estimation in Particle Algorithms for State-Space Models

Chunlin Ji[1]

[1] Kuang-Chi Institute of Advanced Technology, Shenzhen, China

Correspondence: Chunlin Ji, Kuang-Chi Institute of Advanced Technology, Shenzhen 518000, China.
E-mail: chunlin.ji@kuang-chi.org

Abstract

Particle methods, also known as Sequential Monte Carlo, have been ubiquitous for Bayesian inference for state-space models, particulary when dealing with nonlinear non-Gaussian scenarios. However, in many practical situations, the state-space model contains unknown model parameters that need to be estimated simultaneously with the state. In this paper, We discuss a sequential analysis for combined parameter and state estimation. An online learning method is proposed to approach the distribution of the model parameter by tuning a flexible proposal mixture distribution to minimize their Kullback-Leibler divergence. We derive the sequential learning method by using a truncated Dirichlet processes normal mixture and present a general algorithm under a framework of the auxiliary particle filtering. The proposed algorithm is verified in a blind deconvolution problem, which is a typical state-space model with unknown model parameters. Furthermore, in a more challenging application that we call meta-modulation, which is a more complex blind deconvolution problem with sophisticated system evolution equations, the proposed method performs satisfactorily and achieves an exciting result for high efficiency communication.

Keywords: sequential learning, sequential Monte Carlo, Kullback-Leibler divergence, blind deconvolution, meta-modulation

1. Introduction

State-space model, a class of probabilistic graphical model (Koller and Friedman, 2009) that describes the dependence between the unobserved state variable and the observed measurement, is a fundamental model for statistical inference with diversely applications in fields like statistics, econometrics, and information engineering (West and Harrison, 1997; Cappé et al., 2005). A linear or 'weakly' nonlinear state space model with Gaussian noise can be easily solved using Kalman filter and its derivatives. In nonlinear non-Gaussian scenarios, particle methods (also known as Sequential Monte Carlo methods) have been proven to be the most successful approach for the numerical approximation and Bayesian inference of the unknown state, because of their simplicity, flexibility, and ease of implementation (Gordon et al., 1993; Liu and Chen, 1998; Doucet et al., 2000; Liu, 2001; Doucet et al., 2001). In most of previous research, the inference has been focused on filtering or smoothing for the state, with the assumption that the model parameter in the state-space model are known. However, in practical situations, the model may contain unknown parameter that need to be estimated simultaneously with the state, and it might even be the case that the inference of the model parameter is the primary problem of interest.

One early and straightforward way to deal with this problem is to extend the original state to an augmented state that includes the state and the parameter together, and then to apply a standard particle filter to perform inference for both of them. However, this naive approach has been recognized not to be efficient, because the parameter space is not well explored (Kitagawa, 1998; Liu and West, 2001). Consequently, various improve methods has been developed over the past fifteen years (Refer to Kantas et al. (2015) for thoroughly review): maximum likelihood methods have been developed with different Monte Carlo evaluations of the likelihood of the model parameter (Hürzeler and Künsch, 2001; DeJong et al. 2013), and gradient based optimization (Ionides et al., 2006; Ionides et al., 2011) or expectation maximization methods (Andrieu et al., 2005; Cappé, 2009) have been introduced for an on-line or off-line estimation of the model parameter. The maximum likelihood approach generally converges rather slowly, but it may be a good choice for large data sets because of its low complexity; Bayesian methods apply directly to the augmented states and Markov chain Monte Carlo steps are utilized to improve the inference/estimation of the model parameter (Gilks and Berzuini, 2001; Fearnhead, 2002; Andrieu et al., 2010). In the inspiring work by(Liu and West, 2001), an artificial dynamic is introduced to the static model parameter, and a kernel density estimation method is proposed to capture the density of the parameters. The significant idea of these authors is to use a shrinkage strategy for kernel locations and variances inflation, which therefore removes the problem of information loss over time (West, 1993). However, the learning method proposed in this work seems to be ad hoc and weak in explaining the underlying driving strategy. In this paper, a stochastic algorithm is proposed that tunes a flexible proposal mixture distribution to minimize the Kullback-Leibler divergence between the distribution of the unknown model parameter and the distribution of the proposed mixture: this leads to an online learning method for the numerical approximation of the distribution of the model parameter. We derive the sequential learning method by using a truncated Dirichlet processes normal mixture, and present the general algorithm using a proposed learning approach under a framework of the auxiliary particle filtering. The proposed algorithm is verified and compared with other related algorithms in a blind deconvolution problem, which is a typical state-space model with unknown model parameter (Liu and Chen, 1995). Furthermore, a more novel and challenging application is presented, which is essentially a more complex blind deconvolution problem with sophisticated system evolution equation. The proposed method performs satisfactorily in dealing with the demodulation of this system and obtains an exciting result for high efficiency communication.

The paper is organized as follows. In Section 2 we present the general state space model and the auxiliary particle filter algorithm for the Bayesian inference of this model. In Section 3 we introduce the sequential learning method and a general framework for joint parameter and state estimation. In Section 4 we verify the proposed method in a simple blind deconvolution and a more channelling problem in wireless communication. Finally, we summarize our methods in Section 5.

2. Problem Statement and Particle Filter Algorithm

A state-space model can be defined by the following two processes: the Markovian evolution equation, or state equation,

$$\mathbf{x}_t = f(\mathbf{x}_{t-1}, \theta) + w_t \tag{1}$$

defining the transition density $p(\mathbf{x}_t|\mathbf{x}_{t-1}, \theta)$, in which the state vector at time t is $\mathbf{x}_t$, the model parameter vector is θ, $f(\cdot)$ is the system evolution function and w_t is the system noise; and the observation equation

$$\mathbf{y}_t = h(\mathbf{x}_t, \theta) + v_t \tag{2}$$

defining the observation density $p(\mathbf{y}_t|\mathbf{x}_{t-1}, \theta)$, where $h(\cdot)$ is the observation function and v_t is the observation noise. The above expressions covers a very broad class of interesting dynamic models (West and Harrison 1997).

With a known parameter θ the task of sequential Bayesian inference is to estimate the posterior distribution $p(\mathbf{x}_t|\mathbf{y}_{0:t}, \theta)$. However, to apply particle methods, a more general approach is to estimate the sequence of joint posteriors $p(\mathbf{x}_{0:t}|\mathbf{y}_{0:t}, \theta)$ recursively, which leads to the following fundamental recursions: for $t \geq 1$,

$$\begin{aligned}
p(\mathbf{x}_{0:t}|\mathbf{y}_{0:t}, \theta) &= \frac{p(\mathbf{x}_{0:t}, \mathbf{y}_{0:t}|\theta)}{p(\mathbf{y}_{0:t}|\theta)} && (3)\\
&= \frac{p(\mathbf{x}_t, \mathbf{y}_t|\mathbf{x}_{0:t-1}, \mathbf{y}_{0:t-1}, \theta)p(\mathbf{x}_{0:t-1}|\mathbf{y}_{0:t-1}, \theta)p(\mathbf{y}_{0:t-1}|\theta)}{p(\mathbf{y}_t|\mathbf{y}_{0:t-1}, \theta)p(\mathbf{y}_{0:t-1}|\theta)} && (4)\\
&= \frac{p(\mathbf{x}_t, \mathbf{y}_t|\mathbf{x}_{0:t-1}, \mathbf{y}_{0:t-1}, \theta)p(\mathbf{x}_{0:t-1}|\mathbf{y}_{0:t-1}, \theta)}{p(\mathbf{y}_t|\mathbf{y}_{0:t-1}, \theta)} && (5)\\
&= p(\mathbf{x}_{0:t-1}|\mathbf{y}_{0:t-1}, \theta)\frac{p(\mathbf{y}_t|\mathbf{x}_{0:t}, \mathbf{y}_{0:t-1}, \theta)p(\mathbf{x}_t|\mathbf{x}_{0:t-1}, \mathbf{y}_{0:t-1}, \theta)}{p(\mathbf{y}_t|\mathbf{y}_{0:t-1}, \theta)} && (6)\\
&= p(\mathbf{x}_{0:t-1}|\mathbf{y}_{0:t-1}, \theta)\frac{p(\mathbf{y}_t|\mathbf{x}_t, \theta)p(\mathbf{x}_t|\mathbf{x}_{t-1}, \theta)}{p(\mathbf{y}_t|\mathbf{y}_{0:t-1}, \theta)} && (7)
\end{aligned}$$

where $p(\mathbf{y}_t|\mathbf{y}_{0:t-1}, \theta) = \int p(\mathbf{x}_{t-1}|\mathbf{y}_{0:t-1}, \theta)p(\mathbf{x}_t|\mathbf{x}_{t-1}, \theta)p(\mathbf{y}_t|\mathbf{x}_t, \theta)d\mathbf{x}_{t-1:t}$, $p(\mathbf{x}_t|\mathbf{x}_{t-1}, \theta)$ represents the system dynamic in (1), and $p(\mathbf{y}_t|\mathbf{x}_t, \theta)$ refers to the likelihood function obtained in (2). The recursion is initialised with some distribution, for example, $p(\mathbf{x}_0)$.

Particle filtering is a class of importance sampling and resampling techniques designed to give a numerical approximation for the recursions in (7). We present the auxiliary particle filter (APF) (Pitt and Shephard, 1999) here, as it covers a class of particle filter algorithms (Kantas et al., 2015) and is widely used in parameter and state estimation (Liu and West, 2001; Flury and Shephard, 2011). Let the proposal be $q(\mathbf{x}_t, \mathbf{y}_t|\mathbf{x}_{t-1}, \phi) = q(\mathbf{x}_t|\mathbf{y}_t, \mathbf{x}_{t-1}, \phi)q(\mathbf{y}_t|\mathbf{x}_{t-1}, \phi)$, where $q(\mathbf{x}_t|\mathbf{y}_t, \mathbf{x}_{t-1}, \phi)$ is a probability density function which is easy to sample from and $q(\mathbf{y}_t|\mathbf{x}_{t-1}, \phi)$ is a nonnegative function that can be evaluated. The auxiliary particle filter sequentially draws samples from $q(\mathbf{x}_t|\mathbf{y}_t, \mathbf{x}_{t-1}, \phi)$ and calculates the following importance weights,

$$w_t(\mathbf{x}_{t-1:t}) = \frac{p(\mathbf{x}_t|\mathbf{x}_{t-1}, \theta)p(\mathbf{y}_t|\mathbf{x}_t, \theta)}{q(\mathbf{x}_t, \mathbf{y}_t|\mathbf{x}_{t-1}, \phi)}. \tag{8}$$

For $t \geq 1$ the APF algorithm can be summarized as follows,

At iteration $t \geq 1$, for all $i = 1, ..., N$:

Step 1: Sample $\mathbf{x}_t^{(i)} \sim q(\mathbf{x}_t|\mathbf{y}_t, \mathbf{x}_{t-1}^{(i)}, \phi)$ and set $\mathbf{x}_{0:t}^{(i)} = [\mathbf{x}_{0:t-1}^{(i)}, \mathbf{x}_t^{(i)}]$;

Step 2: Compute the weights $\widetilde{w}_t^{(i)} \propto w_t(\mathbf{x}_{t-1:t}^{(i)})q(\mathbf{y}_t|\mathbf{x}_{t-1}^{(i)}, \phi)$, where $\widetilde{w}_t^{(i)}$ is normalized as $\sum_{i=1}^N \widetilde{w}_t^{(i)} = 1$;

Step 3: Resample $\mathbf{x}_{0:t}^{(i)} \sim \sum_{i=1}^N \widetilde{w}_t^{(i)}\delta(\mathbf{x}_{0:t}^{(i)})$.

In many applications, the model parameter θ is unknown and it might even be of interest for inference. A straightforward solution is to define an extended state that includes the state $\mathbf{x}_t$ and the parameter θ together. For example in (Liu and West, 2001), an 'artificial evolution' equation for the model parameter θ is introduced: $\theta_t = \theta_{t-1} + \varepsilon$, where $\varepsilon \sim N(0, \Sigma_t)$ with some specified variance matrix Σ_t. A kernel density estimation method is proposed to capture the density of the parameter θ_t. However, the method of the kernel density estimation is critical in such a problem. As studied by Liu and West (2001), they use a shrinkage strategy for kernel locations and variances estimation in order to removes the over-dispersion of the variances. Although this method works satisfactorily with a careful chosen shrinkage strategy, more efficient approach are still worth for further exploration (Kantas et al., 2015).

3. Sequential Learning and Joint Parameter and State Estimation

3.1 Sequential Learning

A sequential learning method is proposed here to deal with the online learning of the unknown distribution of the model parameter. Assume that the parameter vector θ is subject to some unknown distribution $\pi(\theta)$. We propose to utilize a parametric distribution $q(\theta; \psi)$ with the controlling parameter ψ to approximate the unknown $\pi(\theta)$. To measure the distance between these two distributions, the broadly used Kullback-Leibler divergence (KL-divergence) is employed here. The idea of the proposed sequential learning approach is to tune the parameter ψ by learning from samples of $\pi(\theta)$, and enable $q(\theta; \psi)$ to approximate $\pi(\theta)$ in the sense of minimizing the KL-divergence $\mathcal{D}[\pi(\cdot) \parallel q(\cdot)] = \mathbb{E}_\pi \left[\log \frac{\pi(\cdot)}{q(\cdot)}\right]$. Other criterion like moment matching can also be applied to measure the closeness of two distributions (Ji, 2006), but it is not straightforward to derive the algorithm to find the optimal controlling parameter of the proposal distribution. Under the measurement of KL-divergence, the optimal parameter ψ^* which minimizes $\mathcal{D}\left[\pi(\theta) \parallel q(\theta; \psi)\right]$ can be obtained by finding the root of the derivative of $\mathcal{D}$ (if exists):

$$h(\psi) = -\int \pi(\theta)\frac{\partial}{\partial\psi}\left(\log \frac{\pi(\theta)}{q(\theta; \psi)}\right)d\theta = 0. \tag{9}$$

The closed-form solution of the integral equation (9) is generally intractable, as $h(\psi)$ involves an integral with unknown distribution $\pi(\theta)$. However, suppose that we can obtain samples $\theta^{(1:N)}$ from $\pi(\theta)$, denote $f(\theta, \psi) = \frac{\partial}{\partial\psi}\left(\log \frac{\pi(\theta)}{q(\theta;\psi)}\right)$, then we can numerically evaluate $h(\psi)$ by Monte Carlo integration:

$$\hat{h}(\theta^{(1:N)}; \psi) = \frac{1}{N}\sum_{i=1}^{N} f(\theta^{(i)}, \psi) \tag{10}$$

where $\hat{h}(\theta^{(1:N)}; \psi)$ can be viewed as a noisy 'observation' of $h(\psi)$. One available approach for obtaining roots of $h(\psi) = 0$ when we only have noisy evaluations of $h(\psi)$ is the Stochastic Approximation (SA) algorithm (Kusher and Yin, 1997). The SA algorithm iteratively updates ψ to approximate its optimal values by the following formula:

$$\begin{aligned}\psi_t &= \psi_{t-1} + r_t\left[h(\psi_{t-1}) + \xi_t\right] \\ &= \psi_{t-1} + r_t\,\hat{h}(\theta_{t-1}^{(1:N)}; \psi_{t-1})\end{aligned} \tag{11}$$

where t is an iterative index, $\{\xi_t\}$ is a sequence of 'noise' (thus the Monte Carlo estimation $\hat{h}(\theta_t^{(1:N)}; \psi_t)$ can be interpreted as the ground true $h(\psi_t)$ plus noise ξ_t), and $\{r_t\}$ is a sequence of decreasing step-sizes satisfying $\sum_t r_t = \infty$ and $\sum_t r_t^2 < \infty$. The proof of covergence of this Monte Carlo estimation based sequential learning algorithm is detailed in the appendix.

We assume that $q(\theta; \psi)$ is a truncated Dirichlet process (TDP) normal mixture (Ishwaran and James, 2001; Ji, 2009). Let ψ denotes the set of controlling parameters (V_k, μ_k, Σ_k), a TDP normal mixture can be expressed as, $q(\theta; \psi) = \sum_{k=1}^K \omega_k N(\cdot|\mu_k, \Sigma_k)$, where $\omega_k = V_k \prod_{l=1}^{k-1}(1 - V_l)$ and $V_l \in (0, 1)$ which can be initially drawn from a beta distribution $Beta(\alpha_0,\beta_0)$. The the partial derivative of $\mathcal{D}[\pi(\theta) \parallel q(\theta; \psi)]$ with respect to V_k, μ_k and Σ_k can be derived respectively as follows (refer to Ji (2009) for more details of derivations),

$$h_{V_k}(\theta; \psi) = \int \pi(\theta)\frac{-\sum_{l=k+1}^K V_l \prod\limits_{\iota\le l-1, \iota\ne k}(1 - V_\iota)q(\theta|\mu_l, \Sigma_l) + \prod\limits_{l=1}^{k-1}(1 - V_l)q(\theta|\mu_k, \Sigma_k)}{\sum\limits_{m=1}^{K} \omega_m q(\theta|\mu_m, \Sigma_m)}d\theta \tag{12}$$

$$H_{\mu_k}(\theta;\psi) \propto \int \pi(\theta) \frac{\omega_k q(\theta;\mu_k,\Sigma_k)}{\sum_{m=1}^K \omega_m q(\theta;\mu_m,\Sigma_m)} \times (\theta-\mu_k)\, d\theta \tag{13}$$

$$H_{\Sigma_k}(\theta;\psi) \propto \int \pi(\theta) \frac{\omega_k q(\theta;\mu_k,\Sigma_k)}{\sum_{m=1}^K \omega_m q(\theta;\mu_m,\Sigma_m)} \times \left((\theta-\mu_k)(\theta-\mu_k)^T - \Sigma_k\right) d\theta \tag{14}$$

Given the samples $\Theta_t = \{\theta_t^{(i)}\}_{i=1}^N$ from $\pi(\theta)$, the Monte Carlo approximation of these partial derivatives is

$$H_{V_k}(\Theta_t;\psi) = \frac{1}{N}\sum_{i=1}^N \frac{-\sum_{l=k+1}^K V_l \prod_{\iota\le l-1,\iota\ne k}(1-V_\iota)q(\theta_t^{(i)}|\mu_l,\Sigma_l) + \prod_{l=1}^{k-1}(1-V_l)q(\theta_t^{(i)}|\mu_k,\Sigma_k)}{\sum_{m=1}^K \omega_m q(\theta_t^{(i)}|\mu_m,\Sigma_m)} \tag{15}$$

$$H_{\mu_k}(\Theta_t;\psi) \propto \frac{1}{N}\sum_{i=1}^N \frac{\omega_k q(\theta_t^{(i)};\mu_k,\Sigma_k)}{\sum_{m=1}^K \omega_m q(\theta_t^{(i)};\mu_m,\Sigma_m)} \times \left(\theta_t^{(i)}-\mu_k\right) \tag{16}$$

$$H_{\Sigma_k}(\Theta_t;\psi) \propto \frac{1}{N}\sum_{i=1}^N \frac{\omega_k q(\theta_t^{(i)};\mu_k,\Sigma_k)}{\sum_{m=1}^K \omega_m q(\theta_t^{(i)};\mu_m,\Sigma_m)} \times \left(\left(\theta_t^{(i)}-\mu_k\right)\left(\theta_t^{(i)}-\mu_k\right)^T - \Sigma_k\right) \tag{17}$$

3.2 Joint Parameter and State Estimation

Let us return to the inference of the augmented states under a particle filter framework. Assuming the particle samples $(\mathbf{x}_t^{(i)}, \theta_t^{(i)})$ and weights $w_t^{(i)}$ $(i = 1, ..., N)$ are available to represent the joint posterior $p(\mathbf{x}_t, \theta|\mathbf{y}_t)$, with the sequential learning algorithm to approach the unknown distribution of θ by samples $\theta_t^{(i)}$ $(i = 1, ..., N)$, we now have an extended version of the auxiliary particle filter algorithm, incorporating the parameter and the state estimation together. The resulting general algorithm is presented as follows:

Step 1: At iteration t, the sampled N particles containt $\left\{\mathbf{x}_t^{(i)}, \theta_t^{(i)}, c_t^{(i)}\right\}_{i=1}^N$ (where $c_t^{(i)}$ is the label of the TDP normal mixture component which $\theta_t^{(i)}$ belongs to), with its weights $w_t^{(i)}$. Calculate the auxiliary variable

$$\phi_{t+1}^{(i)} = E(\mathbf{x}_{t+1}|\mathbf{x}_t^{(i)}, \theta_t^{(i)}).$$

Step 2: Sample the auxiliary index from the set $\{1, ..., N\}$ with probabilities proportion to

$$g_{t+1}^{(i)} \propto w_t^{(i)} p(\mathbf{y}_{t+1}|\phi_{t+1}^{(i)}, \theta_t^{(i)});$$

obtain new index set j, and new label set $c_t^{(j)}$.

Step 3: Sample a new parameter vector $\theta_{t+1}^{(j)}$ from the $k = c_t^{(j)}$ normal components of the mixture model

$$\theta_{t+1}^{(j)} \sim N(\cdot|\mu_{k,t}, \Sigma_{k,t}).$$

Step 4: Sample the new state vector $\mathbf{x}_{t+1}^{(j)}$, from the system equation

$$\mathbf{x}_{t+1}^{(j)} \sim p(\cdot|\mathbf{x}_t^{(j)}, \theta_{t+1}^{(j)}).$$

Step 5: Calculate the importance weights

$$w_{t+1}^{(j)} \propto \frac{p(\mathbf{y}_{t+1}|\mathbf{x}_{t+1}^{(j)}, \theta_{t+1}^{(j)})}{p(\mathbf{y}_{t+1}|\phi_{t+1}^{(j)}, \theta_t^{(j)})}$$

Step 6: Update the parameters of the TDP mixture V_k, ω_k, μ_k and Σ_k (for $k = 1, ..., K$) using $\left\{w_{t+1}^{(j)}, \theta_{t+1}^{(j)}\right\}_{j=1}^{N}$ as follows,

$$V_{k,t+1} = V_{k,t} + r_{t+1} H_{V_k}(\Theta_t; \psi_t) \quad (18a)$$

$$\omega_{k,t+1} = V_{k,t+1} \prod_{l=1}^{k-1} (1 - V_{l,t+1}) \quad (18b)$$

$$\mu_{k,t+1} = \mu_{k,t} + r'_{t+1} \sum_{j=1}^{N} w_{t+1}^{(j)} \alpha_{k,t+1}^{(j)} \left(\theta_{t+1}^{(j)} - \mu_{k,t}\right) \quad (18c)$$

$$\Sigma_{k,t+1} = \Sigma_{k,t} + r'_{t+1} \sum_{j=1}^{N} w_{t+1}^{(j)} \alpha_{k,t+1}^{(j)} \left((\theta_{t+1}^{(j)} - \mu_{k,t})(\theta_{t+1}^{(j)} - \mu_{k,t})^T - \Sigma_{k,t}\right) \quad (18d)$$

where $H_{V_k}(\Theta_t; \psi_t) = \sum_{i=1}^{N} w_{t+1}^{(j)} \frac{-\sum_{l=k+1}^{K} V_l \prod_{\iota \le l-1, \iota \ne k} (1-V_\iota) q(\theta_t^{(j)}|\mu_l, \Sigma_l) + \prod_{l=1}^{k-1} (1-V_l) q(\theta_t^{(j)}|\mu_k, \Sigma_k)}{\sum_{m=1}^{K} \omega_m q(\theta_t^{(j)}|\mu_m, \Sigma_m)}$, $\alpha_{k,t+1}^{(j)} = \frac{\omega_{k,t} q(\theta_{t+1}^{(j)}; \mu_{k,t}, \Sigma_{k,t})}{\sum_{m=1}^{K} \omega_{m,t} q(\theta_{t+1}^{(j)}; \mu_{m,t}, \Sigma_{m,t})}$, r_{t+1} and r'_{t+1} are the step-sizes in the stochastic approximation algorithm.

Step 7: If the stopping criterion is satisfied, then stop; otherwise, set $t := t + 1$ and go back to step 1.

3. Simulation Study

3.1 Blind Deconvolution

Blind deconvolution of source signals is a subject that has been widely studied and applied in various fields, such as wireless communications, sonar and radar systems, audio and acoustics, and image processing (Benveniste and Goursat, 1984; Haykin, 1994). Take the wireless communication over a multipath fading channel as an example (Chen and Liu, 2000; Ali et al., 2002), the unknown information signal propagates through the channel which mixes the signal as it passes through a filter. The blind deconvolution problem is to recover the independent sources from the mixed signal without any priori knowledge of the original signal and the conditions of the channel. This problem is a typical state-space model problem, and can be expressed as follows,

$$x_t = \sum_{i=0}^{q} h_i s_{t-i} \quad (19)$$

$$y_t = x_t + \epsilon_t \quad (20)$$

where the information bits $s_t \in \{-1, 1\}$ are independent, identically distributed (i.i.d) samples from a *Bernoulli*(0.5), $[h0, h_1, ..., h_q]^T$ is a set of coefficients representing the channel conditions, q is the number of the multipath, y_t is the received signal for decoding and the noise ϵ are i.i.d. normal with mean zero and variance σ.

In the simulation study, we consider the following settings: the number of particles in the particle filter is 200, the number of components in the TDP normal mixture is 20, the initial values of μ_k are randomly chosen from $(-0.5, 1)$ and Σ_k are $0.1 * \mathbf{I}$ where $\mathbf{I}$ is an identity matrix. The noise variance σ is an important parameter in the blind deconvolution problem, where we use the signal to noise ratios(SNR) to measure the noise level. For the first study, the SNR is 14 dB and the coefficients of the channel are shown in Table 1. We run 10 trials, and each trial has 10,000 information bits. The sequential learning algorithm outputs an online estimation of the coefficients $\theta \triangleq [h_0, h_1, h_2]^T$. At each iteration we estimate the mean value of θ using the particles $\left\{w_t^{(j)}, \theta_t^{(j)}\right\}_{j=1}^{200}$. The resulting trajectories from 1 to 2000 are shown in Figure 1, which demonstrates the convergence rate and stability of the sequential learning algorithm. The mean values of each h_i over a long period (iterations 1,000 to 10,000) are shown in Table 1, which show the high accuracy of our algorithm in estimating the model parameter. The performance of blind deconvlution in the above communication system can be well measured using the bit error rate (BER) of different SNRs (Chen and Liu, 2000; Ali et al., 2002). We compare our proposed method (SL+APF) with a naive bootstrap particle filter (BPF) with augmented state (x, θ), a BPF with known θ, an APF with augmented state (x, θ), an APF with known θ and a matched filter bound (MFB) which represents a reference line for the BER in Gaussian noise channel (Ali et al., 2002). The MFB can be calculated as follows (Ali et al., 2002): $1 - \int_{-\infty}^{p} \frac{1}{\sqrt{2\pi}} e^{-\frac{x^2}{2}} dx$, where $p = \frac{\sqrt{\sum_{i=0}^{q} h_i^2}}{\sigma}$. Figure 2 shows the BERs for different SNRs (in dB). The APF based methods can generally outperform the BPF based methods. The proposed SL+APF method outperforms the APF with augmented states, and the resulting BER is close to the one evaluated by the APF when θ is known. Moreover, APF with augmented states has a apparent higher SNR requirement for low BERs below 10^{-3}, while our proposed method overcomes this drawback because of the efficient learning for channel coefficients.

Table 1. The coefficients of channel response and the estimation via sequential learning algorithm

Coefficients	h_0	h_1	h_2
True value	0.9460	0.2340	-0.1670
Estimation	0.9450	0.2322	-0.1669

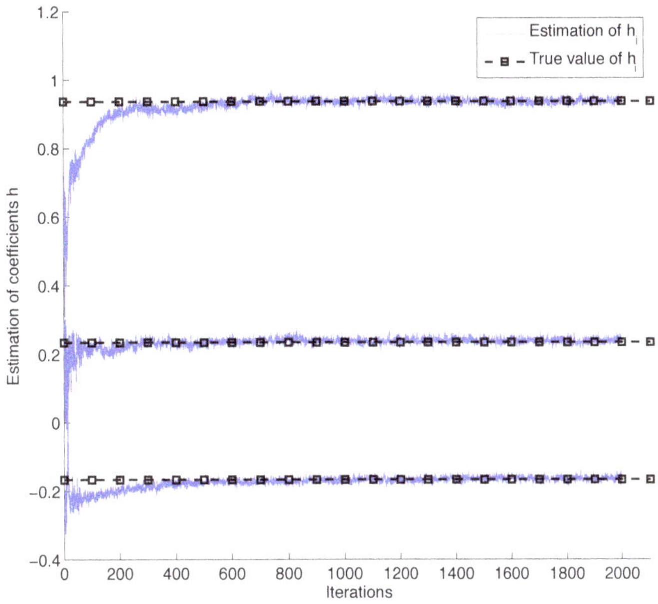

Figure 1. Trajectory plots of the mean value of $\theta = [h_0, h_1, h_2]^T$ estimated using the particles $\left\{w_t^{(j)}, \theta_t^{(j)}\right\}_{j=1}^{200}$ from iteration 0 to 2000. The estimation of each h_i comes close to the true value with in a few hundreds iterations and becomes stable since then.

3.2 Meta-modulation

In this example we consider a realistic state space model with sophisticated system evolution equations. In the above example, only the channel response is considered as the mixing mechanism, which is expressed as the coefficients **h**. Here, we consider a more general situation, where the impulse response consists of all the filters and the channel response during the entire transmission. Let $\mathbf{h}_t = [h_0, ..., h_{q_t}]^T$ represents all the filter responses in the transmitter, such as the pulse shaping filter, the transmitter filter and etc. Let $\mathbf{h}_c = [h_0, ..., h_{q_c}]^T$ be the channel response, and let $\mathbf{h}_r = [h_0, ..., h_{q_r}]^T$ be the response of all the filters in the receiver, such as the receiver filter, the matched filter and etc.

$$x_t = \sum_{q=0}^{Q} h_q s_{t-q} \tag{21}$$

$$y_t = x_t + \epsilon_t \tag{22}$$

where $\mathbf{h} = [h_0, h_2, ..., h_Q]^T$ represents the convolution of all the impulse responses $\mathbf{h} = \mathbf{h}_t \otimes \mathbf{h}_c \otimes \mathbf{h}_r$, and $Q = q_t + q_c + q_r$ is the total number of the filter coefficients.

Let $S = [s_0, s_\tau, s_{2\tau}, ...]^T$ denotes the transmitted i.i.d. bit sequence. In traditional communication system, S will be N times upsampled to $S' = [s_0, 0, ..., 0, s_\tau, 0, ..., 0, s_{2\tau}...]^T$, and then S' passes through a pulse shaping filter $h(n\tau/N)$ $(n = 0, ..., N-1)$ with duration τ. The resulting convolution signal will be

$$\mathbf{x} = [s_0 h((N-1)\tau/N), s_0 h((N-2)\tau/N), ..., s_0 h(0), s_\tau h((N-1)\tau/N), s_\tau h((N-2)\tau/N), ..., s_\tau h(0), ...]^T. \tag{23}$$

To demodulate such a signal, one can apply the matched filter to the received noisy signal **y**, where $\mathbf{y} = \mathbf{x} + \epsilon$, and then decide the value of each s_i by using the signal detection theory. Obviously, such a demodulation is not a blind deconvolution problem. On the other hand, consider the scenario that S is transmitted Q times faster, then S becomes

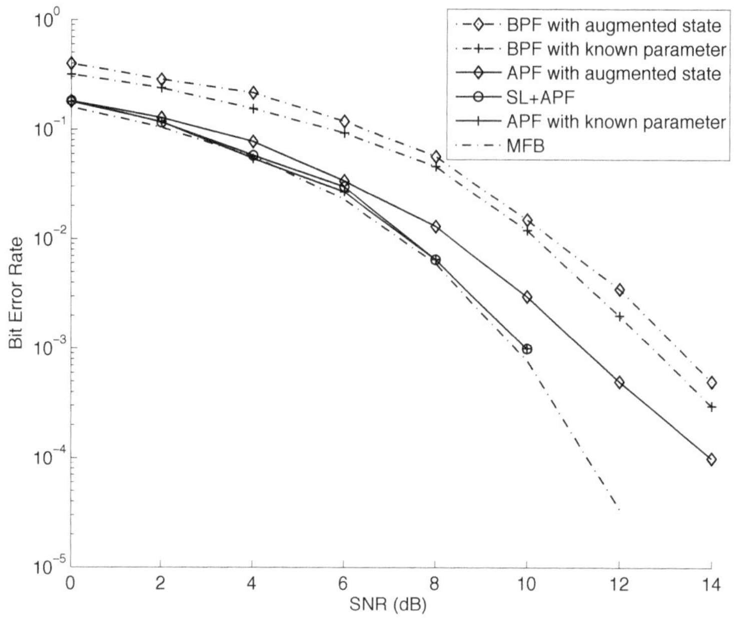

Figure 2. Comparisons of BERs for different methods. Both APF with known parameter and our proposed SL+APF algorithms deconvolute all the 10,000 bits correctly when SNR ≥ 12dB, so no points appear in 12dB and 14dB for these two curves.

$S' = [s_0, s_{\tau/Q}, ..., s_{(Q-1)\tau/Q}, s_\tau, ...]^T$. Let S' passes through a pulse shaping filter $h(q\tau/Q)$ $(q = 0, ..., Q-1)$ with duration τ, then we get a truly convoluted signal

$$\tilde{\mathbf{x}} = [s_0 h((Q-1)\tau/Q), s_0 h((Q-2)\tau/Q) + s_{\tau/Q} h((Q-1)\tau/Q), ..., \sum_{q=0}^{Q} s_{(n-q)\tau/Q} h(q\tau/Q), ...]^T. \tag{24}$$

Consequently this communication system violates the Nyquist intersymbol interference criterion, and traditional matched filter and detection approaches fail to demodulated such signal, however, we can appeal to the blind deconvolution techniques to separate the information bits at the receiver. The notable advantage of this system is that the bandwidth of modulated signal $\tilde{\mathbf{x}}$, which depends mainly on the pulse shaping filter $h(q\tau/Q)$, is comparable with the bandwidth of $\mathbf{x}$ as the pulse duration τ is the same, but the information sequence S has been managed to transmitted Q times faster.

Furthermore, it can be verified that this system essentially belongs to a parallel Gaussian channel model: let the vector $S_{Q\times 1}$ represents Q bits information, $H_{(2Q-1)\times Q}$ be a matrix that each row is the filter coefficients $\mathbf{h}^T$ shifted with one time slot delay, then the received signal $Y_{(2Q-1)\times 1}$ is given by the matrix format $Y = HS + \epsilon$. It is trivial to find that the rank of matrix $H_{(2Q-1)\times Q}$ is Q. Apply a singular value decomposition, $H = U\Lambda V$, where both $U_{(2Q-1)\times(2Q-1)}$ and $V_{Q\times Q}$ are the unitary matrix and $\Lambda_{(2Q-1)\times Q}$ is a diagonal matrix with rank Q, then

$$Y' = \Lambda S' + \epsilon' \tag{25}$$

where $Y' = U^{-1}Y$, $S' = VS$ and $\epsilon' = U^{-1}\epsilon$. According to (25) , this system is equivalent to Q independent parallel channels (Tse and Viswanath 2005; Cover and Thomas, 2006). Consequentially, the transmission rate R of this communication system will be bounded by the capacity of a parallel Gaussian channel model: $R \leq C = \frac{Q}{2}\log_2(1 + \mathrm{SNR}/Q)$ bits per transmission. This high efficient communication strategy is proposed to be named as 'meta-modulation': it utilizes ordinary modulation framework but allows heavy interaction such as convolution in certain domain of the information bits, and therefore format an independent parallel channel to achieve an extraordinary high spectrum efficiency.

In this simulation study, we consider the case that Q is sufficient large, then it is not trivial to demodulate the convolutional signal of (21) particulary some of the coefficients are unknown. For example, the Viterbi algorithm can be modified to deal with deconvolution or blind deconvoultion (Ali et al., 2001; Li, 2014), but it becomes infeasible for large Q as the potential paths in Viterbi algorithm increase exponentially. However, our proposed SL+APF method seems well suite for the deconvolution of this meta-modulation system. The simulation is set as follows, h_t is a Gaussian filter with $q_t = 100$, h_r is also a Gaussian filter with $q_r = 100$, $h_c = [0.925, 0.430, -0.201, -0.117, 0.091]$. Parameter settings of the algorithm are

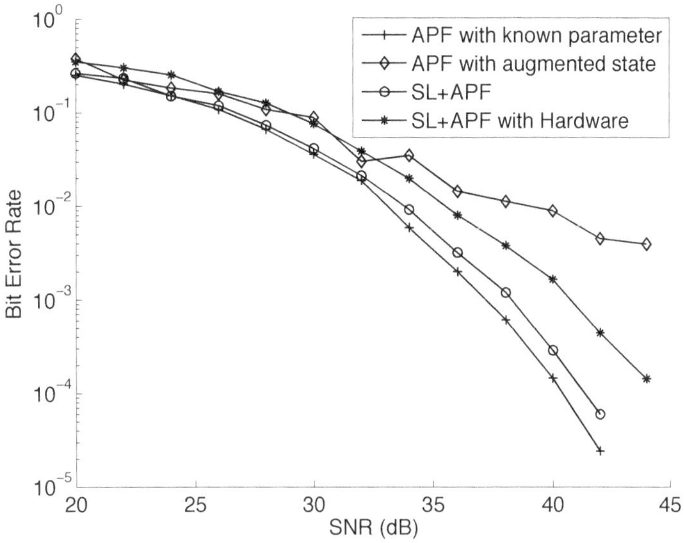

Figure 3. Comparisons of BERs for different methods. Both APF with known parameter and our proposed SL+APF algorithms deconvolute all the 10^5 bits correctly when SNR > 42dB, so no points appear in 44 dB for these two curves.

kept the same with example 3.1 except we simulated 10^5 bits. Coefficients of h_t and h_r are predefined and known, while the channel response h_c are unknown at the receiver. We compare our proposed method with an APF with augmented state and an APF with known channel coefficients. The BER plots in Figure 3 shows that the proposed SL+APF algorithm still performs well, but this time the APF with augmented state method performs very unstable because of the insufficiency in exploration of the unknown channel coefficients. Simulation study also demonstrates that the proposed algorithm achieves the BER < 10^{-5} at SNR=44 dB, which is approximately 4 dB higher than the theoretical required SNR 40.1dB to reach the capacity 200 bits per transmission. We also verify that the meta-modulation works in real-world physical channels, we implemented the communication process on a standard verification system, the universal software radio peripheral (USRP). The received signal is demodulated by the SL+APF algorithm. As shown in Figure 3, the BER becomes slightly higher because of the quantization error in the real communication, however, hardware verification shows the potential in real-world applications. In conclusion, this is an exciting result, because not only such a high spectrum efficiency system has rarely been reported, but also the required SNR for demodulation is moderately low that can be satisfied in practice.

Acknowledgement

The author acknowledge the support by Guangdong Natural Science Funds for Distinguished Young Scholar (No.S2012001253) and Shenzhen Key Laboratory of Data Science and Modeling (CXB201109210-103A).

References

Ali, R. A., Murua, A., Richardson, T., Roy, S. (2002). Evaluation of Sequential Importance Sampling for blind deconvolution via a simulation study. *In Proc. 11th European Signal Processing Conference*, 1-4.

Andrieu, C., Doucet, A., & Holenstein, R. (2010) . Particle Markov chain Monte Carlo methods. *Journal of the Royal Statistical Society. Series B (Statistical Methodology)*, *72*, 269-342. http://dx.doi.org/10.1111/j.1467-9868.2009.00736.x

Andrieu, C., Doucet, A., & Tadić, V. B. (2005). Online parameter estimation in general state-space models. *In Proc. 44th IEEE Conf. on Decision and Control*, 332-337. http://dx.doi.org/10.1109/CDC.2005.1582177

Benveniste, A., & Goursat, M. (1984). Blind equalizers. *IEEE Transactions on Communications*, *32*, 871-883. http://dx.doi.org/10.1109/TCOM.1984.1096163

Cappé, O. (2009). Online sequential Monte Carlo EM algorithm. *In Proc. 15th IEEE Workshop on Statistical Signal Processing*. 37-40. http://dx.doi.org/10.1109/ssp.2009.5278646

Cappé, O., Moulines, E., & Rydén, T. (2005). *Inference in Hidden Markov Models*. Springer, New York.

Chen, R., & Liu, J. S. (2000). Mixture Kalman Filters. *Journal of the Royal Statistical Society. Series B (Statistical Methodology)*, *62*, 493-508. http://dx.doi.org/10.1111/1467-9868.00246

Cover, T. M., & Thomas, J. A. (2006). *Elements of Information Theory*. John Wiley & Sons, Inc.

DeJong, D. N., Liesenfeld, R., Moura, G. V., Richard, J.-F., & Dharmarajan, H. (2013). Efficient likelihood evaluation of state-space representations. *Review of Economic Studies*, *80*, 538-567. http://dx.doi.org/10.1093/restud/rds040

Doucet, A., De Freitas, J. F. G., & Gordon, N. J., eds. (2001). *Sequential Monte Carlo Methods in Practice*. Springer, New York.

Doucet, A., Godsill, S. J., & Andrieu, C. (2000). On sequential Monte Carlo sampling methods for Bayesian filtering. *Statistics and Computing*, *10*, 197-208. http://dx.doi.org/10.1023/A:1008935410038

Fearnhead, P. (2002). Markov chain Monte Carlo, sufficient statistics, and particle filters. *Journal of Computational and Graphical Statistics*, *11*, 848-862. http://dx.doi.org/10.1198/106186002835

Flury, T., & Shephard , N. (2011). Bayesian inference based only on simulated likelihood: Particle filter analysis of dynamic economic models. *Econometric Theory*, *27*, 933-956. http://dx.doi.org/10.1017/S0266466610000599

Gilks, W. R., & Berzuini, C. (2001). Following a moving targetłMonte Carlo inference for dynamic Bayesian models. *Journal of the Royal Statistical Society. Series B (Statistical Methodology)*, *63*, 127-146. http://dx.doi.org/10.1111/1467-9868.00280

Gordon, N. J., Salmond, D. J., & Smith, A. F. M. (1993). Novel approach to nonlinear/non-Gaussian Bayesian state estimation. *IEE Proceedings F, Radar and Signal Processing*, *140*, 107-113. http://dx.doi.org/10.1049/ip-f-2.1993.0015

Haykin, S. (1994). *Blind Deconvolution*. Prentice-Hall, New Jersey.

Hürzeler, M., & Künsch , H. R. (2001). Approximating and maximising the likelihood for a general state-space model. *In Sequential Monte Carlo Methods in Practice. Stat. Eng. Inf. Sci*, 159-175. http://dx.doi.org/10.1007/978-1-4757-3437-9_8

Kantas, N., Doucet, A., Singh S. S., Maciejowski, J., & Chopin, N. (2015). On Particle Methods for Parameter Estimation in State-Space Models. *Statistical Science*, *30*(3), 328-351. http://dx.doi.org/10.1214/14-STS511

Kitagawa, G. (1998). A self-organizing state-space model.*Journal of the American Statistical Association*, *93*, 1203-1215. http://dx.doi.org/10.2307/2669862

Koller, D., & Friedman N. (2009) *Probabilistic Graphical Models*. Cambridge, MA: MIT Press.

Kushner, H. J., & Yin, G. G. (1997), *Stochastic Approximation Algorithms and Applications*. Springer-Verlag, New York.

Ionides, E. L., Bretó, C., & King, A. A. (2006). Inference for nonlinear dynamical systems. *Proc. Natl. Acad. Sci. USA*, *103*, 18438-18443. http://dx.doi.org/10.1073/pnas.0603181103

Ishwaran, H., & James, L. (2001). Gibbs sampling methods for stick-breaking priors, *Journal of the American Statistical Association*, *96*, 161-173. http://dx.doi.org/10.1198/016214501750332758

Ji, C. (2006). *Adaptive Monte Carlo Methods for Bayesian Inference*, MPhil dissertation, University of Cambridge.

Ji, C. (2009). *Advances in Bayesian Modelling and Computation: Spatio-temporal Processes, Model Assessment and Adaptive MCMC*, PhD dissertation, Duke University.

Liu, J., & West, M. (2001). Combined parameter and state estimation in simulation-based filtering. *In Sequential Monte Carlo Methods in Practice*. Springer, New York. http://dx.doi.org/10.1007/978-1-4757-3437-9_10

Liu, J. S. (2001). *Monte Carlo Strategies in Scientific Computing*. Springer, New York.

Liu, J. S., & Chen, R. (1995). Blind deconvolution via sequential imputations. *Journal of the American Statistical Association*, *90*, 567-576. http://dx.doi.org/10.1080/01621459.1995.10476549

Liu, J. S., & Chen, R. (1998). Sequential Monte Carlo methods for dynamic systems. *Journal of the American Statistical Association*, *93*, 1032-1044. http://dx.doi.org/10.1080/01621459.1998.10473765

Olsson, J., Cappé, O., Douc, R., & Moulines, E. (2008). Sequential Monte Carlo smoothing with application to parameter estimation in nonlinear state space models. *Bernoulli*, *14*, 155-179. http://dx.doi.org/10.3150/07-BEJ6150

Pitt, M., & Shephard, N. (1999). Filtering via simulation: Auxiliary particle filters, *Journal of the American Statistical Association*, *94*, 590-599. http://dx.doi.org/10.1080/01621459.1999.10474153

Poyiadjis, G., Doucet, A., & Singh, S. S. (2011). Particle approximations of the score and observed information matrix in state space models with application to parameter estimation. *Biometrika*, *98*, 65-80.

http://dx.doi.org/10.1093/biomet/asq062

Robert, C., & Casella, G. (1999). *Monte Carlo Statistical Methods*, 2nd ed. Springer, New York.

Wang, I. , Chong, E., and Kulkarni, S. (1995). On Equivalence of Some Noise Conditions for Stochastic Approximation Algorithms, *in Proceedings of the 34th Conference on Decision and Control.*

West, M. (1993). Approximating posterior distributions by mixtures, *Journal of Royal Statistical Society 55*, 409-422.

West, M., & Harrison, J. (1997). *Bayesian Forecasting and Dynamic Models*, 2nd ed. Springer, New York.

Appendix: Proofs of Convergence of Monte Carlo SA Algorithm

Stochastic Approximation (SA) is a class of algorithms to finding the roots of possibly non-linear equation $h(\psi) = 0$, in the situation where only noisy measurements of $h(\psi)$ are available. In its simplest form, the Robbins-Monro algorithm is a recursive process as follows,

$$\psi^{(t+1)} = \psi^{(t)} + r^{(t+1)}\zeta^{(t+1)} \tag{26}$$

where $\left\{r^{(t)}, t \geq 1\right\}$ is a sequence of stepsizes which satisfies standard conditions: $\sum_{t=1}^{\infty} r^{(t)} = \infty$ and $\sum_{t=1}^{\infty} \left[r^{(t)}\right]^2 < \infty$ and for any $t \geq 1$, ζ is a noisy measurement of $h(\psi)$:

$$\zeta^{(t+1)} = h(\psi) + \xi^{(t+1)} \tag{27}$$

where $\left\{\xi^{(t)}, t \geq 1\right\}$ is the so called noise sequence.

In our case, we denote $f(\theta;\psi) = \frac{\partial}{\partial\psi}\left(\log \frac{\pi(\theta)}{q(\theta;\psi)}\right)$. Assume we have Monte Carlo samples $\{\theta^{(i)} : i = 1, \ldots, N\}$ from the distribution $\pi(\theta)$, then $h(\psi)$ can be evaluated by its Monte Carlo estimate,

$$\zeta(\psi) = \frac{1}{N}\sum_{i=1}^{N} f(\theta^{(i)};\psi). \tag{28}$$

The *Central Limit Theorem* gives

$$\xi = [\zeta(\psi) - h(\psi)] \rightarrow Norm\left(0, \frac{\sigma^2}{N}\right), \quad \text{as} \quad N \rightarrow \infty \tag{29}$$

which implies that ξ is Gaussian noise, with mean zero and variance $\frac{\sigma^2}{N}$ where $\sigma^2 = \frac{1}{N-1}\sum_{i=1}^{N}(h(\psi) - \zeta(\psi))$ (Robert and Casella, 1999).

By using the iterative stochastic approximation method, we can estimate ψ iteratively by

$$\psi^{(t+1)} = \psi^{(t)} + r^{(t+1)}\zeta\left(\psi^{(t)}\right). \tag{30}$$

Assume that ψ_L is the root of equation $h(\psi) = 0$. We present a theorem to show that $\psi^{(t)} \rightarrow \psi_L$ in probability one as $t \rightarrow \infty$.

Theorem 1. *Consider the following conditions:*

(A1) *By the central limit theorem,* $\xi(\psi) \rightarrow N\left(0, \frac{\sigma^2}{N}\right)$ *in distribution and*

$$\int_{\Theta} h(\theta;\psi)^2 q(\theta;\psi) d\theta < \infty.$$

(A2) Ψ *is an open subset of* R^{n_ψ}. *The mean field* $f : \Psi \rightarrow R^{n_\psi}$ *is continuous and there exists a continuously differentiable function* $w : \Psi \rightarrow [0, \infty)$ *(with the convention* $w(\psi) = \infty$ *when* $\psi \notin \Psi$ *) such that:*

1. *For any* $M > 0$, *the level set* $W_M \equiv \{\psi \in \Psi, w(\psi) \leq M\} \subset \Psi$ *is compact,*
2. *The set of stationary point(s)* $\mathcal{L} \equiv \{\psi \in \Psi, \langle\nabla w(\psi), f(\psi)\rangle = 0\}$ *belongs to the interior of* Ψ,
3. *For any* $\psi \in \Psi$, $\langle\nabla w(\psi), f(\psi)\rangle \leq 0$ *and the closure of* $w(\mathcal{L})$ *has an empty interior.*

(A3) *The sequence* $\{r^{(t)}, t \geq 1\}$ *is non-increasing, positive and*

$$\sum_{t=1}^{\infty} r^{(t)} = \infty \quad \text{and} \quad \sum_{t=1}^{\infty}\left[r^{(t)}\right]^2 < \infty. \tag{31}$$

Assume (A1-3). Then,

$$P\left[\lim_{t\to\infty} d\left(\psi^{(t)}, \mathcal{L}\right) = 0\right] = 1. \tag{32}$$

Proof: the recursion is expressed as follows

$$\psi^{(t+1)} = \psi^{(t)} + r^{(t+1)}\zeta^{(t+1)} = \psi^{(t)} + r^{(t+1)} f(\psi) + r^{(t+1)}\xi^{(t+1)}, \tag{33}$$

where $f(\psi)$ is the function of interest and $\xi^{(t+1)}$ is a random perturbation

$$\xi^{(t+1)} = f(\psi) - \zeta^{(t+1)} \tag{34}$$

$$= \int_{\Theta} h(\theta;\psi)d\theta - \frac{1}{N}\sum_{i=1}^{N} h(\theta^{(i)};\psi). \tag{35}$$

Define $M_n = \sum_{t=1}^{n} r^{(t)}\xi^{(t)}$. Then

$$E\left[M_{n+1}|M_k, k \le n\right] = E\left[\sum_{t=1}^{n+1} r^{(t)}\xi^{(t)}|M_k, k \le n\right] \tag{36}$$

$$= E\left[\sum_{t=1}^{n} r^{(t)}\xi^{(t)}|M_k, k \le n\right] + E\left[r^{(t+1)}\xi^{(t+1)}|M_k, k \le n\right] \tag{37}$$

$$= M_n + r^{(t+1)}E\left[\xi^{(t+1)}\right]. \tag{38}$$

Since $\xi^{(t+1)} = \int_{\Theta} h(\theta;\psi)d\theta - \frac{1}{N}\sum_{i=1}^{N} h(\theta^{(i)};\psi) \to 0$, almost sure (a.s.), as $N \to \infty$, $E\left[M_{n+1}|M_k, k \le n\right] \to M_n$ a.s. or with probability one. Therefore $\{M_n, n \ge 1\}$ is a F-martingale.

Then by the martingale inequality,

$$P\left\{\sup_{n\ge j\ge m}\left|M_j - M_m\right| \ge \mu\right\} \le \frac{E\left|\sum_{i=m}^{n-1} r^{(i)}\xi^{(i)}\right|}{\mu} \tag{39}$$

which implies

$$\lim_{m\to\infty} P\left\{\sup_{j\ge m}\left|M_j - M_m\right| \ge \mu\right\} = 0 \tag{40}$$

so that

$$\lim_{m\to\infty}\left(\sup_{m\le j\le m(n,T)}\left\|\sum_{i=n}^{j} r^{(i)}\xi^{(i)}\right\|\right) = 0 \quad \text{for all} \quad T > 0 \tag{41}$$

where $m(n,T) \equiv \max\left\{k : r^{(n)} + \ldots + r^{(k)} \le T\right\}$. This condition is called Kushner and Clark's condition, which is an important sufficient condition on the noise sequence for the convergence of stochastic approximation algorithms (Wang et al., 1995; Kushner and Yin, 1997). By the theorem in Chapter 5 of Kushner and Yin (1997), we can obtain

$$P\left[\lim_{t\to\infty} d\left(\psi^{(t)}, \mathcal{L}\right) = 0\right] = 1. \tag{42}$$

7

Limit Theorems for Negatively Dependent Fuzzy Set-Valued Random Variables

Li Guan[1]

[1] College of Applied Sciences, Beijing University of Technology, Beijing, China

Correspondence: Li Guan, College of Applied Sciences, Beijing University of Technology, 100 Pingleyuan, Chaoyang District, Beijing, 100124, P.R.China. E-mail: guanli@bjut.edu.cn

Abstract

In this paper, we shall discuss negatively dependent fuzzy set-valued random variables. And at last, we shall prove the limit theorems for rowwise negatively dependent fuzzy set-valued random variables in the sense of d_H^∞, which is the extension of (Guan & Sun, 2014) and (Guan & Wan, 2016).

Keywords: laws of large number, fuzzy set-valued random variables, negatively dependent

1. Introduction

We all know that strong laws of large numbers are one of the most important theories in probability. For independent set-valued random variables, many limit results have been obtained (cf. (Artstein & Vitale, 1975), (Cressie, 1978), (Hiai, 1984), (Taylor & Inoue, 1985), (Puri & Ralescu, 1983)). About the convergence theorems of fuzzy set-valued random variables, Klement *et al* proved the strong laws of large numbers (SLLN) for independent identically distributed (i.i.d) fuzzy set-valued random variables in the sense of d_H^1. Colubi *et al.* obtained the SLLN for i.i.d fuzzy set-valued random variables with respect to d_H^∞ in (Colubi, López-Díaz, Dominguez-Menchero & Gil, 1999), where the underlying space is $\mathbb{R}^d$. Li and Ogura proved the same SLLN as in (Colubi, López-Díaz, Dominguez-Menchero & Gil, 1999) by using a new embedding method (Li, Ogura & Kreinovich, 2002, Theorem 6.2.6). In 1991, Inoue (Inoue, 1991) extended the SLLN of set-valued case given by Taylor and Inoue (Taylor & Inoue, 1985) to the case of only independent fuzzy set-valued random variables in the sense of d_H^1. Li and Ogura (Li & Ogura, 2003) proved the SLLN of (Inoue, 1991) in the sense of d_H^∞.

In practice, the random variables are not always independent. So it is necessary to discuss the limit theorems for dependent random variables. Bozorgnia, Patterson and Taylor discussed the properties for negatively dependent random variables in (Bozorgnia, Patterson, & Taylor, 1993), and proved the laws of large number for negative dependence random variables in (Bozorgnia, Patterson & Taylor, 1992), where the random variables are single-valued. In (Guan, 2014), Guan and Sun discussed the property of set-valued random variables in real space $\mathbb{R}$, and proved the weak convergence theorem for dependent set-vallued random variables in the sense of Hausdorff metric. In (Guan, 2016), Guan and Wan proved the strong law of large number for set-valued dependent random variables in the sense of Hausdorff metric. In 2016, Shen and Guan (Shen & Guan, 2016) proved the strong laws of large numbers for independent fuzzy set-valued random variables, where the underlying space is G_α space. In this paper, we are concerned with the weak and strong limit theorems for dependent fuzzy set-valued random variables in the sense of d_H^∞.

This paper is organized as follows. In section 2, we shall briefly introduce some definitions and basic results of set-valued random variables and fuzzy set-valued random variables. In section 3, we shall give basic definition and discuss the properties of fuzzy set-valued negatively dependent random variables. And at last, we shall prove the weak and strong laws of large numbers for weighted sums of fuzzy set-valued negatively dependent random variables.

2. Preliminaries on Fuzzy Set-Valued Random Variables

Throughout this paper, we assume that $(\Omega, \mathcal{A}, \mu)$ is a complete probability space, $(\mathfrak{X}, \|\cdot\|)$ is a real separable Banach space, $\mathbf{K_k}(\mathfrak{X})$ is the family of all nonempty compact subsets of $\mathfrak{X}$, and $\mathbf{K_{kc}}(\mathfrak{X})$ is the family of all nonempty compact convex subsets of $\mathfrak{X}$.

Let A and B be two nonempty subsets of $\mathfrak{X}$ and let $\lambda \in \mathbb{R}$, the set of all real numbers. We define addition and scalar multiplication by

$$A + B = \{a + b : a \in A, b \in B\}$$

$$\lambda A = \{\lambda a : a \in A\}$$

The Hausdorff metric on $\mathbf{K_k}(\mathfrak{X})$ is defined by

$$d_H(A, B) = \max\{\sup_{a\in A}\inf_{b\in B}\|a-b\|,\ \sup_{b\in B}\inf_{a\in A}\|a-b\|\}$$

for $A,\ B \in \mathbf{K_k}(\mathfrak{X})$. For an A in $\mathbf{K_k}(\mathfrak{X})$, let $\|A\|_{\mathbf{K}} = d_H(\{0\}, A)$.

The metric space $(\mathbf{K_k}(\mathfrak{X}), \mathbf{d_H})$ is complete and separable, and $\mathbf{K_{kc}}(\mathfrak{X})$ is a closed subset of $(\mathbf{K_k}(\mathfrak{X}), \mathbf{d_H})$ (cf. (Li, Ogura & Kreinovich, 2002), Theorems 1.1.2 and 1.1.3). Concerning the concepts and results of set-valued random variables, readers may refer to the book (Li, Ogura & Kreinovich, 2002).

A set-valued mapping $F : \Omega \to K_k(\mathfrak{X})$ is called a set-valued random variable (or a random set) if, for each open subset O of $\mathfrak{X}$, $F^{-1}(O) = \{\omega \in \Omega : F(\omega) \cap O \neq \emptyset\} \in \mathcal{A}$. In fact, set-valued random variables can be defined as mappings from Ω to the family of all closed subsets of $\mathfrak{X}$. Concerning its equivalent definitions, please refer to (Castaing & Valadier, 1977) and (Hiai & Umegaki, 1977).

A set-valued random variable F is called integrably bounded (cf. (Hiai & Umegaki, 1977) or (Li, Ogura & Kreinovich, 2002)) if $\int_\Omega \|F(\omega)\|_{\mathbf{K}} d\mu < \infty$. Let $L^1[\Omega, \mathcal{A}, \mu; \mathbf{K_k}(\mathfrak{X})]$ denote the space of all integrably bounded random variables, and $L^1[\Omega, \mathcal{A}, \mu; \mathbf{K_{kc}}(\mathfrak{X})]$ denote the space of all integrably bounded random variables taking values in $\mathbf{K_{kc}}(\mathfrak{X})$. For $F, G \in L^1[\Omega, \mathcal{A}, \mu; \mathbf{K_k}(\mathfrak{X})]$, $F = G$ if and only if $F(\omega) = G(\omega)$ $a.e.(\mu)$.

For each set-valued random variable F, define $S_F = \{f \in L^1[\Omega; \mathfrak{X}] : f(\omega) \in F(\omega), a.e.(\mu)\}$. The expectation of F, denoted by $E[F]$, is defined as

$$E[F] = \Big\{\int_\Omega f d\mu : f \in S_F\Big\},$$

where $\int_\Omega f d\mu$ is the usual Bochner integral in $L^1[\Omega, \mathfrak{X}]$, the family of integrable $\mathfrak{X}$-valued random variables. This integral was first introduced by Aumann (Aumann, 1965), called Aumann integral in literature.

For each $A \in \mathbf{K}(\mathfrak{X})$, define the support function by

$$s(x^*, A) = \sup_{a\in A} < x^*, a >, \quad x^* \in \mathfrak{X}^*,$$

where $\mathfrak{X}^*$ is the dual space of $\mathfrak{X}$.

Let $\mathbf{S}^*$ denote the unit sphere of $\mathfrak{X}^*$, $C(\mathbf{S}^*)$ the all continuous functions of $\mathbf{S}^*$, and the norm is defined as $\|v\|_C = \sup_{x^*\in S^*}$

From now on, we begin to introduce necessary concepts, notation and basic results on fuzzy set space and fuzzy set-valued random variables.

Let $\mathbf{F_k}(\mathfrak{X})$ be the family of all special fuzzy sets: $v : \mathfrak{X} \to [0, 1]$ satisfying the following conditions:
(1) the 1-level set $v_1 = \{x \in \mathfrak{X} : v(x) = 1\} \neq \emptyset$,
(2) each v is upper semicontinuous, i.e. for each $\alpha \in [0, 1]$, the α level set $v_\alpha = \{x \in \mathfrak{X} : v(x) \geq \alpha\}$ is a closed subset of $\mathfrak{X}$,
(3) the support set $v_{0+} = \mathrm{cl}\{x \in \mathfrak{X} : v(x) > 0\}$ is compact.

A fuzzy set v in $\mathbf{F_k}(\mathfrak{X})$ is called convex if it satisfies

$$v(\lambda x + (1-\lambda)y) \geq \min\{v(x), v(y)\}, \text{ for any } x, y \in \mathfrak{X}, \lambda \in [0, 1].$$

It is known that v is convex if and only if each level set v_α ($\alpha \in (0, 1]$) of v is a convex subset of $\mathfrak{X}$. Let $\mathbf{F_{kc}}(\mathfrak{X})$ be the subset of all convex fuzzy sets in $\mathbf{F_k}(\mathfrak{X})$.

The uniform metric in $\mathbf{F_k}(\mathfrak{X})$, which is an extension of the Hausdorff metric d_H, is often used (cf. Puri & Ralescu, 1986): for $v^1, v^2 \in \mathbf{F_k}(\mathfrak{X})$,

$$d_H^\infty(v^1, v^2) = \sup_{\alpha\in(0,1]} d_H(v_\alpha^1, v_\alpha^2).$$

Let $\|v\|_{\mathbf{F}} =: d_H^\infty(v, I_0) = \sup_{\alpha>0} \|v_\alpha\|_{\mathbf{K}}$, where I_0 is the function taking value one at 0 and zero for all $x \neq 0$. The space $(\mathbf{F_k}(\mathfrak{X}), \mathbf{d_H^\infty})$ is a complete metric space (cf. Li, & Ogura, 1996) but not separable (Li, Ogura & Kreinovich, 2002), Remark 5.1.7). Completeness was first proved by Puri and Ralescu (Puri & Ralescu, 1986) in the case of $\mathfrak{X} = \mathbb{R}^d$, the d-dimensional Euclidean space.

A fuzzy set-valued random variable (or $\mathbf{F_k}(\mathfrak{X})$-valued random variable) is a measurable mapping X from the space $(\Omega, \mathcal{A})$ to the space $(\mathbf{F_k}(\mathfrak{X}), \mathcal{B}(\mathbf{F_k}(\mathfrak{X})))$, where $\mathcal{B}(\mathbf{F_k}(\mathfrak{X}))$ is the Borel σ-field of $\mathbf{F_k}(\mathfrak{X})$ with respect to d_H^∞.

It is well known that for any fuzzy set v, $v_\alpha = \bigcap_{\beta<\alpha} v_\beta$, for every $\alpha \in (0,1]$, but usually $v_\alpha \neq \bigcup_{\beta>\alpha} v_\beta$. We denote $v_{\alpha+} = \text{cl}(\bigcup_{\beta>\alpha} v_\beta)$, for $\alpha \in [0,1)$, which will be used later. Obviously, v_{0+} is the support set of v. Due to the completeness of $(\mathbf{F_k}(\mathfrak{X}), \mathbf{d_H^\infty})$, every Cauchy sequence $\{v^n : n \in \mathbb{N}\}$ has a limit v in $\mathbf{F_k}(\mathfrak{X})$.

A sequence of set-valued random variables $\{F_n : n \in \mathbb{N}\}$ is called to be stochastically dominated by a set-valued random variable F if

$$\mu\{\|F_n\|_\mathbf{K} > t\} \leq \mu\{\|F\|_\mathbf{K} > t\}, \quad t \geq 0, \ n \geq 1.$$

A sequence of fuzzy set-valued random variables $\{X^n : n \geq 1\}$ is called to be stochastically dominated by a fuzzy set-valued random variable X if

$$\mu\{\|X_\alpha^n\|_\mathbf{K} > t\} \leq \mu\{\|X_\alpha\|_\mathbf{K} > t\}, \quad t \geq 0, \ n \geq 1, \alpha \in (0,1].$$

It is obvious that if $\{X^n : n \geq 1\}$ is stochastically dominated by a fuzzy set-valued random variable X, then $\forall\ \alpha \in (0,1]$, $\{X_\alpha^n : n \geq 1\}$ is stochastically dominated by the set-valued random variable X_α. And also for any $\alpha \in (0,1]$, $\{X_{\alpha+}^n : n \geq 1\}$ is stochastically dominated by the set-valued random variable $X_{\alpha+}$ (Guan & Li, 2004).

An $\mathbf{F_k}(\mathfrak{X})$-valued random variable X is called integrably bounded if the real-valued random variable $\|X_{0+}(\omega)\|_\mathbf{K}$ is integrable. Let $L^1[\Omega, \mathcal{A}, \mu; \mathbf{F_k}(\mathfrak{X})]$ be the set of all integrably bounded $\mathbf{F_k}(\mathfrak{X})$-valued random variables and $L^1[\Omega, \mathcal{A}, \mu; \mathbf{F_{kc}}(\mathfrak{X})]$ be the set of all integrably bounded $\mathbf{F_{kc}}(\mathfrak{X})$-valued random variables. Two $\mathbf{F_k}(\mathfrak{X})$-valued random variables $X, Y \in L^1[\Omega, \mathcal{A}, \mu; \mathbf{F_k}(\mathfrak{X})]$ are considered to be identical if for any $\alpha \in [0,1], X_\alpha(\omega) = Y_\alpha(\omega)\ a.e.(\mu)$.

A sequence of $\mathbf{F_k}(\mathfrak{X})$-valued random variables $\{X^n : n \in \mathbb{N}\}$ is called to converge to an $\mathbf{F_k}(\mathfrak{X})$-valued random variable X in the sense of d_H^∞, if $d_H^\infty(X^n(\omega), X(\omega)) \to 0\ a.e.(\mu)\ as\ n \to \infty$.

The expectation of an $\mathbf{F_k}(\mathfrak{X})$-valued random variables X, denoted by $E[X]$, is an element in $\mathbf{F_k}(\mathfrak{X})$ such that for every $\alpha \in (0,1]$,

$$(E[X])_\alpha = cl \int_\Omega X_\alpha d\mu = cl\{E(f) : f \in S_{X_\alpha}\}$$

where the closure is taken in $\mathfrak{X}$ and $S_{X_\alpha} = \{f \in L^1[\Omega; \mathfrak{X}] : f(\omega) \in X_\alpha(\omega), a.e.(\mu)\}$. By virtue of the existence theorem (cf. (cf. Li, & Ogura, 1996), (Li, Ogura & Kreinovich, 2002)), we have an equivalent definition as follows:

$$E[X](x) = \sup\{\alpha \in (0,1] : x \in E[X_\alpha]\}.$$

Furthermore, $(E[\overline{co}X])_\alpha = E[\overline{co}X_\alpha]$ for any $\alpha \in (0,1]$.

3. Negatively Dependent and Main Results

In this section, we will give the definition of negatively dependent for fuzzy set-valued random variables and discuss the properties. Then we will prove the weak and strong limit theorem of weighted sums for fuzzy set-valued negatively dependent random variables in the sense of d_H^∞.

The following definition is Toeplitz sequence, which will be used later.

Definition 3.1 *A double array $\{a_{nk} : n, k = 1, 2, \cdots\}$ of real numbers is said to be a Toeplitz sequence, if*

(i) $\lim_{n\to\infty} a_{nk} = 0$ *for each k;*

(ii) $\sum_{k=1}^{\infty} |a_{nk}| \leq c$ *for each n.*

Now we will recall some concepts of negatively dependent random variables.

Definition 3.2 (cf. Bozorgnia, Patterson & Taylor, 1993)) *A finite family of real-valued random variables $X_1, \cdots, X_n$ is said to be negatively dependent if for all real $x_1, x_2, \cdots, x_n$,*

$$\mu\{X_1 > x_1, \cdots, X_n > x_n\} \leq \prod_{i=1}^{n} \mu\{X_i > x_i\}$$

and

$$\mu\{X_1 \le x_1, \cdots, X_n \le x_n\} \le \prod_{i=1}^{n} \mu\{X_i \le x_i\}.$$

An infinite family of random variables is negatively dependent if every finite subfamily is negatively dependent. The following results are from (Bozorgnia, Patterson & Taylor, 1993) of Bozorgnia et. al., and we will use them in the later.

Lemma 3.3 (cf. Bozorgnia, Patterson & Taylor, 1993) *Let real-valued random variables* $\{X_i : 1 \le i \le n\}$ *be negatively dependent. Then the following are true:*

(i) $E[\prod_{i=1}^{n} X_i] \le \prod_{i=1}^{n} E[X_i]$;

(ii) $Cov(X_i, X_j) \le 0, \quad i \ne j$;

(iii) If $\{g_i : 1 \le i \le n\}$ *be all nondecreasing (or all nonincreasing) Borel functions, then random variables* $g_1(X_1), g_2(X_2), \cdots, g_n(X_n)$ *are negatively dependent random variables.*

Definition 3.4 (cf. (Guan & Wan, 2016) *A finite family of set-valued random variables* $F_1, F_2, \cdots, F_n$ *is said to be negatively dependent if* $s(\cdot, F_1), \cdots, s(\cdot, F_n)$ *is single-valued negatively dependent random variables.*

Now we give the definition of fuzzy set-valued random variables and discuss the property.

Definition 3.5 *A finite family of fuzzy set-valued random variables* $X^1, X^2, \cdots, X^n$ *be said to be negatively dependent if for any* $\alpha \in (0, 1]$, *the set-valued random variables* $X^1_\alpha, X^2_\alpha, \cdots, X^n_\alpha$ *are negatively dependent.*

From the definition 3.5, we can easily obtain the following result.

Theorem 3.6 *Fuzzy set-valued random variables* $X^1, X^2, \cdots, X^n$ *are negatively dependent, then for any* $\alpha \in (0, 1]$, $X^1_{\alpha+}, X^2_{\alpha+}, \cdots, X^n_{\alpha+}$ *are negatively dependent set-valued random variables.*

Proof. By definition 3.5, we know that for any $\alpha \in (0, 1]$, $X^1_\alpha, X^2_\alpha, \cdots, X^n_\alpha$ are negatively dependent set-valued random variables. That means for any $x^* \in \mathfrak{X}^*, \alpha \in (0, 1]$, $s(x^*, X^1_\alpha), s(x^*, X^2_\alpha), \cdots, s(x^*, X^n_\alpha)$ are negatively dependent single-valued random variables. Since $X^n_{\alpha+} = cl(\bigcup_{\beta>\alpha} X^n_\beta)$, take a decreasing sequence α_j which converges to α. Then

$$\lim_{j\to\infty} d_H(X^n_{\alpha_j}, X^n_{\alpha+}) = 0.$$

So we have

$$s(x^*, X^n_{\alpha_j}) \longrightarrow s(x^*, X^n_{\alpha+}), \quad \text{as} \quad j \to \infty.$$

Then by the continuous of probability and lemma 3.3, we have

$$\begin{aligned}
&\mu\Big\{s(x^*, X^1_{\alpha+}) \le x_1, \cdots, s(x^*, X^n_{\alpha+}) \le x_n\Big\} \\
&= \mu\Big\{\lim_j s(x^*, X^1_{\alpha_j}) \le x_1, \cdots, \lim_j s(x^*, X^n_{\alpha_j}) \le x_n\Big\} \\
&\le \lim_j \mu\Big\{s(x^*, X^1_{\alpha_j}) \le x_1, \cdots, s(x^*, X^n_{\alpha_j}) \le x_n\Big\} \\
&\le \lim_j \prod_{i=1}^{n} \mu\Big\{s(x^*, X^i_{\alpha_j}) \le x_i\Big\} \\
&= \prod_{i=1}^{n} \lim_j \mu\Big\{s(x^*, X^i_{\alpha_j}) \le x_i\Big\} \\
&= \prod_{i=1}^{n} \mu\Big\{\lim_j s(x^*, X^i_{\alpha_j}) \le x_i\Big\} \\
&= \prod_{i=1}^{n} \mu\Big\{s(x^*, X^i_{\alpha+}) \le x_i\Big\}
\end{aligned}$$

Similarly, we can prove

$$\mu\Big\{s(x^*, X^1_{\alpha+}) > x_1, \cdots, s(x^*, X^n_{\alpha+}) > x_n\Big\} \leq \prod_{i=1}^{n} \mu\Big\{s(x^*, X^i_{\alpha+}) > x_i\Big\}.$$

Then the result was proved. □

The following limit theorem is a weak convergence result for fuzzy set-valued negatively dependent random variables, which is the extension of (Guan & Sun, 2014). Here we assume the Banach space $\mathfrak{X} = \mathbb{R}$.

Theorem 3.7 *Let $\{X^{nk} : k \geq 1, n \geq 1\}$ be an array of fuzzy set-valued random variables in $\mathbf{F}_k(\mathbb{R}^+)$ which are stochastically dominated by a fuzzy set-valued random variable X, and are pairwise negatively dependent in each row. Let $E[X^{nk}] = I_0$ for all n and k. Let $\{a_{nk}\}$ be an array of nonnegative real numbers such that $\sum_{n=1}^{\infty} a^r_{nk} \leq M$ for all n where $0 < r \leq 1$ and $\max_k a_{nk} \to 0$ as $n \to \infty$. If $E[\|X\|_{\mathbf{F}}] < \infty$, then*

$$\lim_{n\to\infty} \mu\Big\{d^{\infty}_H\Big(\sum_{k=1}^{\infty} a_{nk} X^{nk}, \{0\}\Big) > \varepsilon\Big\} = 0.$$

Proof. From the definition 3.5 and theorem 3.6, we know that $\{X^{nk}_\alpha\}, \{X^{nk}_{\alpha+}\} \in \mathbf{K}_k(\mathbb{R}^+)$ are all rowwise negatively dependent set-valued random variables. Then by theorem 4.1 of (Guan & Sun, 2014), for any $\alpha \in (0, 1]$, $\varepsilon > 0$, we have

$$\lim_{n\to\infty} \mu\Big\{d_H\Big(\sum_{k=1}^{\infty} a_{nk} X^{nk}_\alpha, \{0\}\Big) > \frac{\varepsilon}{2}\Big\} = 0,$$

and

$$\lim_{n\to\infty} \mu\Big\{d_H\Big(\sum_{k=1}^{\infty} a_{nk} X^{nk}_{\alpha+}, \{0\}\Big) > \frac{\varepsilon}{2}\Big\} = 0.$$

Take a finite partition $0 = \alpha_0 < \alpha_1 < \cdots < \alpha_M = 1$, for $\alpha_{k-1} < \alpha < \alpha_k$, by virtue of monotone property of level sets and the formula

$$(\sum_{k=1}^{\infty} a_{nk} X^{nk})_\alpha = \sum_{k=1}^{\infty} a_{nk} X^{nk}_\alpha,$$

it holds that

$$\begin{aligned} d_H\Big((\sum_{k=1}^{\infty} a_{nk} X^{nk})_\alpha, \{0\}\Big) &\leq d_H\Big((\sum_{k=1}^{\infty} a_{nk} X^{nk})_{\alpha_k}, \{0\}\Big) + d_H\Big((\sum_{k=1}^{\infty} a_{nk} X^{nk})_{\alpha_{k-1}+}, \{0\}\Big) \\ &= d_H\Big(\sum_{k=1}^{\infty} a_{nk} X^{nk}_{\alpha_k}, \{0\}\Big) + d_H\Big(\sum_{k=1}^{\infty} a_{nk} X^{nk}_{\alpha_{k-1}+}, \{0\}\Big). \end{aligned}$$

Consequently, we have

$$\begin{aligned} d^{\infty}_H\Big(\sum_{k=1}^{\infty} a_{nk} X^{nk}, \{0\}\Big) &= \sup_{\alpha\in(0,1]} d_H\Big((\sum_{k=1}^{\infty} a_{nk} X^{nk})_\alpha, \{0\}\Big) \\ &\leq \max_{1\leq k\leq M} d_H\Big(\sum_{k=1}^{\infty} a_{nk} X^{nk}_{\alpha_k}, \{0\}\Big) \\ &\quad + \max_{1\leq k\leq M} d_H\Big(\sum_{k=1}^{\infty} a_{nk} X^{nk}_{\alpha_{k-1}+}, \{0\}\Big). \end{aligned}$$

Then we have

$$\begin{aligned} \lim_{n\to\infty} \mu\Big\{d^{\infty}_H\Big(\sum_{k=1}^{\infty} a_{nk} X^{nk}, \{0\}\Big) > \varepsilon\Big\} &\leq \lim_{n\to\infty} \mu\Big\{\max_{1\leq k\leq M} d_H\Big(\sum_{k=1}^{\infty} a_{nk} X^{nk}_{\alpha_k}, \{0\}\Big) > \frac{\varepsilon}{2}\Big\} \\ &\quad + \lim_{n\to\infty} \mu\Big\{\max_{1\leq k\leq M} d_H\Big(\sum_{k=1}^{\infty} a_{nk} X^{nk}_{\alpha_{k-1}+}, \{0\}\Big) > \frac{\varepsilon}{2}\Big\} \\ &= 0. \end{aligned}$$

The result was proved. □

Next, we shall prove the strong convergence theorem, which is the extend of theorem 4.1 of (Guan & Wan, 2016). In (Guan & Wan, 2016), the authors proved the convergence theorem of set-valued random variables in the sense of d_H. Here we extend their result to fuzzy set-valued random variables, and the metric is d_H^∞.

Theorem 3.8 *Let $\{X^{nk}\} \in \mathbf{F}_k(\mathbb{R}^+)$ be an array of rowwise negatively dependent fuzzy set-valued random variables with $E[X^{nk}] = I_0$ and stochastic dominated by a fuzzy set-valued random variable X. If $\max_k a_{nk} = O(n^{-r})$, $r > 0$, then $E[\|X\|_{\mathbf{F}}^{1+1/r}] < \infty$ implies that*

$$\sum_{k=1}^{\infty} a_{nk} X^{nk} \to 0 \ a.e.$$

with respect to the Hausdorff metric d_H^∞.

Proof. From the definition 3.5 and theorem 3.6, we know that $\{X_\alpha^{nk}\}, \{X_{\alpha+}^{nk}\} \in \mathbf{K}_k(\mathbb{R}^+)$ are all rowwise negatively dependent set-valued random variables. Then by theorem 4.1 of (Guan & Wan, 2016), we have

$$\sum_{k=1}^{\infty} a_{nk} X_\alpha^{nk} \to 0 \ a.e.$$

with respect to the Hausdorff metric d_H. And

$$\sum_{k=1}^{\infty} a_{nk} X_{\alpha+}^{nk} \to 0 \ a.e.$$

with respect to the Hausdorff metric d_H.

We can find a finite partition $0 = \alpha_0 < \alpha_1 < \cdots < \alpha_M = 1$, for $\alpha_{k-1} < \alpha < \alpha_k$. By virtue of monotone property of level sets and the formula

$$(\sum_{k=1}^{\infty} a_{nk} X^{nk})_\alpha = \sum_{k=1}^{\infty} a_{nk} X_\alpha^{nk},$$

it holds that

$$\begin{aligned} d_H\Big((\sum_{k=1}^{\infty} a_{nk} X^{nk})_\alpha, \{0\}\Big) &\le d_H\Big((\sum_{k=1}^{\infty} a_{nk} X^{nk})_{\alpha_k}, \{0\}\Big) + d_H\Big((\sum_{k=1}^{\infty} a_{nk} X^{nk})_{\alpha_{k-1}+}, \{0\}\Big) \\ &= d_H\Big(\sum_{k=1}^{\infty} a_{nk} X_{\alpha_k}^{nk}, \{0\}\Big) + d_H\Big(\sum_{k=1}^{\infty} a_{nk} X_{\alpha_{k-1}+}^{nk}, \{0\}\Big). \end{aligned}$$

Consequently, we have

$$\begin{aligned} \sup_{\alpha\in(0,1]} d_H\Big((\sum_{k=1}^{\infty} a_{nk} X^{nk})_\alpha, \{0\}\Big) &\le \max_{1\le k\le M} d_H\Big(\sum_{k=1}^{\infty} a_{nk} X_{\alpha_k}^{nk}, \{0\}\Big) \\ &\quad + \max_{1\le k\le M} d_H\Big(\sum_{k=1}^{\infty} a_{nk} X_{\alpha_{k-1}+}^{nk}, \{0\}\Big). \end{aligned}$$

Since the two terms on the right hand converge to 0 in the sense of d_H, then we can have

$$\limsup_{n\to\infty} \sup_{\alpha\in(0,1]} d_H\Big((\sum_{k=1}^{\infty} a_{nk} X^{nk})_\alpha, \{0\}\Big) = 0$$

The result was proved. □

From theorem 3.8, we can easily get the following two corollaries.

Corollary 3.9 If $\{X^n\}$ is a sequence of independent identical distributed fuzzy set-valued random variables in $\mathbf{F}_k(\mathbb{R}^+)$ with $E[X^1] = I_0$ and $\max_k |a_{nk}| = O(n^{-\gamma}), \gamma > 0$, then $E[\|X^1\|_{\mathbf{F}}^{1+1/\gamma}] < \infty$ implies that

$$\sum_{k=1}^{\infty} a_{nk} X^{nk} \to 0 \ a.e.$$

with respect to the Hausdorff metric d_H^∞.

Corollary 3.10 If $\{X^n\}$ is a sequence of independent fuzzy set-valued random variables in $\mathbf{F}_k(\mathbb{R}^+)$ with $E[X^k] = I_0$ for $k = 1, 2, \cdots$ and stochastically dominated by a fuzzy set-valued random variable X. If $\max_k |a_{nk}| = O(n^{-\gamma}), \gamma > 0$, then $E[||X||_{\mathbf{F}}^{1+1/\gamma}] < \infty$ implies that

$$\sum_{k=1}^{\infty} a_{nk} X^{nk} \to 0 \ a.e.$$

with respect to the Hausdorff metric d_H^∞.

4. Discussion

In this paper, we mainly proved the limit theorems for negatively dependent fuzzy set-valued random variables, where underlying space is $\mathbb{R}+$. When the underlying space is $\{x \leq 0 : x \in \mathbb{R}\}$, then the above results are also true. Since the $\mathbf{K}(\mathfrak{X})$ are not linear space, even the $\mathbf{K}_{kc}(\mathfrak{X})$ are not linear space, the difference of sets do not have good properties. So it is not easy to discuss the properties of negatively dependent and obtain the convergence theorems in general Banach space.

Acknowledgements

The author is extremely grateful to the referees for their valuable comments and suggestions on improvement of the paper. The research was supported by NSFC(11301015,11571024, 11401016).

References

Artstein, Z., & Vitale, R. A. (1975). A strong law of large numbers for random compact sets. *Ann. Probab. 3*, 879-882. http://dx.doi.org/10.1214/aop/1176996275

Aumann, R. (1965). Integrals of set-valued functions. *J. Math. Anal. Appl., 12*, 1-12. http://dx.doi.org/10.1016/0022-247X(65)90049-1

Bozorgnia, A., Patterson, R. F.,& Taylor, R. L., (1992). Limit theorems for dependent random variables. *WCNA'92 proceedings of the first world congress on world congress of nonlinear analysts'92, 2*, 1639-1650.

Bozorgnia, A., Patterson, R. F., & Taylor, R.L. (1993). Limit theorems for ND random variables. *University of Georgia Technical Report.*

Castaing, C., & Valadier, M. (1977). Convex Analysis and Measurable Multifunctions. *Lect. Notes in Math., 580*, Springer–Verlag, Berlin, New York.

Colubi, A., López-Díaz, M., Dominguez-Menchero J. S, & Gil, M. A. (1999). A generalized strong law of large numbers. *Probab. Theory and Rel. Fields, 114*, 401-417.

Cressie, N. (1978). A strong limit theorem for random sets. *Suppl. Adv. in Appl. Probab., 10*, 36-46.

Guan, L., & Li, S. (2004). Laws of large numbers for weighted sums of fuzzy set-valued random variables. *International J. Uncertainty , Fuzziness and Knowlidge-Based Systems, 12*, 811-825. http://dx.doi.org/10.1142/S0218488504003223

Guan, L., & Sun, L. J. (2014). The laws of large numbers for set-valued negatively dependent random variables. *International Journal of Intelligent Technologies and Applied Statistics, 7*(3), 99-112.

Guan, L., & Wan, Y. (2016). A strong law of large numbers for set-valued negatively dependent random variables. *International Journal of Statistics and Probability, 5*(3), 102-112. http://dx.doi.org/10.5539/ijsp.v5n3p102

Hiai, F. (1984). Strong laws of large numbers for multivalued random variables, Multifunctions and Integrands (G. Salinetti, ed.). *Lecture Notes in Math., 1091*, Springer, Berlin, 160-172.

Hiai, F., & Umegaki, H. (1977). Integrals, conditional expectations and martingales of multivalued functions. *Jour. Multivar. Anal., 7*, 149-182.

Inoue, H. (1991). A strong law of large numbers for fuzzy random sets. *Fuzzy Sets and Syst., 41*, 285-291.

Li, S., & Ogura, Y. (1996). Fuzzy random variables, conditional expectations and fuzzy martingales. *J. Fuzzy Math., 4*, 905-927.

Li, S., Ogura, Y., & Kreinovich, V. (2002). Limit Theorems and Applications of Set-Valued and Fuzzy Sets-Valued Random Variables. *Kluwer Academic Publishers.* http://dx.doi.org/10.1007/978-94-015-9932-0

Li, S,& Ogura, Y. (2003). Strong laws of numbers for independent fuzzy set-valued random variables. *Fuzzy Sets and*

Systems, *157*, 2569-2578. http://dx.doi.org/10.1016/j.fss.2003.06.011

Puri, M. L, & Ralescu, D. A. (1983). Strong law of large numbers for Banach space valued random sets. *Ann. Probab.*, *11*, 222–224. http://dx.doi.org/10.1214/aop/1176993671

Puri, M. L, & Ralescu, D. A. (1986). Fuzzy random variables. *J. Math. Anal. Appl.*, *114*, 406-422.

Shen, L. M., & Guan, L. (2016). Strong laws of large numbers foe fuzzy set-valued random variables in G_α space. *Advances in Pure Mathematics*, *6*, 583-592. http://dx.doi.org/10.4236/apm.2016.69047

Taylor, R. L., Daffer, P. Z., & Patterson, R. F. (1985). Limit theorems for sums of exchangeable variables. *Rowman and Allanheld*, Totowa.

Taylor, R. L., & Inoue, H. (1985). A strong law of large numbers for random sets in Banach spaces. *Bull. Instit. Math. Academia Sinica*, *13*, 403-409.

Taylor, R. L., & Inoue, H. (1985) A strong law of large numbers for random sets in Banach spaces. *Bull. Instit. Math. Academia Sinica*, *13*, 403-409.

8

First-passage Time Estimation of Diffusion Processes through Time-Varying Boundaries with an Application in Finance

Imene Allab[1] & Francois Watier[1]

[1] Department of mathematics, Université du Québec à Montréal, Montreal, Canada

Correspondence: Francois Watier, Université du Québec à Montréal, Department of mathematics, P.O. Box 8888, Downtown Station, Montreal, QC., H3C 3P8, Canada. E-mail: watier.francois@uqam.ca

Abstract

In this paper, we develop a Monte Carlo based algorithm for estimating the FPT (first passage time) density of the solution of a one-dimensional time-homogeneous SDE (stochastic differential equation) through a time-dependent frontier. We consider Brownian bridges as well as local Daniels curve approximations to obtain tractable estimations of the FPT probability between successive points of a simulated path of the process. Under mild assumptions, a (unique) Daniels curve local approximation can easily be obtained by explicitly solving a non-linear system of equations.

Keywords: first passage time, stochastic differential equation, Monte Carlo, Daniels curve

1. Introduction

Let X be a time homogeneous one-dimensional diffusion process which is the unique (strong) solution of the following stochastic differential equation:

$$dX(t) = \mu(X(t))\,dt + \sigma(X(t))\,dW(t),\ X(0) = x_0 \tag{1}$$

that is the functions μ and σ satisfy regularity conditions as described for example in Karatzas and Shreve (1991).
If S is a time dependent boundary, we are interested in estimating either the pdf or cdf of the first-passage time of the diffusion process through this boundary that is we will study the following random variable:

$$\tau_S = \inf\{t > 0 | X(t) = S(t)\}$$

In general, there is no explicit expression for the first-passage time density of a diffusion process through a time-varying boundary. To this date, few specific cases provide closed-form formulas for some classes of time-varying boundaries for example when the process is Gaussian (Durbin (1985), Durbin and Williams (1992), DiNardo et al. (2001)) or a Bessel process (Deaconu and Herrmann (2013)). Thus, we mainly rely on simulation techniques to estimate this density for the more general case. Besides standard Monte Carlo estimation, only a few papers suggest alternatives or improvements such as considering upcrossing probabilities in between simulation points (Giraudo and Sacerdote (1999), Giraudo, Sacerdote, and Zucca (2001)).
The main goal of this work is to construct a computationally efficient algorithm that will provide reliable FPT density and probability estimates. The paper is organized as follows. In section 2, we review existing Monte Carlo based techniques. In section 3, we establish the mathematical foundations leading to a novel algorithm. Section 4 is devoted to various examples enabling us to evaluate the algorithm's performance. Section 5 consists of an empirical comparison study of Monte Carlo based methods. Finally, section 6 illustrates the pertinence of FPT estimation in the context of portfolio management.

2. Standard FPT Estimation Methods

2.1 Crude Monte Carlo

This is the simplest and best-known approach based on the law of large numbers. After fixing a time interval, basically we divide the latter into smaller ones, simulate a path of the process along those time points and, if it occurs, note the subinterval where the first upcrossing occurs. Generally, the midpoint of this subinterval forms the estimated first-passage time of this simulated path. We repeat the process a large number of time in order to construct a pdf or cdf estimate of this stopping time.

Usually, we need sizeable amounts of simulations and an extremely fine partition of time intervals in order to have a suitable estimation of first-passage time probabilities. Another drawback of the crude Monte Carlo approach is that it tends to overestimates the true value of the first-passage time since an upcrossing may occur earlier in between simulated points of a complete path as illustrated in figure 1.

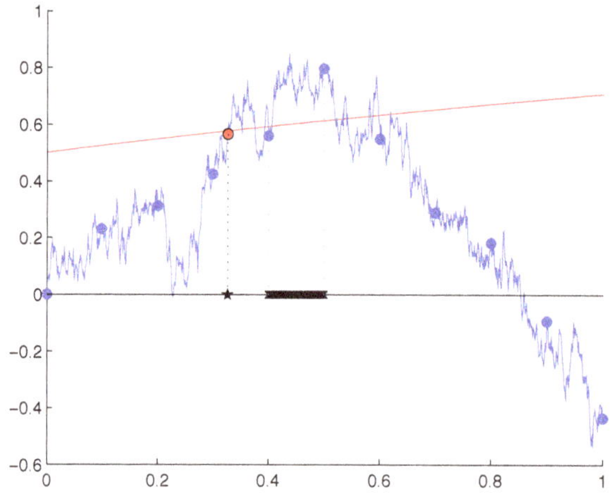

Figure 1. Undetected prior upcrossing through a basic Monte Carlo path simulation

2.2 Monte Carlo Approach with Upcrossings

Instead of continuously repeating the whole Monte Carlo procedure with an even finer interval partition to obtain better estimates, let us see how one could improve on the initial estimates without discarding the original simulated paths.

An astute idea that have been put forward is to ideally obtain the probability law of an upcrossing between simulated points, thus if p_k is the probability of an upcrossing in the time interval $[t_k, t_{k+1}]$, then one would simply generate a value U_k taken from a uniform random variable on $[0, 1]$ and assert that there is an upcrossing if $U_k \leqslant p_k$.

Unfortunately the FPT of a diffusion bridge will more than often not be available for more general diffusion processes, thus we need to propose an adequate estimation of this probability.

For each subinterval, one could simulate paths of tied-down processes to obtain FPT estimates for each subintervals along simulated paths of the process. This approach was used for example in Giraurdo and Sacerdote (1999) and Giraurdo, Sacerdote and Zucca (2001) where Kloeden-Platen numerical schemes were used for path simulations to construct approximation of diffusion bridges.

Another alternative, as first proposed in Strittmatter (1987) is to consider that for a small enough interval, the diffusion part of the process should remain fairly constant and then look at a Brownian bridge approximation of the diffusion bridge and exploit known results on the FPT of Brownian bridges.

For the Brownian bridge to constitute an efficient approximation, the diffusion part $\sigma(X_t)$ of the SDE (stochastic differential equation) should be constant. Therefore one should apply a Lamperti transform on both the original process and the frontier as described for example in Iacus (2008).

Indeed , define

$$F(x) = \int^{x} \frac{1}{\sigma(u)} du \tag{2}$$

and apply it on the original process and the time-varying boundary. Assuming that F is one-to-one, then the original problem is equivalent to finding the FPT density of

$$\tau_{S^*} = \inf \{t > 0 | Y(t) = S^*(t)\}$$

where the new boundary is given by $S^*(t) = F(S(t))$ and, by Itô's formula, the diffusion process Y follows the dynamics

$$dY(t) = b(Y(t))\, dt + dW(t), \;\; Y(0) = F(x_0) \tag{3}$$

where

$$b(t, Y(t)) = \frac{\mu\left(F^{-1}(Y(t))\right)}{\sigma\left(F^{-1}(Y(t))\right)} - \frac{1}{2}\frac{\partial \sigma}{\partial x}\left(F^{-1}(Y(t))\right)$$

Another advantage of opting for the Lamperti transformed FPT problem in this form is that one can readily make use of results from Downes and Borovkov (2008) on bounds on the FPT pdf.

Following the Brownian bridge approximation, we need to compute the upcrossing probabilities through the new boundary S^*. Dating back to Durbin's (1985) paper, we know that the FPT density of continuous Gaussian processes through a time-varying boundary must satisfy a Volterra type integral equation of the second kind. Oftentimes, we cannot explicitly solve this equation and therefore we must rely on some numerical approximation. For example, Durbin and Williams (1992) proposed an expansion series where each term involve computing multiple integrals while Di Nardo et al. (2001) proposed a more tractable iterative scheme to numerically solve the equation.

3. Local Daniels Curve Approximation Approach

While still favoring Brownian bridge approximations of the diffusion bridges after a Lamperti transform as described previously, instead of attempting to approximate the FPT of the brownian bridge through the original boundary using a costly computational method, we could instead introduce carefully chosen successive local curve approximations of the given frontier.

A first attempt would be to consider piecewise linear approximations of the time-varying boundary since explicit formulas are readily available in this case. Although computationally fast, since it does not take into account local curvatures, this technique may sometimes underestimate (locally convex) or overestimate (locally concave) the Brownian bridge FPT upcrossing probability and therefore should be used for a fine partition of the time interval.

In our approach, we propose to focus on local Daniels curve approximations of the time-varying boundary. Indeed, the more flexible three-parameter Daniels curve of the form

$$S(t) = \frac{\alpha}{2} - \frac{t}{\alpha}\ln\left[\frac{\beta + \sqrt{\beta^2 + 4\gamma e^{-\alpha^2/t}}}{2}\right] \tag{4}$$

also produces explicit tractable formulas of the first passage time probability, thus one would typically get a better approximation of the true probability p_k.

First we must show, under mild assumptions, that a (unique) Daniels curve approximation can easily be obtained by simply taking the endpoints of the segment and the value at midpoint (or another point of our choosing). Thus the following proposition will provide useful.

Proposition 1. *Let* $[0, \Delta t]$ *be a time interval, consider the points* $(0, a)$, $(\Delta t/2, b)$,$(\Delta t, c)$ *and set* $A = e^{\frac{2a^2}{\Delta t}}$ $B = e^{-\frac{4ab}{\Delta t}}$ *et* $C = e^{-\frac{2ac}{\Delta t}}$. *If*

$$\frac{1}{A^2} < \frac{B}{C} < \frac{1}{A} + \frac{\sqrt{A^2-1}}{A^2} \tag{5}$$

then there is a unique Daniels curve (4) passing through the three points with parameters

$$\alpha = 2a,\ \beta = \frac{A\left(A^4B^2 - C^2\right)}{A^3B - C},\ \gamma = A^4C^2 - \beta A^3 C \tag{6}$$

Proof. The set of points generate the following non-linear system of equations:

$$\begin{aligned} \frac{\alpha}{2} &= a \\ \frac{\alpha}{2} - \frac{\Delta t}{2\alpha}\ln\left[\frac{\beta + \sqrt{\beta^2 + 4\gamma e^{-2\alpha^2/\Delta t}}}{2}\right] &= b \\ \frac{\alpha}{2} - \frac{\Delta t}{\alpha}\ln\left[\frac{\beta + \sqrt{\beta^2 + 4\gamma e^{-\alpha^2/\Delta t}}}{2}\right] &= c \end{aligned} \tag{7}$$

obviously the first equation gives $\alpha = 2a > 0$, while simple algebraic manipulations on the last two equations lead us to solve the following linear system

$$\begin{aligned} \beta e^{\frac{\alpha^2-2\alpha b}{\Delta t}} + \gamma e^{-\frac{2\alpha^2}{\Delta t}} &= e^{\frac{2\alpha^2-4\alpha b}{\Delta t}} \\ \beta e^{\frac{\alpha^2-\alpha b}{\Delta t}} + \gamma e^{-\frac{\alpha^2}{\Delta t}} &= e^{\frac{\alpha^2-2\alpha b}{\Delta t}} \end{aligned}$$

which can be rewritten in the form

$$\begin{aligned} \beta A^6 B + \gamma &= A^8 B^2 \\ \beta A^3 C + \gamma &= A^4 C^2 \end{aligned}$$

since $A^6B - A^3C = A^3\left(A^3B - C\right) > 0$ then there exists a unique solution given by:

$$\begin{aligned}\beta &= \frac{A\left(A^4B^2 - C^2\right)}{A^3B - C}\\ \gamma &= A^4C^2 - \beta A^3C\\ &= A^7BC\left[\frac{C - AB}{A^3B - C}\right]\end{aligned}$$

this would constitute the solution to the original system provided that $\beta > 0$ and $\lim_{t\to\Delta t}\beta^2 + 4\gamma e^{-\alpha^2/t} > 0$.

Notice first that $\frac{1}{A^2} < \frac{B}{C} \Rightarrow A^4B^2 - C^2 > 0$ and therefore $\beta > 0$ and if furthermore $\frac{B}{C} \leqslant \frac{1}{A}$ then $\gamma \geqslant 0$ and clearly $\lim_{t\to\Delta t}\beta^2 + 4\gamma e^{-\alpha^2/t} > 0$ is satisfied. So if we assume now that $\frac{B}{C} > \frac{1}{A}$ then $\gamma < 0$, thus we need to verify that $\beta^2 + \frac{4\gamma}{A^2} > 0$ which is the case since

$$\begin{aligned}\beta^2 + \frac{4\gamma}{A^2} &= \frac{A^2\left(A^4B^2 - C^2\right)^2}{(A^3B - C)^2} + 4A^5BC\left[\frac{C - AB}{A^3B - C}\right]\\ &= \frac{A^2}{(A^3B - C)^2}\left(\left(A^4B^2 - C^2\right)^2 + 4A^3BC(C - AB)\left(A^3B - C\right)\right)\\ &= \frac{A^2}{(A^3B - C)^2}\left(A^4B^2 + C^2 - 2A^3BC\right)^2\end{aligned}$$

The final step is to make sure that it solves the original system. Substituting back in system (7), (where only positive square roots are involved), we see that is the case only if $A^4B^2 + C^2 - 2A^3BC < 0$, or equivalently

$$\left(\frac{B}{C} - \frac{1}{A}\right)^2 < \frac{A^2 - 1}{A^4}$$

which is verified through (5). □

Next, by applying Theorem 3.4 of Di Nardo et al. (2001) to the special case of Brownian bridges, the following proposition provides us with an explicit expression for the FPT upcrossing probabilities.

Proposition 2. *Consider a Brownian bridge $W^{0,\Delta t}$ define on an time interval $[0,\Delta t]$ and S a Daniels curve defined by (4) where $\alpha,\beta > 0$, $\gamma \in R$ and $\lim_{t\to\Delta t}\beta^2 + 4\gamma e^{-\alpha^2/t} > 0$, if $\tau_S = \inf\left\{0 \leqslant t \leqslant \Delta t | W^{0,\Delta t}(t) = S(t)\right\}$ then*

$$P(\tau_S \leqslant \Delta t) = \beta e^{-\frac{\alpha^2}{2\Delta t}} + \gamma e^{-\frac{2\alpha^2}{\Delta t}}$$

Finally, proposition 1 and 2 suggest that once a path of the diffusion process is produced, for each pair of simulated points on a subinterval of length Δt, one simply applies a linear translation of the line segment between these points onto the interval $[0,\Delta t]$ while translating the corresponding boundary segment.

This naturally leads us to the following algorithm:

[Step 1:] Apply the Lamperti transform (2) to the original diffusion process (1) and frontier S to obtain the new process (3) and boundary $S^* = F(S(t))$

[Step 2:] Select a time interval $[T_l, T_u]$ and construct a partition $T_l = t_0 < t_1 < \ldots < t_n = T_u$

[Step 3:] Initialize FPT vector counter to $\tau := \{0,\ldots,0\}$

[Step 4:] Initialize path counter to $k := 1$

WHILE k is less than N the number of desired paths DO the following:

[Step 5:] Simulate a path of the process $\{Y(t_1),\ldots,Y(t_n)\}$

[Step 6:] Initialize subinterval counter to $i := 1$

WHILE i is less than n the number of desired subintervals DO the following:

[Step 7:] IF $Y(t_i) \geqslant S^*(t_i)$ THEN set i^{th} FPT vector component to $\tau_i := \tau_i + 1$ and path counter to $k := k+1$, GO TO Step 5

[Step 8:] Set $\Delta := t_i - t_{i-1}$, $a := S(t_{i-1}) - Y(t_{i-1})$, $b := S\left(\frac{t_{i-1}+t_i}{2}\right) - Y\left(\frac{t_{i-1}+t_i}{2}\right)$, $c := S(t_i) - Y(t_i)$, finally set A, B, C, α, β and γ as in (6) of proposition 2

[Step 9:] IF $\frac{1}{A^2} < \frac{B}{C} < \frac{1}{A} + \frac{\sqrt{A^2-1}}{A^2}$ THEN set $c_1 := \beta$, $c_2 := \gamma$ll IF $\frac{1}{A^2} \geq \frac{B}{C}$ THEN set $c_1 := 0$, $c_2 := A^4C^2$
IF $\frac{B}{C} \geq \frac{1}{A} + \frac{\sqrt{A^2-1}}{A^2}$ THEN set $c_1 := 2AC$ $c_2 := -A^4C^2$

[Step 10:] Set probability upcrossing to $p := c_1 e^{-\frac{\alpha^2}{2\Delta}} + c_2 e^{-\frac{2\alpha^2}{\Delta}}$

[Step 11:] Generate a value U taken from a uniform random variable

[Step 12:] IF $U \leq p$ THEN set i^{th} FPT vector component to $\tau_i := \tau_i + 1$ and path counter to $k := k + 1$, GO TO Step 5, ELSE set $i := i + 1$, GO TO Step 7

Note that step 9 includes extreme cases where the middle point of the frontier in a subinterval may not be reached by a Daniels curve, thus we use the closest curve possible.

4. Numerical Examples

We will focus our examples on diffusion processes which paths can be simulated exactly. Therefore with known results on FPT density and bounds, it will allow us to better visualize the approximation error due essentially to the algorithm.

Example 1 Consider the following Ornstein-Uhlenbeck process and time varying boundary

$$\begin{aligned} dX(t) &= (1.0 - 0.5X(t))\,dt + dW(t), \; X(0) = 1.6 \\ S(t) &= 2.0\,(1.0 - \sinh(0.5t)) \end{aligned}$$

This diffusion process is a Gauss-Markov process and according to Di Nardo et al. (2001) the chosen boundary allows us to obtain an explicit FPT density given by

$$f(S(t), t) = \frac{0.2}{\sinh(0.5t)}\varphi_0(S(t))$$

where φ_0 is the probability density function of the Ornstein-Uhlenbeck process starting at $X(0) = 1.6$.

Figure 2 compares the true FPT density with the empirical density histogram obtained through our algorithm using a time step discretization of 0.01 of the time interval [0, 1] and 10 000 simulated paths. Furthermore, the algorithm gives us a FPT probability estimate of 0.9619 over the whole interval compared to the true value of 0.9608 representing a relative error of about 0.12%.

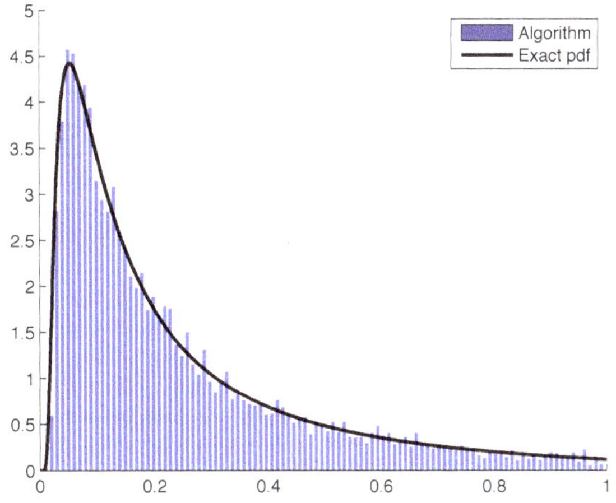

Figure 2. Ornstein-Uhlenbeck FPT pdf through the boundary $2.0\,(1.0 - \sinh(0.5t))$

Example 2 Consider the following squared Bessel process and time-varying boundary

$$\begin{aligned} dX(t) &= 2dt + 2\sqrt{X(t)}dW(t), \; X(0) = 0 \\ S(t) &= 2t(t+1)\ln\left(\frac{t+1}{t}\right) \end{aligned}$$

By applying the Lamperti transform to both the process and boundary we obtain respectively

$$dY(t) = \frac{1}{2Y(t)}dt + dW(t), \; Y(0) = 0$$
$$S^*(t) = \sqrt{2t(t+1)\ln\left(\frac{t+1}{t}\right)}$$

This transformed diffusion process is the Bessel process and, according to Deaconu and Herrmann (2013) the chosen boundary allows us to obtain an explicit FPT density given by.

$$f(t) = \frac{t^t}{(t+1)^{t+1}} \ln\left(\frac{t+1}{t}\right)$$

Figure 3 compares the true FPT density with the empirical density histogram obtained through our algorithm using a time step discretization of 0.01 of the time interval $[0, 1]$ and 10000 simulated paths. Furthermore, the algorithm gives us a FPT probability estimate of 0.7432 over the whole interval compared to the true value of 0.7500 representing a relative error of about 0.91%.

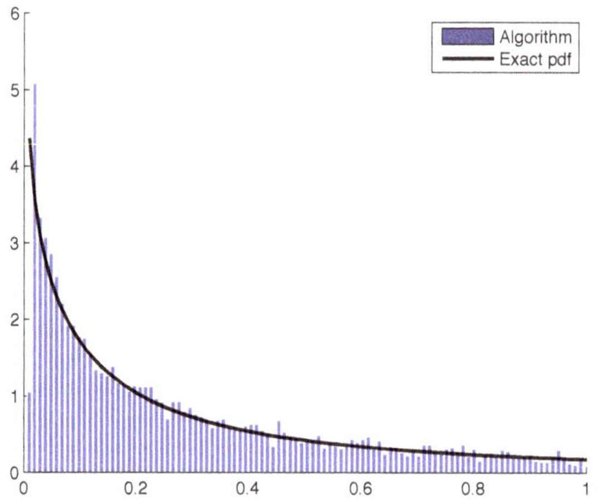

Figure 3. Squared Bessel process FPT pdf through the boundary $2t(t+1)\ln\left(\frac{t+1}{t}\right)$

In this case, it is worth noting that since the true density does not satisfy the initial condition $S(0) > X(0)$, therefore to increase the algorithm's accuracy near the origin one should envisage a finer partition around this point.

Example 3 Consider the following geometric Brownian process and linear boundary

$$dX(t) = 5.0X(t)\,dt + 2.5X(t)\,dW(t), \; X(0) = 0.5$$
$$S(t) = 1.0 + 2.0t$$

By applying the Lamperti transform to both the process and boundary we obtain respectively

$$dY(t) = 0.75dt + dW(t), \; Y(0) = 0.4\ln(0.5)$$
$$S^*(t) = 0.4\ln(1.0 + 2.0t)$$

As in example 1, this transformed diffusion process is a Gauss-Markov process and, although the new frontier does not allow an explicit FPT density, using the deterministic algorithm in Di Nardo et al. (2001) with a 0.01 time step discretization of the time interval $[0, 1]$, we can obtain a reliable approximation.

Figure 4 compares the Di Nardo FPT density approximation with the empirical density histogram obtained through our algorithm using the same time step discretization with 10 000 simulated paths.

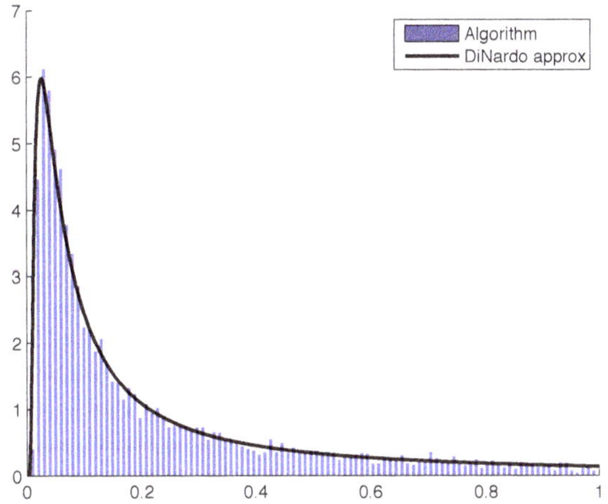

Figure 4. Geometric Brownian Motion FPT pdf through the boundary $1.0 + 2.0t$

Example 4 Consider the modified Cox-Ingersoll-Ross process and linear boundary

$$\begin{aligned} dX(t) &= -0.5X(t)\,dt + \sqrt{1 + X(t)^2}dW(t), \; X(0) = 0 \\ S(t) &= 0.3 + 0.2t \end{aligned}$$

By applying the Lamperti transform to both the process and boundary we obtain respectively

$$\begin{aligned} dY(t) &= -\tanh(Y(t))\,dt + dW(t), \; Y(0) = 0 \\ S^*(t) &= \operatorname{arcsinh}(0.3 + 0.2t) \end{aligned}$$

As opposed to the preceding examples, this diffusion process has neither an explicit expression nor probability law however using Beskos and Roberts' (2005) exact algorithm we can simulate exact sample paths. Although an explicit FPT density is not available, using results of Downes and Borovkov (2008) we can, in this case, obtain the following lower and upper bounds:

$$\begin{aligned} f_L(S(t), t) &= \frac{1}{t}\left(S(t) - \frac{0.2t}{\sqrt{1.09}}\right)\frac{e^{-0.5t}}{\cosh(S(t))}\varphi_W(S(t)) \\ f_U(S(t), t) &= \frac{1}{t}\left(S(t) - \frac{0.2t}{\sqrt{1.25}}\right)\frac{e^{0.5t}}{\cosh(S(t))}\varphi_W(S(t)) \end{aligned}$$

Figure 5 compares the FPT bounds with the empirical density histogram obtained through our algorithm using a 0.01 time step discretization of the time interval [0, 1] starting initially with 15000 simulations and obtaining 11768 valid paths through the exact algorithm.

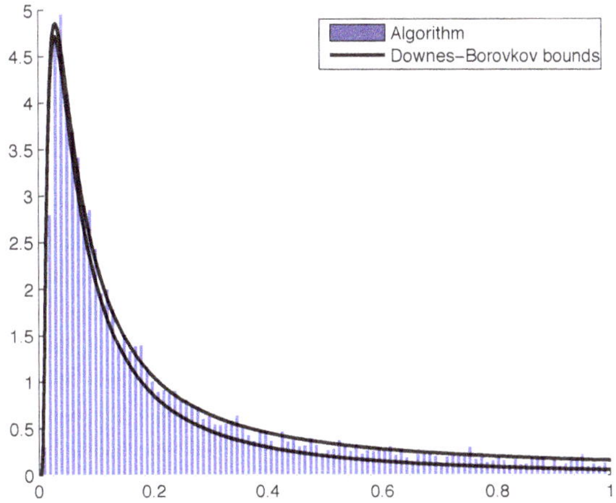

Figure 5. Modified CIR process FPT pdf through the boundary $0.3 + 0.2t$

5. Empirical Efficiency Comparison of Monte Carlo Based Techniques

As of now, essentially three different approaches exist in estimating the FPT probabilities of a general time-homogeneous one-dimensional diffusion process through a time-varying boundary, all of them are Monte Carlo based and involve estimating upcrossing probabilities of a diffusion bridge in between simulation points.

While Giraurdo, Sacerdote and Zucca (2001) suggest Monte Carlo subsimulations of these diffusion bridges, Giraurdo and Sacerdote (1999) and the present authors propose first a Brownian bridge approximation then followed by different deterministic numerical estimations of the upcrossing probabilites.

We will focus our comparison study on the total FPT probability of the Ornstein-Uhlenbeck process of our first numerical example. This choice is mainly motivated by the fact that both the diffusion process and associated diffusion bridges can be simulated exactly and furthermore the FPT pdf of the diffusion process has a known closed-form expression.

Since here we are only interested in comparing the total FPT probability on the time interval [0,1], once the original simulated paths are drawn only paths for which no barrier crossings were detected will require upcrossing probability estimations.

5.1 Comparison with Simulated Diffusion Bridges Approach

Basically, once standard Monte Carlo path simulations are drawn, we fix the number m of time steps discretization of all subintervals as well as the number M of simulated diffusion bridge paths in each subinterval. In Giraurdo, Sacerdote and Zucca's (2001) paper they devise an elaborate scheme for a general diffusion bridge through Kloeden-Platen approximation schemes. In our case an Ornstein-Uhlenbeck bridge can directly be simulated (see Bladt and Sorensen (2014)), thus we will use the later since we will gain in accuracy as well as computing time. Using $N = 10000$ simulated paths with time steps of length $\Delta = 0.01$, we obtained the following results:

	m	M	$P(\tau_S \leqslant 1.0)$	Relative error	Time
O.-U. bridge approx	10	10	0.9596	0.1227 %	2.13 s
		100	0.9593	0.1539 %	2.79 s
	100	10	0.9607	0.0082 %	2.74 s
		100	0.9615	0.0750 %	6.22 s
	1000	10	0.9620	0.1271 %	13.12 s
		100	0.9620	0.1271 %	86.48 s
Daniels boundary approx	n/a	n/a	0.9619	0.1167 %	0.58 s

We notice that while roughly both methods give us the same relative errors for the FPT estimate, our method produces a reasonable estimate at least four times faster. Although the Ornstein-Uhlenbeck bridge simulations would give us eventually the true upcrossing probabilities with larger values of M and m it seems there would be a heavy price to pay in regards to computing time.

5.2 Comparison with FPT Fdf Approximation of Brownian Bridges Approach

Again we start with the usual Monte Carlo path simulations but, following the Brownian bridge approach, we need a FPT probabilty approximation of an upcrossing in between simulation points. Giraurdo and Sacerdote (1999) proposed an iterative scheme to produce m estimated points of the true FPT pdf in each subinterval, it is based on approximating the Volterra integral equation associated with the FPT pdf. As a standard for comparison, since our method makes use of an estimation based on the values of only three points taken from the boundary in each subinterval, we will do the same for their method. Therefore three-point Simpson's rules will estimate the upcrossing probabilities. Using each time $N = 10000$ simulated paths, we obtained the following results:

Here we achieved a relative error of less than 0.2% as fast as about one second of computing time while the three-point FPT pdf approximation needed as much as 80 seconds to achieve the same level of accuracy. Of course one could increase the number of FPT pdf estimation points in order to achieve better accuracy but the immediate drawback seems to be generating sensibly greater computing time.

6. Application in Portfolio Management

Consider a risky strategy on an horizon $[0, T]$, the investor may encounter a specific instant t when the amount of wealth $x(t)$ be sufficient enough so that he may, at this point, safely reinvest all of his money in a simple bank account with constant interest rate r and the resulting terminal wealth $x(T)$ will attain his financial goal z. So we consider the following

	Time step Δ	$P(\tau_S \leqslant 1.0)$	Relative error	Time
FPT pdf approx	1	0.9405	2.1086 %	0.15 s
Daniels boundary approx		0.9575	0.3402 %	0.20 s
FPT pdf approx	0.1	0.9544	0.6671 %	0.70 s
Daniels boundary approx		0.9604	0.0384 %	1.07 s
FPT pdf approx	0.01	0.9580	0.2892 %	8.00 s
Daniels boundary approx		0.9600	0.0852 %	11.37 s
FPT pdf approx	0.002	0.9595	0.1310 %	81.68 s
Daniels boundary approx		0.9603	0.0509 %	97.40 s

first passage time random variable:

$$\tau_z = \inf\left\{0 \le t \le T | x(t)e^{r(T-t)} = z\right\}$$

and we naturally want to compute the probability $P(\tau_z \le T)$ of such an event. If $x_0 > 0$ is his initial wealth then we will assume that $z > x_0 e^{rT}$ so that the investor cannot achieve his financial goal by simply placing his initial investment in a bank account.

In order to investigate this goal-achieving problem, we must first define a mathematical setting for the dynamics of the financial market. We will consider here the celebrated Black-Scholes model that we next describe. The first asset is a bank account whose price at time t, $S_0(t)$, is the solution to the following ODE (ordinary differential equation):

$$dS_0(t) = rS_0(t)\,dt$$

The next asset consist of a stock whose price $S_1(t)$ at time t are the solutions to the following SDE (stochastic differential equation):

$$dS_1(t) = S_1(t)\left[bdt + \sigma dW\right]$$

where $\{W(t), t \geq 0\}$ is a standard Brownian motion and $b > r$.

Let $u(t)$, $0 \le t \le T$ be a financial strategy (or portfolio) where $u(t)$ is the amount placed in the stock. If we assume that all strategies $u(t)$ are self-financed (no outside injection of funds to the investors) and with no transaction costs then the wealth dynamic at time t is given by the following SDE:

$$dx(t) = (rx(t) + \sigma\theta u(t))\,dt + u(t)\,\sigma dW(t)$$

where $\theta = \frac{b-r}{\sigma}$.

Finally, among all the possible strategies, we will focus on the one generated by a family of stochastic control problems defined by

$$\min VAR(x(T)) \text{ subject to } E(x(T)) = z$$

These are known as mean-variance problems and are considered the cornerstone of modern portfolio management.

In the case where the portfolio is unconstrained, the optimal wealth process is given by:

$$x(t) = y_0 e^{\left(r-\frac{3}{2}\theta^2\right)t-\theta W(t)} + (\gamma - z)\,e^{-r(T-t)}$$

where

$$y_0 = \frac{x_0 - ze^{-rT}}{1 - e^{-\theta^2 T}}$$

$$\gamma = \frac{z - x_0 e^{\left(r-\theta^2\right)T}}{1 - e^{-\theta^2 T}}$$

Simple calculation shows that

$$\begin{aligned}\tau_z &= \inf\left\{0 \le t \le T | x(t)e^{r(T-t)} = z\right\} \\ &= \inf\left\{0 \le t \le T | W(t) = \theta T - \frac{3}{2}\theta t\right\}\end{aligned}$$

The computation of the probability $P(\tau_z \leq T)$, is reduced to the calculation of the probability of the first passage time of a Brownian motion through a straight line, more precisely the probability is given by:

$$P(\tau_z \leq T) = \Phi\left(\frac{1}{2}\theta\sqrt{T}\right) + e^{3\theta^2 T}\Phi\left(-\frac{5}{2}\theta\sqrt{T}\right)$$

Surprinsingly, as first noted by Li and Zhou (2006), the first passage time probability does not depend on either the initial wealth x_0 or targeted wealth z.

In the case where the optimization problem is subject to a non-negativity (no bankruptcy) constraint on the wealth process, the optimal wealth process has a more complex expression, according to Bielecki et al. (2005) it is given by

$$x(t) = \lambda e^{-r(T-t)} f_t(\theta W(t))$$

where

$$f_t(Z) = \Phi(-d_t^-(y_t(Z))) - \frac{y_t(Z)}{\lambda}e^{-r(T-t)}\Phi(-d_t^+(y_t(Z)))$$

$$d_t^+(y) = \frac{\ln\left(\frac{y}{\lambda}\right) + \left(r + \frac{1}{2}\theta^2\right)(T-t)}{\theta\sqrt{T-t}}$$

$$d_t^-(y) = d_t^+(y) - \theta\sqrt{T-t}$$

$$y_t(Z) = \mu e^{(\theta^2 - 2r)T} e^{(r - \frac{3}{2}\theta^2)t} e^{-Z}$$

and $\lambda > z$ and $\mu > 0$ are Lagrange multipliers obtained by solving the nonlinear system of equations:

$$\begin{aligned} \lambda\Phi\left(d_0^-\left(\frac{\lambda^2}{\mu}\right)\right) - \mu e^{-(r-\theta^2)T}\Phi\left(d_0^-\left(\frac{\lambda^2}{\mu}\right) - \theta\sqrt{T}\right) &= x_0 e^{rT} \\ \lambda\Phi\left(d_0^+\left(\frac{\lambda^2}{\mu}\right)\right) - \mu e^{-rT}\Phi\left(d_0^-\left(\frac{\lambda^2}{\mu}\right)\right) &= z \end{aligned}$$

Unfortunately, an explicit form for the corresponding goal-achieving probability as in the unconstrained case appears improbable. However, observe that for for each fixed time $t \in [0, T]$ there exists an inverse function f_t^{-1} therefore

$$\begin{aligned} \tau_z &= \inf\left\{0 \leq t \leq T | x(t)e^{r(T-t)} = z\right\} \\ &= \inf\left\{0 \leq t \leq T | \theta W(t) = f_t^{-1}\left(\frac{z}{\lambda}\right)\right\} \end{aligned}$$

thus the goal-achieving probability problem can be reduced to the FPT probability of a Brownian motion through an implicitly defined time-varying boundary and consequently we can make use of our Monte Carlo based algorithm to obtain a good estimate.

As in Bielecki's and al. (2005) numerical example, let $r = 0.06$, $b = 0.12$, $\sigma = 0.15$ thus $\theta = 0.40$ be the parameters of the Black-Scholes model. For a fixed time horizon of $T = 1$ and initial wealth $x_0 = 1$, figure 6 illustrates the estimated first passage-time probabilities of safe reinvestment for increasing values of wealth objective $z > x_0 e^{rT}$ in the wealth constraint case which we can compare to the constant probability value obtained for the unconstrained mean-variance strategy.

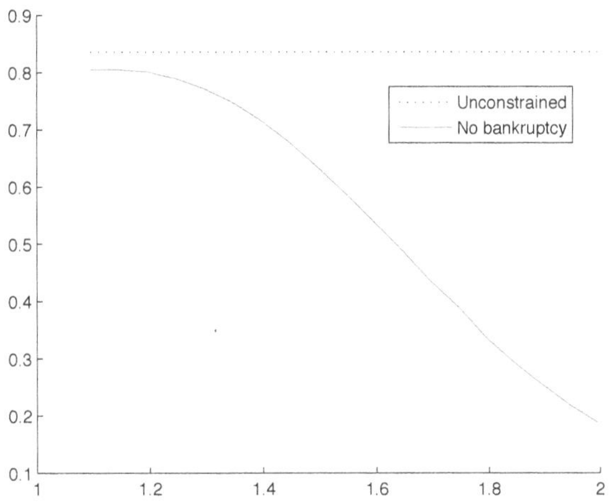

Figure 6. Comparison of goal-achieving FPT probabilities of mean-variance strategies

Clearly we conclude in our example that, in the case of mean-variance strategies under a no-bankruptcy restriction, the FPT probability is a decreasing function of the targeted wealth.

5. Conclusion

In this paper, we proposed an innovative Monte Carlo based algorithm for estimating the first-passage time density of a one-dimensional diffusion process through a deterministic time-dependent boundary. After considering a Lamperti transform, the main idea relied on a piecewise Daniels curve approximation of the boundary as well as Brownian bridge approximations of diffusion bridges. Through specific examples we illustrated the efficiency of our method by comparison with either known theoretical results or an existing deterministic method. We followed with an empirical comparison of accuracy and speed of our method with other Monte Carlo based methods. We concluded with a practical application pertaining to a portfolio management problem. Finally, an interesting question would be to see if our methodology could be extended to other types of processes with continuous paths such as superdiffusion and subdiffusion processes.

Acknowledgements

The authors wish to extend their gratitude to anonymous referees for their suggestions which improved the present paper. This research is supported by the Natural Sciences and Engineering Research Council of Canada (NSERC).

References

Beskos, A., & Roberts, G. O. (2005). Exact simulation of diffusions, *Annals of Applied Probability*, *15*(4), 2422-2444. http://dx.doi.org/10.1214/105051605000000485

Bielecki, T. R. et al. (2005). Continuous-time Mean-variance Portfolio Selection with Bankruptcy Prohibition, *Mathematical Finance*, *15*(2), 213-244. http://dx.doi.org/10.1111/j.0960-1627.2005.00218.x

Bladt, M., & Sorensen, M. (2014). Simple simulation of diffusion bridges with application to likelihood inference for diffusions,*Bernoulli*, *20*(2), 645-675. http://dx.doi.org/10.3150/12-BEJ501

Deaconu, M., & Herrmann, S. (2013). Hitting Time for Bessel Process - Walk on Moving Spheres Algorithm (WOMS), *Annals of Applied Probability*, *23*(6), 2259-2289. http://dx.doi.org/10.1214/12-AAP900

Di Nardo, E. et al. (2001). A Computational Approach to First Passage Time Problems for Gauss-Markov Processes, *Advances in Applied Probability*, *33*, 453-482. http://dx.doi.org/10.1239/aap/999188324

Downes, A.N., & Borovkov, K. (2008). First Passage Densities and Boundary Crossing Probabilities for Diffusion Processes, *Methodology and Computing in Applied Probability*, *10*, 621-644. http://dx.doi.org/10.1007/s11009-008-9070-x

Durbin, J. (1985). The First-Passage Density of a Continuous Gaussian Process to a General Boundary, *Journal of Applied Probability*, *22*(1), 99-122. http://dx.doi.org/10.2307/3213751

Durbin, J., & Williams, D. (1992). The First-Passage Density of the Brownian Motion Process to a Curved Boundary, *Journal of Applied Probability*, *29*(2), 291-304. http://dx.doi.org/10.2307/3214567

Giraudo, M.T., & Sacerdote, L. (1999). An improved technique for the simulation of first passage times for diffusion processes, *Communications in Statistics - Simulation and Computation*, *28*(4), 1135-1163. http://dx.doi.org/10.1080/03610919908813596

Giraudo, M.T., Sacerdote, L., & Zucca, C. (2001). A Monte Carlo Method for the Simulation of First Passage Time of Diffusion Processes, *Methodology and Computing in Applied Probability*, *3*, 215-231. http://dx.doi.org/10.1023/A:1012261328124

Iacus, S.M. (2008). *Simulation and Inference for Stochastic Differential Equations*, Springer, New York.

Karatzas, I., & Shreve, S. (1991). *Brownian Motion and Stochastic Calculus*, Springer, New York.

Li, X., & Zhou, X.Y. (2006). Continuous-time Mean-variance Efficiency: The 80% Rule, *Annals of Applied Probability*, *16*(4), 1751-1763. http://dx.doi.org/10.1023/10.1214/105051606000000349

Strittmatter, W. (1987). Numerical simulation of the mean first passage time, Preprint, University Freiburg.

9

Parameter Estimation of Shared Frailty Models Based on Particle Swarm Optimization

Oykum Esra Askin[1], Deniz Inan[2] & Ali Hakan Buyuklu[1]

[1] Department of Statistics, Yildiz Technical University, Istanbul, Turkey

[2] Department of Statistics, Marmara University, Istanbul, Turkey

Correspondence: Oykum Esra Askin, Department of Statistics, Yildiz Technical University, Istanbul, Turkey. E-mail: oykumesra@gmail.com

Abstract

Standard survival techniques such as proportional hazards model are suffering from the unobserved heterogeneity. Frailty models provide an alternative way in order to account for heterogeneity caused by unobservable risk factors. Although vast studies have been done on estimation procedures, Evolutionary Algorithms (EAs) haven't received much attention in frailty studies. In this paper, we investigate the estimation performance of maximum likelihood estimation (MLE) via Particle Swarm Optimization (PSO) in modelling multivariate survival data with shared gamma frailty. Simulation studies and real data application are performed in order to assess the performance of MLE via PSO, quasi-Newton and conjugate gradient method.

Keywords: frailty models; unobserved heterogeneity; particle swarm optimization; maximum likelihood estimation

1. Introduction

The general concept of survival models is to investigate the impact of risk factors on individual failure times. However, it's impossible to include all important factors into the model. This is due to the lack of information on individual level or the researcher may not be aware of the importance of the factor, or even the existence of it (Hougaard, 1991). Several authors studied the consequences of ignoring the existence of unobserved heterogeneity caused by unobservable risk factors such as Aalen (1988), Lancaster (1979, 1990), Henderson and Oman (1999) and Van den Berg (2001).

In order to account for unobserved heterogeneity, the frailty term was first introduced by (Vaupel, Manton, & Stallard, 1979) for univariate survival data. They indicated that the individual hazard is the product of two terms: an individual level frailty and a baseline hazard function. The multivariate generalization was then introduced by Clayton (1978). The proposed model was a random effect model, which is an extension of a proportional hazards (PH) approach. In a shared frailty model, lifetimes of a group of observations in the same cluster (e.g., individuals in a family) are related to each other (Aalen, Borgan, & Gjessing, 2008). Each cluster shares the same level of frailty. In other words, the observations in a cluster share the same unobservable risks such as genetic structures. The variance of this common frailty is a measure of dependence among lifetimes within a cluster.

There exists a huge amount of literature on estimating parameters according to information of the distributional form of the baseline hazard function. If a parametric form is not assumed for the baseline hazard, different estimation methods are proposed for the semi-parametric frailty models (Clayton & Cuzick, 1985; Gill, 1985; Clayton, 1991; McGilchrist & Aisbett, 1991; Klein, 1992; Nielsen, Gill, Andersen, & Srensen, 1992; McGilchrist, 1993; Rondeau, Commenges, & Joly, 2003; Therneau, Grambsch, & Pankratz, 2003). Also, important works relevant with Bayesian framework were offered by Sinha and Dey (1997), Ibrahim, Chen and Sinha (2001) and Hanagal and Pandey (2015). On the other hand, when the baseline hazard is represented parametrically, model parameters can be estimated with MLE via derivative-based methods such as Newton Raphson, quasi-Newton and conjugate gradient. In spite of having useful properties, these techniques have some drawbacks. The main drawback is that the starting point should be chosen with a value close to a global optimum because they can easily get stuck in a local optimum at multiple solution space. The poor choice of an initial value for any model parameter could lead to finding an estimate that is far away from the optimal solution. When the number of model parameters is too large, choosing suitable initial values becomes harder and researchers are never sure whether the resulting solution is a local or global. Also, biased parameter estimates can be obtained due to the heavy censoring or small sample size.

PSO is a member of stochastic search family inspired by some natural phenomena to solve complicated and high-dimensional optimization problems. It is a global search strategy that has properties of being robust, fast converge and easy implemented for linear/non-linear functions. In contrast with the derivative-based methods, it is less sensitive to the selection of initial parameter values. Besides, it gives reliable estimates even though the number of parameters being estimated is oversized. Recently, PSO has become an important estimation technique for censored data (see Wang & Huang, 2014). However, to the best of our knowledge, there is no work that uses PSO in frailty models. The main purpose of this study is to present the estimation performance of PSO in modelling multivariate survival data with shared frailty. Two simulation studies with different parameter settings are conducted to evaluate the performances of MLE via PSO, quasi-Newton and conjugate gradient methods. It is shown that PSO is an efficient method to obtain maximum likelihood estimates (MLEs) even though the following data characteristics exist: (i) small number of clusters (ii) large number of observations in a cluster (iii) large number of censored observations (heavy censored). The remainder of paper is organized as follows. Section 2 provides a general concept of shared frailty models. Section 3 provides the basic steps of standard PSO method. Section 4 outlines the MLE via PSO procedure and demonstrates its performance using two simulation studies. Section 5 provides a real data application to confirm the efficiency of procedure.

2. Shared Frailty Models

Suppose that there are N clusters and each cluster i has n_i observations ($i = 1, ..., N$). t_{ij} is the observed failure time of $j\,th$ ($j = 1, ..., n_i$) observation in $i\,th$ cluster. Under a right censoring scheme, $t_{ij} = min(c_{ij}, t^*_{ij})$ where t^*_{ij} is the failure time and c_{ij} is the censoring time. Here, t^*_{ij} and c_{ij} are independent random variables. The observed censoring indicator δ_{ij} is equal to 1 if $t^*_{ij} < c_{ij}$, and 0 otherwise. Conditional on frailty z_i (> 0) and X_{ij}, the hazard function of $i\,th$ cluster has the form

$$h(t_{ij} \backslash X_{ij}, z_i) = z_i h_0(t_{ij}) \exp(\beta' X_{ij}), \tag{1}$$

where $h_0(.)$ is the baseline hazard function, X_{ij} is a vector of observed covariates for the $j\,th$ observation and β is a vector of regression parameters. It is assumed that survival times (in cluster i) are conditionally independent with respect to the frailty. This common frailty is the cause of dependence among lifetimes of within a cluster (Wienke, 2011). The frailties,z_i, are i.i.d. variables with the common probability density function $g(z_i)$.

Let $(t_{i1}, t_{i2}, ..., t_{in_i})$ denote n_i survival times of $i\,th$ cluster. The conditional survival function in that cluster can be expressed such as (for a given $Z_i = z_i$; $t_{ij} > 0$; $j = 1, 2, ..., n_i$),

$$S(t_{i1}, ..., t_{in_i} \backslash z_i, X_i) = S(t_{i1} \backslash z_i, X_{i1}) ... S(t_{in_i} \backslash z_i, X_{in_i}) = \exp\left(-z_i \sum_{j=1}^{n_i} \Lambda_0(t_{ij}) e^{\beta' X_{ij}}\right), \tag{2}$$

where $\Lambda_0(.) = \int_0^{t_{ij}} h_0(s)ds$ is the common cumulative baseline hazard. Here, $S(.\backslash.) = \exp(-z_i \Lambda_0(t_{ij}) e^{\beta' X_{ij}})$ denotes the survival function of jth observation conditional on frailty. The conditional likelihood function of ith cluster has the form

$$L_i(\psi, \beta \backslash z_i) = \prod_{j=1}^{n_i} h(t_{i1}, , ..., t_{in_i} \backslash z_i, X_i)^{\delta_{ij}} S(t_{i1}, ..., t_{in_i} \backslash z_i, X_i), \tag{3}$$

where ψ and β are the vector of baseline hazard parameters and regression parameters, respectively. Unconditional likelihood for observed data can be obtained by integrating Eq.(3) with respect to frailty terms ($i = 1, 2, ..., N$; $j = 1, 2, ..., n_i$),

$$L(\psi, \theta, \beta) = \prod_{i=1}^{N} \int_0^{\infty} \prod_{j=1}^{n_i} (z_i h_0(t_{ij}; \psi) e^{\beta' X_{ij}})^{\delta_{ij}} \exp(-z_i \Lambda_0(t_{ij}; \psi) e^{\beta' X_{ij}}) g(z_i) dz_i \tag{4}$$

Various studies have been done on the choice of distribution of frailty random variables. While some authors use continuous distributions such as Gamma (Clayton, 1978; Vaupel et al., 1979), inverse Gaussian (Hougaard, 1984; Whitmore & Lee, 1991; Hanagal & Sharma, 2015), log-normal (McGilchrist & Aisbett, 1991) and positive stable (Hougaard, 1986), other authors use discrete distributions (Caroni, Crowder, & Kimber, 2010; Ata & Ozel, 2012). However, the Gamma distribution is the most common and widely used in literature for determining the frailty effect, which acts multiplicatively on the baseline hazard (Wienke, 2011). Due to its computational convenience, one parameter Gamma distribution (with mean 1 and variance θ) is used as the frailty distribution. The probability density function of one parameter Gamma distribution is as follows

$$g(z) = \frac{z^{1/\theta-1}\exp(-z/\theta)}{\Gamma(1/\theta)\theta^{1/\theta}}. \tag{5}$$

The larger value of θ indicates the greater degree of heterogeneity among lifetimes within a cluster. When Gamma distribution is degenerate at value 1, Eq.(1) reduces to the standard PH model. Hence, we can conclude that θ takes the value 0 and lifetimes within clusters are independent.

Under the concept of gamma frailty, unconditional survival function for cluster i is obtained by integrating conditional survival function over the Gamma distribution and can be written as

$$S(t_{i1},...,t_{in_i}\backslash X_i) = \int S(t_{i1},...,t_{in_i}\backslash z_i, X_i)g(z_i)dz_i = \left[1+\theta\sum_{j=1}^{n_i}\Lambda_0(t_{ij})e^{\beta' X_{ij}}\right]^{-1/\theta}. \tag{6}$$

The unconditional hazard function of corresponding cluster is,

$$h(t_{i1},...,t_{in_i}\backslash X_i) = \frac{\left(\prod_{j=1}^{n_i} h_0(t_{ij})\right)e^{\sum_{j=1}^{n_i}\beta' X_{ij}}}{\left[1+\theta\sum_{j=1}^{n_i}\Lambda_0(t_{ij})e^{\beta' X_{ij}}\right]}. \tag{7}$$

Once the parametric form of baseline hazard is specified, the unconditional likelihood function can be easily derived (Wienke, 2011). Taking into account of all clusters and denoting the number of observed events in each cluster as $d_i = \sum_{j=1}^{n_i}\delta_{ij}$, one might write the following unconditional likelihood function as (Klein, 1992; Duchateau & Janssen, 2008; Wienke, 2011),

$$L(\psi,\theta,\beta) = \prod_{i=1}^{n}\frac{\Gamma(D_i+1/\theta)\prod_{j=1}^{n_i}\left(h_0(t_{ij};\psi)e^{\beta' X_{ij}}\right)^{\delta_{ij}}}{\theta^{1/\theta}\Gamma(1/\theta)\left(1/\theta+\sum_{j=1}^{n_i}\Lambda_0(t_{ij};\psi)e^{\beta' X_{ij}}\right)^{1/\theta+D_i}}, \tag{8}$$

where $i = 1,...,N$ and $j = 1,...,n_i$. MLEs of model parameters can be obtained by maximizing Eq.(8).

3. Estimation Method: PSO

PSO is a member of Evolutionary Algorithms (EAs) and can be used to overcome some limitations of derivative-based methods. It was first developed by Kennedy and Eberhart (1995) as a population-based method. It is easy to implement for complex and high-dimension, linear/non-linear functions. PSO simultaneously searches a global optimum using more than one candidate solutions in different regions of multiple solution space. Therefore, the problem of being stuck at a local optimum is almost avoided (Aladag, Yolcu, Egrioglu, & Dalar, 2012). Overall, PSO is not affected by the existence of large numbers of unknown parameters as much as derivative-based methods. The algorithm has its own parameters: particle size, inertia weight, acceleration coefficients and velocity vector.

A search in solution space is similar to bird food-finding behavior in a flock. In contrast with the derivative-based methods, the PSO algorithm starts with many candidate solutions (particles). The population of particles is called a swarm, and each particle has a location vector P_m and a velocity vector V_m. A particle flies thought the solution space in search for a better position and adjusts its own position with reference to the values of personel best (Pbest) and neighbors global best (Gbest). The basic steps standard PSO are given as follows (Egrioglu, Yolcu, Aladag, & Kocak, 2013):

[Step 1]: In d-dimensional search space, stochastically generate a position and a velocity for each particle and then store as the vectors $\vec{P}_m = \{p_{m,1},...,p_{m,k},...,p_{m,d}\}$ and $\vec{V}_m = \{v_{m,1},...,v_{m,k},...,v_{m,d}\}$, respectively. For $m = 1,...,L$ and $k = 1,...,d$; $p_{m,k}$ denotes the $k\,th$ position of $m\,th$ particle and L is the total number of particles in a swarm.

[Step 2]: Let $\vec{P}best_m$ represents the vector of best position of $m\,th$ particle, and $\vec{G}best$ is the particle which has the best fitness function value, found so far by all particles. According to the fitness function, determine $\vec{P}best_m$ and $\vec{G}best$ that are respectively given by (9) and (10),

$$\vec{P}best_m = \{p_{m,1}^{best},...,p_{m,k}^{best},...,p_{m,d}^{best}\} \tag{9}$$

$$\vec{G}best = \{p_{g,1}^{gbest}, ..., p_{g,k}^{gbest}, ..., p_{g,d}^{gbest}\} \tag{10}$$

[Step 3]: Update velocities and obtain particles by using following formulas given below,

$$v_{m,k}^{new} = \omega v_{m,k} + c_1 r_1 (p_{m,k}^{best} - p_{m,k}) + c_2 r_2 (p_{g,k}^{gbest} - p_{m,k}) \tag{11}$$

$$p_{m,k} = p_{m,k} + v_{m,k}^{new} \tag{12}$$

where c_1 and c_2 are the positive acceleration coefficients (learning factors); ω is an inertia parameter, and r_1 and r_2 are uniform random variables which are generated within $[0, 1]$. Generally, both the values of (c_1, c_2) are taken fixed at 2 and $\omega = 0.5 + rand/2$ changes in each iteration (Hu, Eberhart, & Shi, 2004).

[Step 4]: Stop when pre-specified iteration number is met. $\vec{G}best$ is the vector of optimal solution. Otherwise, return to Step 2.

The Gelman-Rubin (G-R) statistic, modified by Prasad and Souradeep (2012), is used in order to be sure that pre-specified iteration number satisfy the convergence for each position related to unknown model parameter. Let T represents the total iteration number. $p_{m,k}$ is the trajectory of mth particle of kth position. The variances of within particles (W) and between particles (B) are calculated for position k with the following formulas (13) and (14),

$$W_k = \frac{1}{L}\sum_{m=1}^{L} \sigma_{m,k}^2, \tag{13}$$

where $\sigma_{m,k}^2 = \frac{1}{T-1}\sum_{j=1}^{T}(p_{m,k}^{(j)} - \bar{p}_{m,k})$ and $\bar{p}_{m,k} = \sum_{j=1}^{T} p_{m,k}^{(j)}$.

$$B_k = \frac{T}{L-1}\sum_{m=1}^{L}(\bar{p}_{m,k} - \bar{\bar{p}}_k)^2, \tag{14}$$

where $\bar{\bar{p}}_k = \frac{1}{L}\sum_{m=1}^{L}\bar{p}_{m,k}$.

Using W_k and B_k, the variance of the stationary distribution is calculated as, $\hat{V}_k = [(W_k(T-1))/T] + [B_k/T]$. An estimated potential scale reduction factor (G-R statistic) $\hat{R}_k = \sqrt{\hat{V}_k/W_k}$ monitors the convergence ability of performed total iteration. For the value of close to 1 suggests a good convergence achieved for the corresponding parameter k.

4. Simulation Study

In this section, the estimation performance of MLE via PSO is investigated by two scenarios. In the first simulation study, the performance of PSO is compared with the quasi-Newton method, and then in the second one it is compared with the conjugate gradient method.

For both of simulation studies, the baseline cumulative hazard function is assumed to be Weibull as $\Lambda_0(.) = \lambda t^p$. In the first simulation study, Weibull parameters are set to $(\psi = \lambda, p) = (0.2, 1.5)$ and the frailty variables are taken as Gamma distributed randoms with variance θ= 1.5. Three observed covariates $x^{(1)}$, $x^{(2)}$ and $x^{(3)}$ are used and randomly generated from Bernoulli distribution with success probability 0.1, 0.5 and 0.9, respectively. Regression parameters are assumed to be β_1=β_2=β_3= 1. In the second simulation study, Weibull parameters are set to $(\psi = \lambda, p) = (0.5, 1)$. The variance of frailty is taken as θ= 1. Two observed covariates, $x^{(1)}$ and $x^{(2)}$, are randomly generated from Binomial distribution having the same success probability 0.5 and the number of trials are taken as 2 and 3, respectively. Regression parameters are set to be β_1=β_2= 0.5.

The model for the failure times $(t_{i1}, ..., t_{in_i})$ is considered as,

$$F(t_{ij} \backslash z_i, X_i) = 1 - \exp\left[-z_i \sum_{j=1}^{n_i} \Lambda_0(t_{ij}) \exp(\beta X_{ij})\right]$$

where $\beta X_{ij} = \beta_1 x_{ij}^{(1)} + \beta_2 x_{ij}^{(2)} + \beta_3 x_{ij}^{(3)}$ in the first simulation and $\beta X_{ij} = \beta_1 x_{ij}^{(1)} + \beta_2 x_{ij}^{(2)}$ in the second simulation. Following the study of Yu (2006), the multivariate survival data are generated as follows:

(1) frailty variables are generated from Gamma distribution with mean 1 and variance θ. (2) for the jth observation in ith cluster, uniform random variable is generated as $u_{ij} \sim U(0,1)$. Also censoring time c_{ij} is generated from an exponential distribution $exp[\kappa]$ with parameter κ to create expected censoring rates.

(3) the failure times are generated by,

$$t_{ij}^{*} = \left(\frac{-\log(u_{ij})}{z_i \lambda \exp(\beta X_{ij})} \right)^{1/p} \tag{15}$$

(4) the observed failure times are $t_{ij} = min(c_{ij}, t_{ij}^{*})$. The observed censoring indicator δ_{ij} is equal to 1 if $t_{ij}^{*} < c_{ij}$, and 0 otherwise.

For both of simulations, different size of clusters ($n_i = 2, 3, 4$) and different number of clusters ($N = 20, 30, 40$) are taken into account. Three censoring rates are chosen as 5%, 20% and 40% which represent light, medium and heavy censoring, respectively. Simulations are repeated 100 times. Mean absolute percentage error (MAPE), bias and standard error (SE) vectors are computed for all cases. Let η shows the vector of real parameters and $\hat{\eta}$ shows the vector of MLEs. In the first simulation study $\hat{\eta} = (\hat{\beta}_1, \hat{\beta}_2, \hat{\beta}_3, \hat{\lambda}, \hat{p}, \hat{\theta})$, and in the second one $\hat{\eta} = (\hat{\beta}_1, \hat{\beta}_2, \hat{\lambda}, \hat{p}, \hat{\theta})$. To show the results of 54 different cases less complicated, we report the mean of parameters' MAPEs (mMAPE), biases (mBias) and SEs (mSE) given below,

$$mMAPE(\hat{\eta}) = \frac{\frac{1}{100} \sum_{k=1}^{d} \sum_{r=1}^{100} \left| \frac{(\hat{\eta}_r - \eta)}{\eta} \right|^{(k)}}{d}, \tag{16}$$

$$mSE(\hat{\eta}) = \frac{\sum_{k=1}^{d} \left(\sqrt{\frac{1}{99} \sum_{r=1}^{100} (\hat{\eta}_r - \eta)^2} \right)^{(k)}}{d}, \tag{17}$$

$$mBias(\hat{\eta}) = \frac{\sum_{k=1}^{d} \left(\sqrt{(E(\hat{\eta}) - \eta)^2} \right)^{(k)}}{d}, \tag{18}$$

where $\hat{\eta}_r$ is the vector of MLEs obtained in rth repeat and $E(\hat{\eta}) = \sum_{r=1}^{100} \hat{\eta}_r / 100$. Besides these measures, the mean of log-likelihood values obtained in each repeat are calculated (mLog.Lik.). The steps of MLE via PSO are given as follows:

[Step 1]: PSO parameters are selected as $(\omega, c_1, c_2, L) = (0.5 + rand/2, 2, 2, 30)$. T is taken as 1000. In the first simulation study the solution space is 6-dimensional and in the second, it is 5-dimensional. Positions of each particle and their corresponding velocities are generated randomly from uniform distribution in the range of [0,3] and [0,0.5], respectively. Positions of a particle in the swarm for the first and second simulation are illustrated in Figure 1 and 2, respectively.

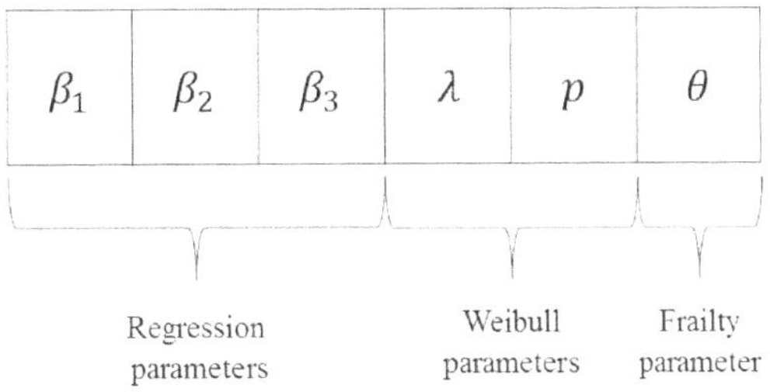

Figure 1. Positions of a particle in Simulation 1

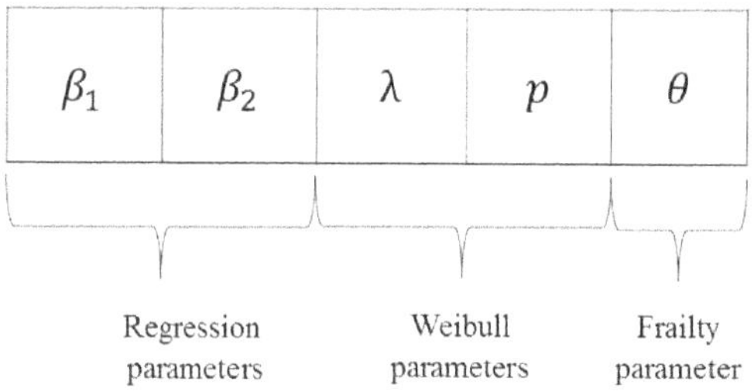

Figure 2. Positions of a particle in Simulation 2

[Step 2]: The unconditional likelihood function given in Eq.(8) is chosen as the fitness function of PSO. For all particles, use the fitness function to calculate the fitness values. Determine $\vec{P}best_m$ and $\vec{G}best$, then update the velocity and and position values for each particle using formulas (11) and (12).

[Step 3]: If the termination criteria is met, then stop and go to Step 4. Otherwise return to Step 2.

[Step 4]: If the G-R statistics of all parameter estimations are satisfactory, $\vec{G}best$ is the vector of MLEs found by using PSO (it is observed that T=1000 is sufficient for converge). To perform the MLE with either the quasi-Newton or conjugate gradient methods, randomly choose a particle from the initial swarm and use it as the initial vector.

Table 1. Simulation Results for Model 1

Cases			Methods							
			PSO				Quasi-Newton			
δ	n_i	N	mMAPE	mSE	mBias	mLog.Lik	mMAPE	mSE	mBias	mLog.Lik
5%	2	20	0.751	0.816	0.076	-66.645	1.639	0.871	0.256	-131.932
		30	0.586	0.506	0.024	-98.473	1.623	0.861	0.250	-209.767
		40	0.542	0.421	0.028	-134.224	1.577	0.839	0.250	-289.869
	3	20	0.576	0.499	0.046	-94.536	1.755	0.855	0.280	-166.141
		30	0.478	0.351	0.013	-143.818	1.703	0.856	0.265	-246.152
		40	0.454	0.305	0.020	-190.658	1.593	0.847	0.253	-341.636
	4	20	0.507	0.368	0.019	-123.813	1.788	0.875	0.276	-239.448
		30	0.460	0.305	0.030	-184.511	1.530	0.866	0.240	-415.138
		40	0.424	0.257	0.011	-249.729	1.813	0.855	0.281	-506.310
20%	2	20	0.844	0.964	0.080	-50.705	1.865	0.881	0.304	-131.768
		30	0.631	0.599	0.054	-76.061	1.631	0.878	0.250	-206.824
		40	0.561	0.485	0.024	-101.141	1.764	0.828	0.275	-206.835
	3	20	0.572	0.537	0.024	-73.551	1.681	0.878	0.257	-149.540
		30	0.530	0.432	0.025	-112.645	1.637	0.860	0.257	-226.845
		40	0.498	0.354	0.017	-149.404	1.737	0.875	0.276	-284.043
	4	20	0.552	0.461	0.024	-95.459	1.673	0.849	0.268	-167.023
		30	0.501	0.351	0.036	-140.327	1.538	0.848	0.244	-265.062
		40	0.444	0.278	0.011	-193.216	1.675	0.873	0.263	-439.114
40%	2	20	0.898	1.041	0.093	-34.374	1.739	0.853	0.269	-97.330
		30	0.779	0.814	0.078	-51.931	1.602	0.873	0.252	-126.279
		40	0.694	0.669	0.074	-71.023	1.595	0.848	0.257	-168.676
	3	20	0.665	0.732	0.059	-49.059	1.817	0.855	0.282	-106.891
		30	0.557	0.482	0.023	-76.464	1.747	0.891	0.268	-162.363
		40	0.506	0.399	0.022	-104.568	1.621	0.878	0.251	-251.969
	4	20	0.566	0.497	0.029	-64.371	1.665	0.858	0.262	-154.974
		30	0.516	0.403	0.028	-96.147	1.553	0.868	0.235	-199.620
		40	0.471	0.336	0.013	-134.302	1.746	0.863	0.276	-262.420

Table 2. Simulation Results for Model 2

Cases			Methods							
			PSO				Conjugate Gradient			
δ	n_i	N	mMAPE	mSE	mBias	mLog.Lik	mMAPE	mSE	mBias	mLog.Lik
5%	2	20	0.442	0.461	0.038	-43.164	1.643	0.848	0.377	-212.481
		30	0.392	0.303	0.023	-66.949	1.644	0.851	0.379	-214.597
		40	0.372	0.267	0.021	-90.148	1.609	0.880	0.362	-395.706
	3	20	0.384	0.269	0.023	-65.337	1.541	0.849	0.351	-210.397
		30	0.366	0.201	0.011	-98.386	1.642	0.841	0.389	-337.160
		40	0.347	0.181	0.006	-129.349	1.645	0.842	0.394	-424.842
	4	20	0.383	0.253	0.018	-83.467	1.498	0.845	0.339	-242.526
		30	0.354	0.176	0.009	-127.168	1.486	0.861	0.333	-307.569
		40	0.351	0.148	0.006	-171.539	1.630	0.860	0.374	-518.609
20%	2	20	0.466	0.541	0.040	-31.532	1.657	0.837	0.396	-133.936
		30	0.394	0.320	0.038	-46.877	1.714	0.881	0.400	-180.451
		40	0.375	0.251	0.017	-62.962	1.578	0.830	0.362	-234.702
	3	20	0.396	0.322	0.024	-42.867	1.735	0.861	0.406	-211.654
		30	0.364	0.242	0.018	-67.121	1.689	0.891	0.396	-245.108
		40	0.371	0.215	0.014	-89.857	1.565	0.843	0.357	-275.511
	4	20	0.391	0.285	0.019	-54.476	1.723	0.851	0.414	-173.668
		30	0.364	0.206	0.018	-85.803	1.721	0.874	0.404	-287.870
		40	0.357	0.178	0.009	-117.091	1.693	0.858	0.402	-370.715
40%	2	20	0.546	0.707	0.062	-16.255	1.614	0.871	0.369	-105.767
		30	0.432	0.390	0.045	-26.241	1.631	0.877	0.374	-132.810
		40	0.399	0.300	0.032	-37.778	1.647	0.856	0.380	-162.449
	3	20	0.433	0.366	0.041	-25.335	1.612	0.875	0.367	-111.564
		30	0.382	0.259	0.021	-37.975	1.569	0.874	0.356	-159.544
		40	0.367	0.222	0.013	-51.761	1.593	0.828	0.379	-208.918
	4	20	0.384	0.294	0.027	-32.055	1.557	0.864	0.350	-144.924
		30	0.366	0.213	0.013	-51.601	1.663	0.866	0.389	-198.873
		40	0.349	0.174	0.015	-67.963	1.628	0.875	0.373	-256.160

The results are shown in Table 1 and Table 2. As observed in the tables, MLE via PSO has lower mMAPE and mBias values for all cases. For all cases except two, the values of mSE are lower in the first simulation. In most cases, as the number of clusters increase, a remarkable reduction in mMAPE, mSE and mBias for PSO is seen. However, the quasi-Newton or conjugate gradient methods do not provide the same pattern. For the same level of N and δ, when the size of clusters increase, mMAPE, mSE and mBias decrease in all cases of MLE via PSO except one (in this case, mBias remains constant). It should also be noted that increasing censoring rate has lower impact on the values of mMAPE, mSE and mBias in MLE via PSO procedure.

Also, mLog.Lik values of PSO are quite better than mLog.Lik values of other two methods. To illustrate change of Log.Lik values in each repeat, randomly two cases are chosen among 54 different cases. Figures 3 and 4 show the cases (n_i = 4; N = 40; δ =20%) and (n_i = 4; N = 40; δ =5%) from first and second simulation study, respectively. As seen in figures, PSO takes the higher and more consistent Log.Lik values compared with derivative-based methods.

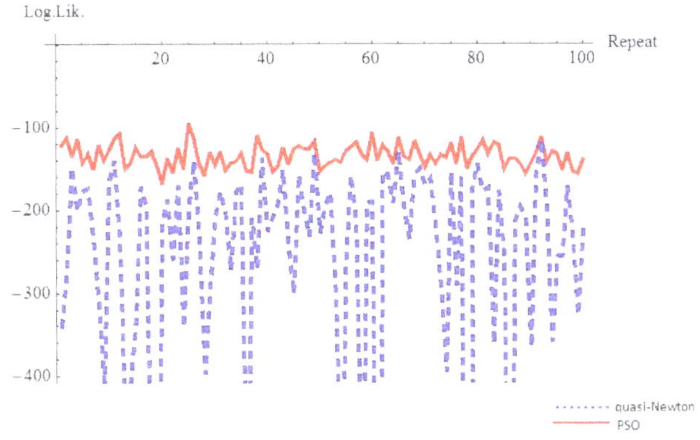

Figure 3. Change of Log.Lik values for case (n_i = 4; N = 40; δ =20%)

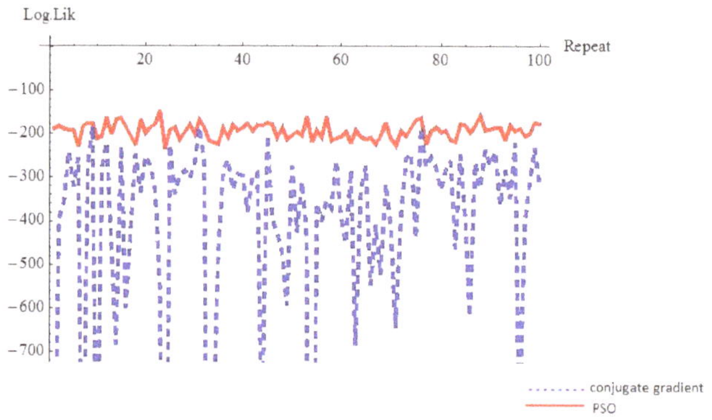

Figure 4. Change of Log.Lik values for case (n_i = 4; N = 40; δ =5%)

5. Real Data Application

The kidney infection data (McGilchrist & Aisbett, 1991) has been widely used in frailty studies. The data set consists of first and second recurrence times of infection for 38 kidney patients using a portable dialysis machine. The event of interest is the occurrence of an infection at the point of insertion of the catheter. The catheter is removed when the infection is observed. Then, the infection is cleared up and the catheter is reinserted for the second time. Each lifetime ($t_{i1}, t_{i2} > 0$) represents the time elapsed (in days) from the catheter insertion to infection. Censoring occurs when the catheter is removed for any reasons except the infection.

We first investigate the goodness of fit of the kidney infection data. Weibull and exponential probability plots (P-P) are shown in Figures 5 and 6. According to P-P plots, Weibull fits the data better than the exponential distribution. Kolmogorov-Smirnov (KS) tests suggest that Weibull is an appropriate distribution on the choice of baseline hazard. Table 3 provides the K-S statistics and associated p-values.

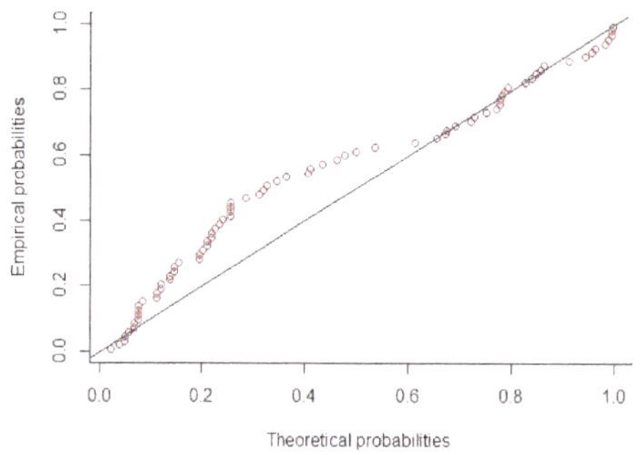

Figure 5. Exponential P-P plot for kidney infection data

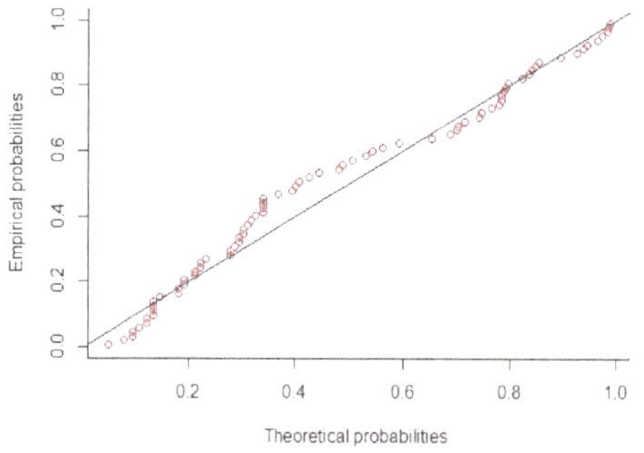

Figure 6. Weibull P-P plot for kidney infection data

Table 3. Kolmogorov-Smirnov statistics and p-values

	MLEs	K-S	p-value
Exponential	$\psi = (0.009)$	0.2239	0.0009
Weibull	$\psi = (0.011, 0.805)$	0.1205	0.2201

Two primary risk factors are included in the model: age and sex of patient (0=male, 1=female). The hazard function of $i\,th$ patient conditional on Gamma shared frailty at time $t_{ij} > 0$ takes the form ($i = 1, ..., 38; j = 1, 2$),

$$h(t_{ij} \backslash z_i, X_{ij}) = z_i(\lambda p t_{ij}{}^{p-1}) \exp(\beta_1 x_{ij}^{(age)} + \beta_2 x_{ij}^{(sex)}) \tag{19}$$

The results of MLE with PSO, quasi-Newton and conjugate gradient are reported in Table 4. Standard errors of parameters (Se) are obtained from the inverse of observed information matrix $I(\beta, \psi, \theta,) = -H(\beta, \psi, \theta)$, where $H(\beta, \psi, \theta)$ denotes the Hessian matrix. As observed in Table, lower standard errors and higher Log.Lik. values are obtained from PSO. Also, the G-R converge statistics of PSO calculated for 5 parameter estimates are found satisfactory (that is, equal to or less than 1.1).

Table 4. Application results

	PSO			Q uasi-Newton		C onjugate Gradient	
Parameter	Value	Se	G-R statistic	Value	Se	Value	Se
β_1	0.007	0.025	0.999	-0.088	0.042	-0.088	0.039
β_2	-2.118	0.577	0.999	0.923	2.066	0.923	1.960
λ	0.011	0.024	1.019	0.281	0.136	0.281	0.128
p	1.293	0.321	1.003	1.146	0.395	1.146	0.375
θ	0.510	0.297	1.030	0.848	0.626	0.846	0.590
Log. Lik.	-332.488			-364.718		-364.776	

Note: The initials selected randomly for quasi-Newton and conjugate gradient are (0.257,0.929,0.307,1.180,0.098)

5. Conclusion

In this paper, the estimation performance of MLE via PSO is investigated against the estimation performance of MLE via quasi-Newton and conjugate gradient methods for the shared Gamma frailty model. Based on two different simulation studies, PSO procedure outperforms the two derivative-based methods for most cases based on all comparison criteria. Also it is shown that smaller biases are obtained under the situation of heavy censoring. MLE via PSO is also implemented to a real data set and it is shown that this procedure can be preferred for estimating the parameters of parametric shared frailty models.

Acknowledgements

Collate acknowledgements in a separate section at the end of the article before the references. List here those individuals who provided help during the research (e.g., providing language help, writing assistance or proof reading the article, etc.).

References

Aalen, O. O. (1988). Heterogeneity in survival analysis. *Statistics in Medicine, 7*(11), 1121-1137. http://dx.doi.org/10.1002/sim.4780071105

Aalen, O., Borgan, O., & Gjessing, H. (2008). *Survival and event history analysis: A process point of view.* New York: Springer.

Aladag, C., Yolcu, U., Egrioglu, E., & Dalar, A. (2012). A new time invariant fuzzy time series forecasting method based on particle swarm optimization. *Applied Soft Computing, 12*(10), 3291-3299. http://dx.doi.org/10.1016/j.asoc.2012.05.002

Ata, N., & Ozel, G. (2012). Survival functions for the frailty models based on the discrete compound Poisson process. *Journal of Statistical Computation and Simulation, 83*(11), 2105-2116. http://dx.doi.org/10.1080/00949655.2012.679943

Caroni, C., Crowder, M., & Kimber, A. (2010). Proportional hazards model with discrete frailty. *Lifetime Data Analysis, 16*(3), 374-384. http://dx.doi.org/10.1007/s10985-010-9151-3

Clayton, D. (1978). A model for association in bivariate life tables and its application in epidemiological studies of familial tendency in chronic disease incidence. *Biometrika, 65*(1), 141-151. http://dx.doi.org/10.1093/biomet/65.1.141

Clayton, D. (1991). A Monte Carlo method for Bayesian inference in frailty models. *Biometrics, 47*(2), 467-485. http://dx.doi.org/10.2307/2532139

Clayton, D., & Cuzick, J. (1985). Multivariate generalizations of the proportional hazards model. *Journal of the Royal Statistical Society Series A, 148*(2), 82-117. http://dx.doi.org/10.2307/2981943

Duchateau, L., & Janssen, P. (2008). *The frailty model.* New York: Springer.

Egrioglu, E., Yolcu, U., Aladag, C. H., & Kocak, C. (2012). An ARMA type fuzzy time series forecasting method based on Particle Swarm Optimization. *Mathematical Problems in Engineering, 2013*, 3291-3299. http://dx.doi.org/10.1155/2013/935815

Gill, R. D. (1985). Discussion of multivariate generalizations of the proportional hazards model, by Clayton and Cuzick. *Journal of the Royal Statistical Society Series A, 148*(2), 82-117. http://dx.doi.org/10.2307/2981943

Hanagal, D., & Pandey, A. (2015). Gamma frailty models for bivariate survival data. *Journal of Statistical Computation and Simulation, 85*(15), 3172-3189. http://dx.doi.org/10.1080/00949655.2014.958086

Hanagal, D., & Sharma, R. (2015). Analysis of bivariate survival data using shared Inverse Gaussian frailty model. *Communications in Statistics-Theory and Methods, 44*(7),1351-1380. http://dx.doi.org/10.1080/03610926.2013.768663

Henderson, R., & Oman, P. (1999). Effect of frailty on marginal regression estimates in survival analysis. *Journal of the Royal Statistical Society Series B, 61*(2),367-379. http://dx.doi.org/10.1111/1467-9868.00182

Hougaard, P. (1984). Life table methods for heterogeneous populations: Distributions describing the heterogeneity. *Biometrika, 71*, 75-83. http://dx.doi.org/10.2307/2336399

Hougaard, P. (1986). A class of multivariate failure time distributions. *Biometrika, 73*(3), 671-678. http://dx.doi.org/10.1093/biomet/73.3.671

Hougaard, P. (1991). Modelling heterogeneity in survival data. *Journal of Applied Probability, 28*, 695-701. http://dx.doi.org/10.2307/3214503

Hu, X., Eberhart, R., & Shi, Y. (2004). Recent advances in particle swarm. In Proceedings of the IEEE Congress on Evolutionary Computation (pp. 90-97), Oregon, Portland.

Ibrahim, J., Chen, M., & Sinha, D. (2001). *Bayesian survival analysis.* New York: Springer.

Kennedy, J., & Eberhart, R. (1995). Particle Swarm Optimization. In Proceedings of the IEEE International Conference on Neural Networks (pp. 1942-1948), Piscataway, NJ.

Klein, J. P. (1992). Semiparametric estimation of random effects using the Cox model based on the EM algorithm. *Biometrics, 48*, 795-806. http://dx.doi.org/10.2307/2532345

Lancaster, T. (1979). Econometric methods for the duration of unemployment. *Econometrica, 47*(4), 939-956. http://dx.doi.org/10.2307/1914140

Lancaster, T. (1990). *The econometric analysis of transition data.* Cambridge: Cambridge U. Press.

McGilchrist, C. A., & Aisbett, C. W. (1991). Regression with frailty in survival analysis. *Biometrics, 47*, 461-466. http://dx.doi.org/10.2307/2532138

McGilchrist, C. A. (1993). REML estimation for survival models with frailty. *Biometrics, 49*, 221-225. http://dx.doi.org/10.2307/2532615

Nielsen, G. G., Gill, R. D., Andersen, P. K., & Sørensen, T.I.A. (1992). A counting process approach to maximum likelihood estimation in frailty models. *Scandinavian Journal of Statistics, 19*, 25-43. http://www.jstor.org/stable/4616223

Prasad, J., & Souradeep, T. (2012). Cosmological parameter estimation using particle swarm optimization (PSO). *Phys.Rev.D., 85*(12), 1-13. http://dx.doi.org/10.1103/PhysRevD.85.123008

Rondeau, V., Commenges, D., & Joly, P. (2003). Maximum penalized likelihood estimation in a gamma-frailty model. *Lifetime Data Analysis, 9*(2),139-153. http://dx.doi.org/10.1023/A:1022978802021

Sinha, D., Dey, & K. (1997). Semiparametric Bayesian analysis of survival data. *Journal of the American Statistical Association, 92*(439), 1195-1212. http://dx.doi.org/10.2307/2965586

Therneau, T. M., Grambsch, P. M., & Pankratz, V. S. (2003). Penalized survival models and frailty. *Journal of Computational and Graphical Statistics, 12*, 156-175. http://dx.doi.org/10.1198/1061860031365

Van den Berg, G. J. (2001). Duration models: Specification, identification and multiple duration. *Handbook of Econo-*

metrics, 5, 3381-3460. http://dx.doi.org/10.1016/S1573-4412(01)05008-5

Vaupel, J. W., Manton, K. G., & Stallard, E. (1979). The impact of heterogeneity in individual frailty on the dynamics of mortality. *Demography, 16*, 439-454. http://dx.doi.org/10.2307/2061224

Wang, F., & Huang, P. (2014). Implementing particle swarm optimization algorithm to estimate the mixture of two Weibull parameters with censored data. *Journal of Statistical Computation and Simulation, 84*(9), 1975-1989. http://dx.doi.org/10.1080/00949655.2013.778992

Whitmore, G. A., & Lee, M. L. T. (1991). A multivariate survival distribution generated by an inverse Gaussian mixture of exponentials. *Technometrics, 33*, 39-50. http://dx.doi.org/10.2307/1269006

Wienke, A. (2011). *Frailty models in survival analysis*. Boca Raton: Chapman & Hall/CRC Biostatistics Series.

Yu, B. (2006). Estimation of shared gamma frailty models by a modified EM algorithm. *Computational Statistics and Data Analysis, 50*(2), 463-474. http://dx.doi.org/10.1016/j.csda.2004.08.010

10

A Directional Bayesian Significance Test for Equality of Variances

Michael B Brimacombe[1]

[1] Department of Biostatistics, KUMC, Kansas City, Kansas, USA

Correspondence: Michael Brimacombe, Department of Biostatistics, KUMC, Kansas City, Kansas, USA 66103. E-mail: mbrimacombe@kumc.org

Abstract

A directional Bayesian pure significance test for the equality of variances is developed. The approach is based on the assessment of observed departure conditioned on the direction of departure in multivariate models. The resulting one-dimensional directional distribution is easily interpreted. Normality is not required. Robustness of prior selection is discussed focusing on directional properties of the multivariate prior. Several examples are considered.

Keywords: Bartlett test, Bayesian directional test, test of variances.

1. Introduction

Statistical significance tests for the equality of variation across treatment groups are often used in experimental settings where ANOVA based analysis is required. This is an issue for example in toxicological and genetic applications (Gastwirth et al. 2009). As shown in Box (1954), the robustness of the one-way ANOVA overall F-test to non-normality is dependent on the degree of inequality among the group variances. Further, in moderate size samples, the p-value calculation may not be accurate (Draper and Smith, 1998) and variance-stabilizing transformations may be required to develop a more stable result. In settings where homogeneity of variation is to be formally tested, the commonly applied Bartlett test (Bartlett, 1937) which is a slight modification of a likelihood ratio test is often used, but is sensitive to the assumption of normality. The non-parametric Levene test can provide a more robust test with the trade-off of lower power. See for example Miller (1997).

Extensions of the Bartlett and Levene tests of homogeneity have been developed, often through modification of the L_2 norm in the ANOVA setting and modifications of the assumed error distribution. The work in Gastwirth et al. (2009) reviews several such approaches where typically bootstrap sampling methods and absolute value based (L_1 norm) measures of departure are used to extend standard tests. Earlier work in this regard can be found for example in Box and Tiao (1973) where Bayesian inference is developed for variance components, tests of homogeneity, and the modeling of variation in linear models with random effects.

With regard to likelihood based approaches for the modeling and testing of variation in a linear model, likelihood and marginal likelihood approaches are given in Harville (1977) which reviews restricted maximum likelihood methods (REML) based on error contrasts. This involves considering the transformation $w = Ay$ where y is the original response and A is a selected orthogonal projection matrix. The restricted likelihood for w resulting from this projection can be used to analyze parameters in the Σ variance-covariance matrix as the mean parameter μ is orthogonalized out of the likelihood. Harville (1974) points out the usefulness of the REML approach in a Bayesian setting. If the prior selection for mean and variance component parameters is independent, the joint posterior resulting from using the restricted likelihood for inference regarding Σ is equivalent to ignoring prior information for the mean parameter and using all the data to make inferences for Σ.

In the setting of one-way ANOVA the assumption of homogeneity of variation across treatments underlies application of the standard global F-test for differences among treatment averages. To test this assumption by examining the null hypothesis $H_0 : \sigma_1^2 = \cdots = \sigma_k^2$ in the model $y_{li} \sim N(\theta_i, \sigma_i^2), i = 1, ..., k; l = 1, ..., n_i$ where $\theta_i = \mu + \tau_i$, the Bartlett test uses a statistic of the form;

$$M = -\sum_{i=1}^{k} v_i(\log s_i^2 - \log s_p^2)$$

where s_i^2 is the sample variance for each treatment group $s_p^2 = (1/v)\sum_{i=1}^{k} v_i s_i^2$, the pooled variance, $v_i = (n_i - 1)$ and $v = \sum_{i=1}^{k} v_i$. Under the null hypothesis M is distributed as a χ_{k-1}^2 random variable. The related p-value calculations are standard in many statistical packages.

In this paper we develop an analogue of the standard Bartlett significance test in a directional Bayesian framework. The pure significance test developed assesses departure of the hypothesized null parameter value from the posterior mode where the posterior distribution is conditioned on the direction of departure. The approach is general and does not require normality as an assumption, though this is assumed here for comparative purposes with the Bartlett test. This measure provides an assessment of the observed discrepancy of the predicted value or posterior mode from the hypothesized null value. Its application to the multi-parameter hypothesis test $H_0 : \sigma_1^2 = \cdots = \sigma_k^2$ is examined in detail. The sensitivity of the test to prior and likelihood selection is defined. Several examples are considered.

2. Method

1. Bayesian Approach

The Bayesian approach to statistical inference is based on the posterior density function given by;

$$p(\theta|data) = c \cdot p(\theta) \cdot L(\theta|\, data)$$

where $L(\theta|\, data)$ is the observed likelihood function, $p(\theta)$ is the prior density function for θ and c denotes the constant of integration. The Bayesian approach to hypothesis testing is based on developing empirical measures of support for null parameter values typically expressed in terms of posterior odds ratios or Bayes factors (Bernardo and Smith, 1994). The posterior odds ratio for example comparing a specified value θ_0 to a general alternative θ can be written

$$\frac{p(\theta_0|data)}{1 - p(\theta_0|data)} \Big/ \frac{p(\theta|data)}{1 - p(\theta|data)} \tag{1}$$

which provides a relative measure of the support given to the competing hypotheses within the context of the joint posterior. Typically $\theta = \widehat{\theta}$,the posterior mode, is taken as a reference value to interpret the resulting odds ratio. If θ is a vector of parameters, some initial marginalisation of parameters not of interest may be necessary.

1.1 Bayesian Bartlett Test

A Bayesian test of variances that is a version of the Bartlett test is given in Box and Tiao (1973), p. 132. Assume the samples $(y_{11}, \ldots, y_{1n_1}), \ldots, (y_{k1}, \ldots, y_{kn_k})$ of $n_1, \ldots, n_k$ independent observations, respectively, are drawn from the Normal populations $N(\theta_i, \sigma_i^2)$, where $i = 1, \ldots, k$. If both the means θ_i and variances σ_i^2 are unknown and it is assumed that *a priori* θ_i and $\log \sigma_i$ are approximately independent and locally uniform it follows that the joint posterior for $\sigma_1^2, \ldots, \sigma_k^2$ is given by

$$p(\sigma_1^2, \ldots, \sigma_k^2|\mathbf{y}) = \prod_{i=1}^{k} c_i(\sigma_i^2)^{-[\nu_i/2+1]} exp(-\frac{\nu_i s_i^2}{2\sigma_i^2}) \tag{2}$$

for $\sigma_i^2 > 0,\ i = 1, \ldots, k$. Here $c_i = [\Gamma(\upsilon_i/2)]^{-1}(\nu_i s_i^2/2)^{\nu_i/2}$, $\nu_i = n_i - 1$, s_i^2 is the standard deviation for the ith sample.

To compare variation across the k samples any (k - 1) linearly independent contrasts in $\log \sigma_i$ can be used. Following the development in Box and Tiao (1973) we can define

$$\psi_i = \log \sigma_i^2 - \log \sigma_k^2$$

for $i = 1, \ldots, (k - 1)$. Changing variables in (2), the $(k - 1)$- dimensional joint posterior for $\psi_1, \ldots, \psi_{k-1}$ can be expressed as;

$$p(\psi\,|\,\mathbf{y}) \;=\; c \cdot (a_1 e^{-\psi_1})^{\nu_1/2} \cdots (a_{k-1} e^{-\psi_{k-1}})^{\nu_{k-1}/2} \tag{3}$$

$$\cdot [1 + a_1 e^{-\psi_1} + \cdots + a_{k-1} e^{-\psi_{k-1}}]^{-\nu/2} \tag{4}$$

where $i = 1, \ldots, k - 1$, $a_i = \nu_i s_i^2 / \nu_{k-1} s_{k-1}^2$, $\nu = \sum_{i=1}^{k-1} \nu_i$ and $-\infty < \psi_i < \infty$. The mode of this distribution is given by $\psi = \widehat{\psi}$, where $\hat{\psi}_i = \log s_i^2 - \log s_k^2$ for $i = 1, \ldots, (k - 1)$.

The null hypothesis $\sigma_i^2 = \cdots = \sigma_k^2$ can be rewritten as $\psi = \mathbf{0}$. Box and Tiao (1973) use a Likelihood Ratio type pivotal quantity in a Bayesian context to develop a χ_{k-1}^2 based approximate testing procedure. The presence of $\psi = \mathbf{0}$ in the approximate 95% credible region is interpreted as evidence that the null hypothesis of homogeneity is supported by the observed data.

2. Directional Bayesian Significance Testing

The Bayesian approach to testing hypotheses is an examination of the observed posterior probability based weight observed for each potential θ value of interest. There is no assumption of a true value, rather a comparison of relative weights for potential θ values. Typically a posterior modal value for θ, $\widehat{\theta}$, can be compared to specific θ values of interest using the posterior odds ratios or Bayes factor.

The directional approach applied here emphasizes the observed magnitude of departure of the hypothesized parameter value θ_0 from a central posterior value $\hat{\theta}$, typically the mode of the posterior distribution. This is viewed in terms of two distinct elements; the direction of departure and the magnitude of departure. The direction of departure out from the central modal value $\hat{\theta}$ is defined along the unit vector $\mathbf{d} = (\theta_0 - \hat{\theta})/||\theta_0 - \hat{\theta}||$. This is formally conditioned upon as other directions are not of immediate interest in assessing $(\theta_0 - \hat{\theta})$. The observed magnitude of departure, $r_0 = ||\theta_0 - \hat{\theta}||$ is then evaluated under the resulting one-dimensional local directional distribution.

Geometrically, this involves examining the posterior distribution along the span of the unit vector $\mathbf{d}$ out from the mode, subject to an initial change of variable $\theta \rightarrow (r, \mathbf{d})$, including a Jacobian factor. The resulting conditional distribution, $p_r(r|\mathbf{d})$ provides tail areas that can be evaluated in terms of a *one-dimensional* integral using $p_r(r|\mathbf{d})$. As noted above, the joint posterior density can be written as;

$$p(\theta|\mathbf{y}) = c \cdot p(\theta) \cdot L(\theta|\mathbf{y}) \tag{5}$$

where $p(\theta)$ is the prior distribution for θ, c is a constant of integration and $L(\theta|\mathbf{y})$ is the observed likelihood function. The value for θ is taken to lie in the parameter space $\Omega \subset \Re^k$. There may be nuisance parameters involved in the initial specification of the model. We assume for simplicity of exposition that nuisance parameters have been integrated out.

Some assumptions regarding the form of the joint posterior for θ are useful; θ must be a continuously valued vector of reals defined in a parameter space Ω which is also a vector space and $p(\theta\,|\,\mathbf{y})$ is assumed to be unimodal. The support for $p(\theta|y)$, i.e., $\{\theta \in \Omega : p(\theta\,|\,\mathbf{y}) > 0\}$, is assumed here to be a convex set with the linear span of the vector $\mathbf{d} = (\theta_0 - \hat{\theta})/||\theta_0 - \hat{\theta}||$, $\mathbf{L}\{\mathbf{d}\}$ entirely within Ω. The vector $\mathbf{d}$ is a unit vector lying on S^{k-1}, the unit sphere in $(k-1)$ dimensions. In cases where the convexity is not global, the region of Ω about θ_0 should contain $\mathbf{L}\{\mathbf{d}\}$. The hypothesized parameter value $\boldsymbol{\theta_0}$ should not lie on the boundary of Ω.

With $\widehat{\theta}$ as the mode of the joint posterior, let $r = ||\theta - \hat{\theta}||$ and change variables $\theta \rightarrow (r, \mathbf{d})$. The Jacobian is proportional to $r^{k-1}\, dr\, d\mathbf{d}$. Conditioning upon $\mathbf{d}$, the resulting conditional distribution $p_r(r\,|\,\mathbf{d})$ is proportional to the initial joint distribution up to a multiplicative constant c and is given by

$$p_{r|\mathbf{d}}(r) = c \cdot p(r\mathbf{d} + \hat{\theta}|y) \cdot r^{k-1}.$$

In the specific case where $\mathbf{p}(\theta\,|\,\mathbf{y})$ is a rotationally symmetric distribution satisfying the above assumptions then $\hat{\theta}$ will be the mean of the posterior density.

To assess the global hypothesis $H_0 : \theta = \theta_0$, let $r_0 = ||\theta_0 - \hat{\theta}||$. A conditional tail area in regard to the null hypothesis can then be defined. This is given by the "Directional Posterior Level of Significance" (*DPLS*);

$$DPLS\,(\theta_0) = \frac{\int_{r_0}^{\infty} p(r\mathbf{d} + \hat{\theta}|\,\mathbf{y})\, r^{k-1}\, dr}{\int_0^{\infty} p(r\mathbf{d} + \hat{\theta}|\,\mathbf{y})\, r^{k-1}\, dr} \tag{6}$$

and the "significance" here reflects the underlying distance measure and can be assessed without reference to a specific Type I error level. This is the tail area of the conditional distribution for r given the unit directional vector $\mathbf{d}$. $DPLS\,(\theta_0)$ is the (conditional) probability of θ lying at or beyond θ_0, in the direction specified by $\mathbf{d}$. Note that this measure of departure is made in terms of r, a scalar measure of the distance from $\boldsymbol{\theta_0}$ to the central value $\hat{\theta}$. It provides a simple summary measure of the (conditional) weight to be accorded the null hypothesis $H_0 : \theta = \theta_0$.

To further interpret results in terms of the number of standard errors found in the magnitude of departure, the mean and variance for $p_{r|\mathbf{d}}(r)$ are generally available. If we define $K(m) = \int_0^{\infty} p(r\mathbf{d} + \hat{\theta} \mid \mathbf{y})\, r^{m-1} d\tau$, it follows that $E[r \mid \mathbf{d}] = K(m+1)/K(m)$ and $Var[r \mid \mathbf{d}] = Var[\tau \mid \mathbf{d}] = K(m+2)/K(m) - (K(m+1)/K(m))^2$. The approach has been applied to multi-parameter testing problems in econometrics (Brimacombe, 1996) and the development of diagnostics for asymptotic convergence in logistic regression (Brimacombe, 2016).

2.1 Rotational Symmetry

In the specific case of a rotationally symmetric posterior distribution about $\hat{\theta}$, the unit vector $\mathbf{d}$ defined above is uniformly distributed on the unit sphere S^{k-1} and the $DPLS\,(\theta_0)$ value is equivalent to an unconditional Bayes significance level, namely,

$$\int_A p(\theta|\mathbf{y})\, d\theta$$

where $A \subset \Omega$ is the set of θ values at or beyond $\boldsymbol{\theta_0}$. Rotational symmetry implies that the tail area in question is equivalent in any chosen direction. In general, if $p(\theta\,|\,\mathbf{y})$ is of the form $p(||\theta - \hat{\theta}||^2|\,\mathbf{y})$, it is trivial to show that the $DPLS(\theta_0)$ is independent of **d**. For example, a multivariate normal model with assumed homogeneity and independence has this property. Note that differences in the *DPLS* value for various directions out from the posterior mode can be used to assess the degree of asymmetry in the joint posterior distribution. This is examined in more detail below.

3. Results

2.2 Example 1: Box and Tiao (1973)

We discuss the application of the *DPLS* measure in the context of an example given in Box and Tiao (1973), p.136. An experiment is conducted to examine similarity in the variation of three distributions. The summary statistics are given by;

$$s_1^2 = 52.785, s_2^2 = 34.457, s_3^2 = 66.030, \ \nu_1 = \nu_2 = \nu_3 = 30, \ \nu = 90.$$

Assuming normality and locally non-informative priors, the joint posterior distribution of the two contrasts, $\psi_1 = log\ \sigma_1^2 - log\ \sigma_3^2$ and $\psi_2 = log\ \sigma_2^2 - log\ \sigma_3^2$ can be written;

$$\begin{aligned} p(\psi_1, \psi_2 | \mathbf{y}) &= \frac{\Gamma(45)}{[\Gamma(15)]^3}(.79941e^{-\psi_1})^{15}(.52184e^{-\psi_2})^{15} \\ &\cdot(1 + .79941e^{-\psi_1} + .52184e^{-\psi_2})^{-45} \end{aligned} \quad (7)$$

where $-\infty < \psi_i < \infty$, for i = 1,2. The mode is given by $\widehat{\psi} = (\hat{\psi}_1, \hat{\psi}_2) = (-.2239, -.6504)$. Approximate 75, 90 and 95 per cent Highest Posterior Density (H.P.D.) regions are given in Box and Tiao (1973), p. 138 with the null hypothesis $\psi_0 = (0, 0)$ lying within the 90% region, providing limited support for the null hypothesis of homogeneity. The joint posterior density is given in Figure 1.

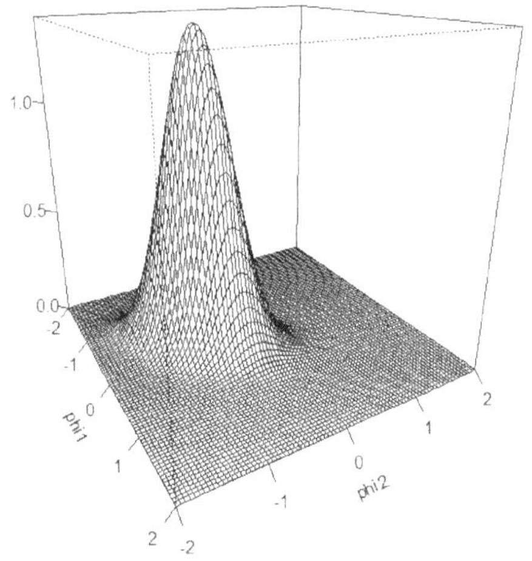

Figure 1. Joint Posterior Density

To derive a directional Bayesian Bartlett test here we initially change variables $\psi \rightarrow (r, \mathbf{d})$ with Jacobian proportional to $r^{2-1}\, dr\, d\mathbf{d}$, $r > 0$ to obtain $p_{r|\mathbf{d}}(r)$ as above. We may also further change variables with $\tau = r^2$ to obtain a chi-square related distribution. The resulting conditional distribution $p_\tau(\tau\,|\,\mathbf{d})$ in the direction defined by the null hypothesis $\psi_0 = (0, 0)$ out from the mode is given by;

$$\begin{aligned} p_{\tau|\mathbf{d}}(\tau) &= c \cdot p(\tau^{1/2}\mathbf{d} + \hat{\psi}) = \\ &= c \cdot (.79941e^{-\tau^{1/2}d_1 - \widehat{\psi}_1})^{15}(.52184e^{-\tau^{1/2}d_2 - \widehat{\psi}_2})^{15} \\ &\cdot[1 + .79941e^{-\tau^{1/2}d_1 - \widehat{\psi}_1} + .52184e^{-\tau^{1/2}d_2 - \widehat{\psi}_2}]^{-45} \cdot \tau^{2/2-1} \end{aligned}$$

This is re-normed and shown in Figure 2.

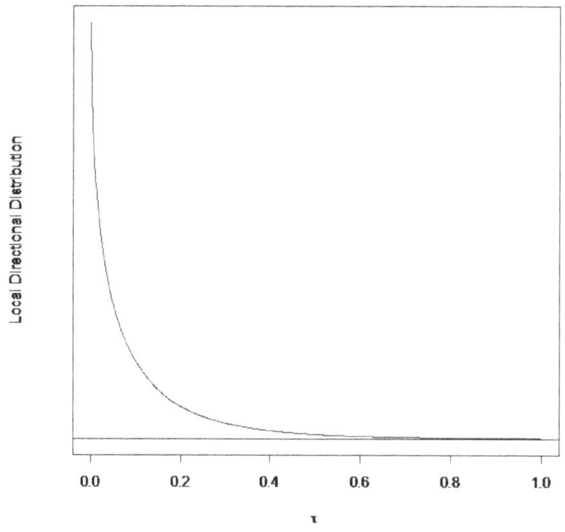

Figure 2. Directional Conditional Distribution ($H_0 : \psi_0 = (0, 0)$)

The direction of departure related to the null is given by $\mathbf{d}_0 = (\hat{\psi} - \mathbf{0})/||\hat{\psi} - \mathbf{0}|| = (0.3255, 0.9455)$ and the observed magnitude of departure $r_0 = ||\psi_0 - \widehat{\psi}|| = 0.6879$ or $\tau_0 = (0.6879)^2 = 0.4732$. The related local tail area for the null hypothesis $\psi_0 = (0, 0)$ is given by;

$$DPLS\,(\psi_0) = \frac{\int_{\tau_0}^{\infty} p_{\tau|\mathbf{d}}(\tau)\, d\tau}{\int_0^{\infty} p_{\tau|\mathbf{d}}(\tau) d\tau} = 1 - .9872 = 0.0128$$

and can be interpreted as providing little support for the null hypothesis of equal variances.

2.3 Example 2: Simulation 1

This example gives a simulated dataset where the Bartlett and Levine tests do not agree. The data and summary statistics are given in Table 1.

Table 1. Test for Equal Variances. 95% Bonferroni Confidence Intervals for Standard Deviations

Group	n	Lower Bound	Standard Deviation	Upper Bound
1	24	0.99	1.34	2.04
2	18	0.89	1.26	2.09
3	28	1.63	2.18	3.19

Group	Data
Group 1 (n=24)	5.37, 5.80, 4.70, 5.70, 3.40, 8.60, 7.48, 5.77, 7.15, 6.49, 4.09, 5.94, 6.38, 9.24, 5.66, 4.53, 6.51, 7.0, 6.20, 7.04, 4.82, 6.73, 5.26, 5.21
Group 2 (n=18)	3.96, 3.04, 5.28, 3.4, 4.1, 3.61, 6.16, 3.22, 7.48, 3.87, 4.27, 4.05, 2.40, 5.81, 4.29, 2.77, 4.4, 4.45
Group 3 (n=28)	5.37, 10.6, 5.02, 14.30, 9.9, 4.27, 5.75, 5.03, 5.74, 7.85, 6.82, 7.9, 8.36, 5.72, 6.0, 4.75, 5.83, 7.3, 7.52, 5.32, 6.05, 5.68, 7.57, 5.68, 8.91, 5.39, 4.4, 7.13

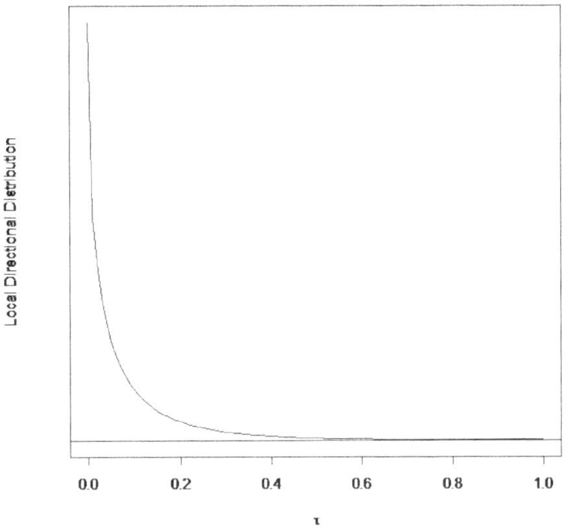

Figure 3. Directional Conditional Distribution ($H_0 : \psi_0 = (0, 0)$)

Here the Bartlett test has a p-value of 0.015 and the Levine test a p-value of 0.23, providing different inferential conclusions. The direction of departure related to the null is given by $\mathbf{d}_0 = (\hat{\psi} - \mathbf{0})/||\hat{\psi} - \mathbf{0}|| = (0.6641, 0.7476)$ and the magnitude of departure by $r_0 = ||\psi_0 - \widehat{\psi}|| = 0.6305$ or $\tau_0 = (0.6305)^2 = 0.3975$. The conditional distribution $p_\tau(\tau \,|\, \mathbf{d})$ for the null hypothesis $\psi_0 = (0, 0)$ is given by;

$$\begin{aligned} p_{\tau|\mathbf{d}}(\tau) &= c \cdot (a_1 e^{-\tau^{1/2} d_1 - \widehat{\psi}_1})^{23} (a_2 e^{-\tau^{1/2} d_2 - \widehat{\psi}_2})^{17} \\ &\quad \cdot [1 + a_1 e^{-\tau^{1/2} d_1 + \widehat{\psi}_1} + a_2 e^{-\tau^{1/2} d_2 + \widehat{\psi}_2}]^{-34} \end{aligned}$$

where c is the constant of integration. This is shown in Figure 3. The exact test tail area for this hypothesis is given by;

$$DPLS(\psi_0) = \frac{\int_{\tau_0}^{\infty} p_{\tau|\mathbf{d}}(\tau)\, d\tau}{\int_0^{\infty} p_{\tau|\mathbf{d}}(\tau) d\tau} = 0.03186$$

which supports mild rejection of the null hypothesis of equal variances, agreeing here with the standard Bartlett test.

2.4 Example 3: Simulation 2

The approach here extends easily into multiple treatment group settings. A simulated example examining variation across five groups ($n_i = 20, i = 1, ..., 5$) is developed with the fifth group being an increasing outlier across each of three simulations. This gives a joint posterior density for ψ in four dimensions. Summary statistics and related tail areas are given in Table 2 (data not shown). Figure 4 shows the three respective conditional densities.

Table 2. Comparison of Five Plasma Groups

	s_1 (95% interval)	s_2 (95% interval)	s_3 (95% Interval)	s_4 (95% Interval)	s_5 (95% interval)	τ_0	DPLS	Levine	Bartlett
Simulation1	1.19 (.83,1.98)	.99 (.69,1.65)	.66 (.46,1.09)	.75 (.53,1.26)	1.23 (.87, 2.06)	0.722	.202	.014	.030
Simulation2	1.19 (.83,1.98)	.99 (.69,1.65)	.66 (.46,1.09)	.75 (.53,1.26)	2.79 (1.96,4.65)	2.06	.089	.0001	.0001
Simulation3	1.19 (.83,1.98)	.99 (.69,1.65)	.66 (.46,1.09)	.75 (.53,1.26)	3.72 (2.61,6.20)	2.55	.013	.0001	.0001

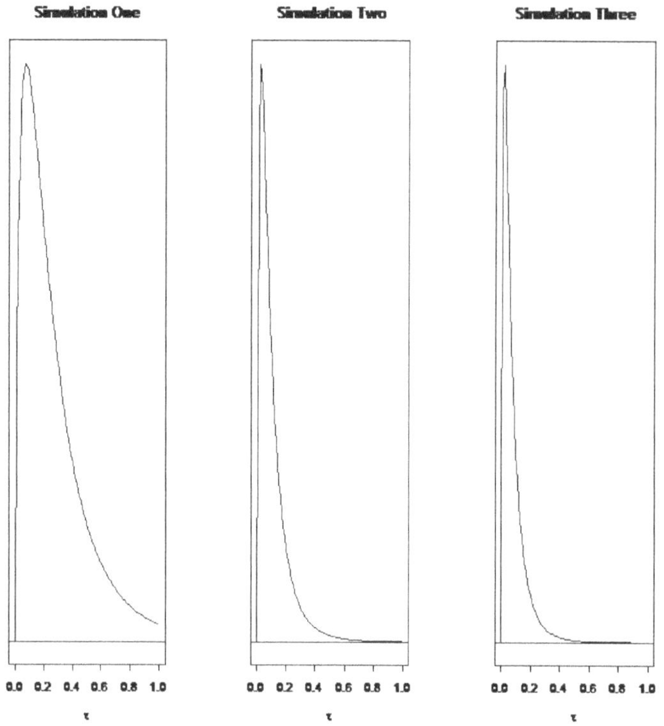

Figure 4. Directional Conditional Distributions for Three Simulations ($H_0 : \psi_0 = (0, 0, 0, 0)$)

The Bartlett and Levine tests here give significant p-values for all three datasets. Standard Bonferroni corrected confidence intervals shown in Table 2 give a large overlap of the outlier group confidence interval (group 5) with the other four group intervals in the first two datasets but none in the third. Note that while the sample size per group is relatively small, the number of groups and overall sample size lead to potential over-sampling. The conditional distribution $p_\tau(\tau \mid \mathbf{d})$ for the null hypothesis $\psi_0 = (0, 0, 0, 0)$ is given by;

$$\begin{aligned} p_{\tau|\mathbf{d}}(\tau) &= c \cdot (a_1 e^{-\tau^{1/2} d_1 - \widehat{\psi}_1})^{20} (a_2 e^{-\tau^{1/2} d_2 - \widehat{\psi}_2})^{20} (a_3 e^{-\tau^{1/2} d_3 - \widehat{\psi}_3})^{20} (a_4 e^{-\tau^{1/2} d_4 - \widehat{\psi}_4})^{20} \\ &\quad \cdot [1 + a_1 e^{-\tau^{1/2} d_1 + \widehat{\psi}_1} + a_2 e^{-\tau^{1/2} d_2 + \widehat{\psi}_2} + a_3 e^{-\tau^{1/2} d_3 - \widehat{\psi}_3} + a_4 e^{-\tau^{1/2} d_4 - \widehat{\psi}_4}]^{-50} \cdot \tau^{4/2-1} \end{aligned}$$

where again c is the constant of integration. See Table 2 for results. The directional approach here is relatively conservative, with clear rejection of the null supported only in the third simulated dataset where the outlier group has no overlap with the other groups in terms of standard Bonferroni corrected confidence intervals.

3. Robustness

3.1 Prior Robustness

Any application of the Bayesian perspective in relation to modeling should discuss potential effects of prior selection. As the analysis here conditions upon $\mathbf{d}$ to obtain the conditional distribution $p_r(r \mid \mathbf{d})$, the robustness of the analysis to choice of prior density may also be defined directionally. While examining the behavior of $p(r|\mathbf{d})$ for a variety of chosen priors is a possibility, the directional and one-dimensional nature of the *DPLS* approach allows for geometric insight. In particular, a chosen prior will have limited effect on the resulting *DPLS* value if it is relatively flat in the direction $\mathbf{d}$, directionally non-informative. This effect can be visualized by plotting the one-dimensional directionally conditioned prior component $p(r\mathbf{d} + \hat{\theta})$ as function of $r > 0$. These can be compared for a set of potential prior densities.

When selecting a multivariate prior for the covariance matrix Σ several approaches are standard. A basic approach is to use a normal-gamma distribution for the (μ, Σ) set of parameters and express this as a conditional distribution $p(\mu|\Sigma)$ times a marginal distribution $p(\Sigma)$. A simpler approach is to assume prior independence. In the *REML* setting where we assume that prior belief regarding μ and Σ are independent, it is possible to use the marginal likelihood based on error contrasts. Harville (1974) showed that the *REML* based marginal posterior for Σ is independent of the mean parameter μ, simplifying the prior selection process. Prior choices can be extended to other multivariate densities. Note that a hierarchical setting may be useful in modeling variation (Daniels, 1999) but is not considered in detail here.

If the prior density is viewed in isolation, directionalized by applying the polar transformation, then we obtain $c \cdot p(r\mathbf{d} + \hat{\theta}) \cdot$

r^{k-1} as the conditional prior density in the direction $\mathbf{d}$ where $\hat{\theta}$ is the prior mode. If a non-informative prior is interpreted as being fairly constant as a function of the parameter $r > 0$ we can plot this one-dimensional directionalized prior as a function of r and examine its behavior. We can express this more directly as;

$$\begin{aligned} c \cdot p(r\mathbf{d} + \hat{\theta}) \cdot r^{k-1} &= C \\ (c/C) \cdot p(r\mathbf{d} + \hat{\theta}) \cdot r^{k-1} &= 1 \\ \log(c/C) + \log p(r\mathbf{d} + \hat{\theta}) + \log r^{k-1} &= 0 \\ -\log p(r\mathbf{d} + \hat{\theta}) &= [\log(c/C) + (k-1)\log r] \end{aligned}$$

Plotting $-\log p(r\mathbf{d} + \hat{\theta})$ versus $\log r$ allows us to see if a straight line regression with slope $(k-1)$ is obtained. If yes, the chosen prior $p(\cdot)$ may be viewed as directionally non-informative. This can be examined in relation to other directions $\mathbf{d}$ if required and across a set of potential prior densities. A similar directional examination of the posterior density comparing actual and large sample likelihood functions can be found in Brimacombe (2016).

4. Discussion

A directional Bayesian pure significance test procedure is developed here for testing the hypothesis of equal variances. It is based on the tails areas of a conditional Bayesian directional distribution defined in relation to the specified null hypothesis and reflects the departure of the null value from the observed posterior mode. The approach does not require an assumption of normality. Where non-informative priors are employed, the approach reflects local properties of the likelihood function. Note that issues related to the power of the pure significance test are not addressed here.

The use of a directional approach allows for a unique perspective regarding prior selection and prior related robustness. The stability of the posterior can be assessed in the specific direction of the null hypothesis, examining how the prior is affecting the relevant directional properties of the posterior density. While related here to the Bartlett test, the directional approach can be applied more widely and provides a one-dimensional conditional approach to higher dimensional multiparameter significance testing problems in general. Applications to MANOVA and generalized linear models will be found elsewhere. Note that standard Markov Chain Monte Carlo based methods may be applied to determining the required tail area in more detailed settings especially where nuisance parameters must initially be averaged out.

References

Bartlettt, M. S. (1937). Properties of Sufficiency and Statistical Tests. Proc. Roy. Soc. London, Series A, *160*, pp. 268-282. http://dx.doi.org/10.1098/rspa.1937.0109

Bernardo, J., & Smith, A. F. M. (1994). Bayesian Theory. *John Wiley and Sons, Inc.*, New-York.

Boos D. D., & Brownie, C. (2004). Comparing Variances and Other Measures of Dispersion. *Statistical Science*, *19*(4), p. 571–578. DOI 10.1214/088342304000000503.

Box, G. E. P. (1954). Some Theorems on Quadratic Forms Applied in the Study of Analysis of Variances Problems I: Effect of Inequality of Variance in the One-Way Classification. *Ann. Math. Statist.,25*, 290-302. http://dx.doi.org/10.1214/aoms/1177728786

Box, G. E. P., & Tiao, G. C. (1973). Bayesian Inference in Statistical Analysis. *Addison-Wesley*, Reading, Mass.

Brimacombe, M. (1996). Conditional Bayesian Significance Levels: Econometric Applications, in Advances in Econometrics Vol. 11 (Part A): Bayesian Computational Methods and Applications, eds. R. Carter Hill, T.B. Fomby, JAI Press, Greenwich.

Brimacombe, M. (2016). Large sample convergence diagnostics for likelihood based inference: Logistic regression. *Statistical Methodology 33*, 114–130. http://dx.doi.org/10.1016/j.stamet.2016.08.001

Draper, N. R., & Smith, H. (1998). Applied Regression Analysis, 3rd ed. *John Wiley and Sons*, Inc. New York. http://dx.doi.org/10.1002/9781118625590

Daniels, M. (1999). A Prior for the Variance in Hierarchical Models. *Canadian Journal of Statistics*, *27*, 569-580. http://dx.doi.org/10.2307/3316112

Gastwirth, J. L., Gel, Y. R., & Miao, W. (2009). The Impact of Levene's Test of Equality of Variances. *Statistical Theory and Practice Statistical Science*, *24*(3), p. 343-360. http://dx.doi.org/10.1214/09-STS301.

Harville, D. A. (1977). Maximum Likelihood Approaches to Variance Component Estimation and to Related Problems. *Journal of the American Statistical Association*, *72*(358), 320-338. http://dx.doi.org/10.1080/01621459.1977.10480998

11

New Flexible Regression Models Generated by Gamma Random Variables with Censored Data

Elizabeth M. Hashimoto[1], Gauss M. Cordeiro[2], Edwin M.M. Ortega[3] & G.G. Hamedani[4]

[1] Departamento Acadêmico de Matemática, UTFPR, Londrina, Brazil

[2] Departamento de Estatística, UFPE, Recife, Brazil

[3] Departamento Ciências Exatas, ESALQ-USP, Piracicaba, Brazil

[4] Departament of Mathematics, Statistics and Computer Science, Milwaukee, USA

Correspondence: Edwin M.M. Ortega, Departamento de Cilncias Exatas, USP, Av. Pdua Dias 11 - Caixa Postal 9, 13418-900, Piracicaba-S£o Paulo, Brazil. E-mail: edwin@usp.br

Abstract

We propose and study a new log-gamma Weibull regression model. We obtain explicit expressions for the raw and incomplete moments, quantile and generating functions and mean deviations of the log-gamma Weibull distribution. We demonstrate that the new regression model can be applied to censored data since it represents a parametric family of models which includes as sub-models several widely-known regression models and therefore can be used more effectively in the analysis of survival data. We obtain the maximum likelihood estimates of the model parameters by considering censored data and evaluate local influence on the estimates of the parameters by taking different perturbation schemes. Some global-influence measurements are also investigated. Further, for different parameter settings, sample sizes and censoring percentages, various simulations are performed. In addition, the empirical distribution of some modified residuals are displayed and compared with the standard normal distribution. These studies suggest that the residual analysis usually performed in normal linear regression models can be extended to a modified deviance residual in the proposed regression model applied to censored data. We demonstrate that our extended regression model is very useful to the analysis of real data and may give more realistic fits than other special regression models.

Keywords: censored data, gamma-Weibull distribution, regression model, residual analysis, sensitivity analysis

1. Introduction

The CAPES *(Coordenação de Aperfeiçoamento de Pessoal de Nível Superior)* is responsible for assessing every three years, postgraduate courses in the country with the goal of keeping the courses with a level of excellence and contribute to the training of researchers (Horta and Moraes, 2005) in Brazil. According to CAPES, one of the requirements is that the master programs are done in two years and doctoral courses in four years (Moreira et al., 2010). For this reason, the programs graduate has studied how to encourage students before or within the stipulated time. To determine the best strategy to take, a particular program postgraduate noted the time it took his doctoral students to hold, according to sex and age at first registration.

One way to study the effect of these explanatory variables (gender and age of the students) on the completion time is through a location-scale regression model, also known as a model of accelerated lifetime. These models consider that the response variable belongs to a family of distributions characterized by a parameter location and scale parameter. Further details on this class of regression models can be found in Cox and Oakes (1984), Kalbfleisch and Prentice (2002) and Lawless (2003).

However, for the case of parametric models, it is assumed that the time until the end of the PhD is sampled from a continuous distribution. In the context of survival analysis, some distributions have been used to analyze censored data. For example, Leiva et al. (2007) conducted a diagnostic study in a log-Birnbaum-Saunders regression model. Carrasco et al. (2008) defined a regression model considering a modification of the Weibull distribution. Ortega et al. (2011) proposed the beta-Weibull regression model and Silva et al. (2011) proposed the log-Burr XII regression model. Thus, using the same approach adopted in this work, a distribution obtained from a generated gamma family will be expressed in the form of models belonging to the location-scale models. In this way, we can study the influence of explanatory variables on the completion time of doctoral students.

Regression models can be proposed in different forms in survival analysis. Among them, the location-scale regression model (Lawless, 2003) is distinguished and it is frequently used in clinical trials. In this paper, we propose a location-scale regression model based on the *log-gamma Weibull* (LGW) distribution. We consider a classic analysis for the LGW regression model. The inferential part was carried out using asymptotic distribution of the maximum likelihood estimators (MLEs), which, in situations when the sample is small, may present difficult results to be justified. As an alternative to classic analysis we explore the use of bootstrap method for survival times analysis as a feasible alternative. After modeling, it is important to check assumptions in the model and to conduct a robustness study in order to detect influential or extreme observations that can cause distortions in the results of the analysis. Numerous approaches have been proposed in the literature to detect influential or outlying observations. On the other hand, when using case deletion, all information from a single subject is deleted at once and, therefore, it is hard to tell whether that subject has any influence on a specific aspect of the model. A solution for the earlier problem can be found in a quite different paradigm, being a local influence approach where one again investigates how the results of an analysis are changed under small perturbations in the model, and where these perturbations can be specific interpretations. We develop a similar methodology to detect influential subjects in LGW regression models with censored data. Further, we compare two residuals to assess departures from the error assumptions and to detect outlying observations in the LGW regression models with censored observations. For different parameter settings, sample sizes and censoring percentages, various simulation studies are performed and the empirical distribution of each residual is displayed and compared with the standard normal distribution.

In Section 2, we perform a brief review on the GW and LGW distributions and derive the quantile function (qf) of the last distribution. Some of the mathematical properties of the LGW model such as the ordinary and incomplete moments, mean deviations, Bonferroni and Lorenz curves and generating function are investigated in Section 3. In Section 4, we consider a brief study on the GW distribution and present certain the characterizations of LGW distribution. In Section 5, we obtain the MLEs and the estimates based on a bootstrap method and provide some results from simulation studies for the LGW regression model with censored data. In Section 6, we use diagnostic measures considering three perturbation schemes and case-deletion in the LGW regression model with censored observations. Section 7 deals with the definition and discussion of the residuals and presents useful comments on the results from various simulation studies. In Section 8, a real data set is analyzed for illustrative purposes. Finally, we offer some conclusions in Section 9.

2. The Log-gamma-Weibull Distribution

The art of proposing generalized distributions has attracted theoretical and applied statisticians due to their flexible properties. Most of the distributions used to describe real data are chosen for the following reasons: a physical or statistical theoretical argument to explain the mechanism of the generated data, a model that has previously been used successfully and an appropriate model whose empirical fit is good to data. The Weibull distribution is a very popular distribution for modeling lifetime data and for modeling phenomenon with monotone failure rates. When modeling monotone hazard rates, the Weibull distribution may be an initial choice because of its negatively and positively skewed density shapes. This distribution has cumulative distribution function (cdf) (for $t > 0$) given by

$$G(t;\alpha,\lambda) = 1 - \exp\Big[-\Big(\frac{t}{\lambda}\Big)^{\alpha}\Big], \qquad (1)$$

where $\alpha > 0$ is a shape parameter and $\lambda > 0$ a scale parameter. The probability density function (pdf) corresponding to (1) is given by

$$g(t;\alpha,\lambda) = \frac{\alpha}{\lambda^{\alpha}}\, t^{\alpha-1} \exp\Big[-\Big(\frac{t}{\lambda}\Big)^{\alpha}\Big]. \qquad (2)$$

We write $T \sim \mathrm{GW}(\alpha,\lambda)$ for a random variable T having the pdf (2).

There has been an increased interest in defining new univariate continuous distributions by introducing one additional shape parameter to the baseline distribution. The extra parameter has been proved useful in some cases to explore tail properties and to improve the goodness-of-fit of the new distribution. In fact, Zografos and Balakrishnan (2009) and Ristic, and Balakrishnan (2012) proposed a family of univariate distributions generated by one-parameter gamma random variables. Based on any baseline cdf $G(t)$, they defined the *gamma-G family* with pdf $f(t)$ and cdf $F(t)$ (for $\phi > 0$) given by

$$f(t) = \frac{1}{\Gamma(\phi)}\,\{-\log[1-G(t)]\}^{\phi-1}\, g(t) \qquad (3)$$

and

$$F(t) = \frac{\gamma(-\log[1-G(t)],\phi)}{\Gamma(\phi)} = \frac{1}{\Gamma(\phi)}\int_0^{-\log[1-G(t)]} u^{\phi-1}\,\mathrm{e}^{-\mathrm{u}}\mathrm{du},$$

respectively, where $g(t) = dG(t)/dt$, $\Gamma(a) = \int_0^{\infty} u^{a-1}\,\mathrm{e}^{-\mathrm{u}}\mathrm{du}$ is the gamma function, and $\gamma(a,z) = \int_0^{z} u^{a-1}\,\mathrm{e}^{-\mathrm{u}}\mathrm{du}$ is the incomplete gamma function. The gamma-G family has the same parameters of the G distribution plus one additional shape parameter $\phi > 0$. For any specified G distribution, we can generate the associated gamma-G distribution. Recently Nadarajah *et al.* (2015) provide a comprehensive treatment of general mathematical properties of gamma-G distributions. For $\phi = 1$, the G distribution is a basic exemplar of the gamma-G distribution with a continuous crossover towards cases with different shapes (for example, a particular combination of skewness and kurtosis).

In this context, we define the *gamma Weibull* (GW) density function by inserting (1) and (2) in equation (3). So, we obtain

$$f(t) = \frac{\alpha}{\lambda^{\phi\alpha}\,\Gamma(\phi)}\,t^{\phi\alpha-1}\exp\left[-\left(\frac{t}{\lambda}\right)^{\alpha}\right], \quad t > 0, \tag{4}$$

where $\phi > 0$ and $\alpha > 0$ are shape parameters and $\lambda > 0$ is a scale parameter. By combining the gamma-G and Weibull distributions, we obtain the generalized gamma distribution (Stacy, 1962). The applications of the GW distribution can be directed to model insurance data, tree diameters, software reliability, extreme value observations in floods, carbon fibrous composites, firmware system failure, reliability prediction and fracture toughness, among others.

To obtain a distribution that belongs to the location-scale model, we consider the transformation of the random variable $Y = \log(T)$, which has the *log-gamma Weibull* (LGW) distribution. Thus, considering the pdf (4) and the transformations $\alpha = \sigma^{-1}$ and $\lambda = e^{\mu}$, the LGW distribution (for $y \in R$) reduces to

$$f(y) = f(y;\phi,\mu,\sigma) = \frac{1}{\sigma\Gamma(\phi)}\exp\left[\frac{\phi(y-\mu)}{\sigma} - \exp\left(\frac{y-\mu}{\sigma}\right)\right], \tag{5}$$

where $\phi > 0$, $\sigma > 0$ and $\mu \in R$. Therefore, Y follows the LGW distribution. Plots of the pdf (5) for selected parameter values are displayed in Figure 1. These plots show great flexibility for different values of the shape parameter ϕ with $\mu = 0$ and $\sigma = 1$. If Y is a random variable having pdf (5), we write $Y \sim \mathrm{LGW}(\phi,\mu,\sigma)$.

If $T \sim \mathrm{GW}(\phi,\alpha,\lambda)$, then $Y = \log(T) \sim \mathrm{LGW}(\phi,\mu,\sigma)$. The survival function corresponding to (5) becomes

$$S(y) = S(y;\phi,\mu,\sigma) = 1 - \frac{1}{\Gamma(\phi)}\gamma\left[\exp\left(\frac{y-\mu}{\sigma}\right),\phi\right]. \tag{6}$$

The simulation of Y is very easy: if V is a gamma random variable with shape parameter ϕ and unit scale parameter then $Y = \mu + \sigma\,\log(V)$, will have the LGW density function (5).

Let $z = Q^{-1}(a,u)$ be the inverse function of $Q(a,z) = 1 - \gamma(a,x)/\Gamma(a) = u$, see http:// functions.wolfram.com/ Gamma-BetaErf/ InverseGammaRegularized/ for details. The asymptotes of $z = Q^{-1}(a,u)$ can be determined using known properties of $Q^{-1}(a,u)$. Using http:// functions.wolfram.com/ GammaBetaErf/ InverseGammaRegularized/ 06/ 02/ 01/, we can write $z = Q^{-1}(a,u)$ as one can see that

$$z = Q^{-1}(a,u) = \left[-(1-a)\,W_{-1}\left(-\frac{(1-u)^{1/(a-1)}\Gamma(a)^{1/(a-1)}}{a-1}\right)\right]$$

as $u \to 0$, where $W_{-1}(\cdot)$ denotes the product log function. Further, using http:// functions.wolfram.com/ GammaBetaEr-f/InverseGammaRegularized/06/01/03/0001/, we have

$$z = Q^{-1}(a,u) = \left[-(1-u)^{1/a}\,\Gamma(a+1)^{1/a} + \frac{(1-u)^2\,\Gamma(a+1)^{2/a}}{(a+1)}\right] + O((1-u)^3).$$

Further, inverting $1 - S(y) = u$, we obtain the qf of Y as

$$y = \mu + \sigma\,\log\left[Q^{-1}(a,u)\right]. \tag{7}$$

We define the standardized random variable $Z = (Y-\mu)/\sigma$ having pdf given by

$$\pi(z;\phi) = \frac{1}{\Gamma(\phi)}\exp\{\phi\,z - \exp(z)\}, \qquad z \in R. \tag{8}$$

The special case $\phi = 1$ corresponds to the log-Weibull (LW) (or extreme-value) distribution and, for $\sigma = 1$, we obtain the log-gamma-exponential (LGE) model.

3. Mathematical Properties

In this section, we derive explicit expressions for the ordinary and incomplete moments, generating function and Bonferroni and Lorenz curves for the LGW distribution. Its mathematical properties are not difficult to be implemented in applications because of the computational and analytical facilities available in programming softwares like MATHEMATICA and MAPLE that can easily tackle the problems involved in computing the special functions in these properties.

3.1 Moments

The need for necessity and the importance of moments in any statistical analysis especially in applied work is obvious. Some of the most important features and characteristics of a distribution can be studied through moments (e.g. tendency, dispersion, skewness and kurtosis). The rth raw moment of Y can be expressed as

$$\mu'_r = E(Y^r) = \frac{1}{\Gamma(\phi)} \sum_{i=0}^{r} \binom{r}{i} \sigma^i \mu^{r-i} \int_0^{\infty} [\log(u)]^i \, u^{\phi-1} \, e^{-u} du, \tag{9}$$

where $u = \exp\left(\frac{y-\mu}{\sigma}\right)$. Further, based on a result by Prudnikov et al. (1986, equation 2.6.21.1), we can rewrite (9) as

$$\mu'_r = \frac{1}{\Gamma(\phi)} \sum_{i=0}^{r} \binom{r}{i} \sigma^i \mu^{r-i} \frac{\partial^i \Gamma(\phi)}{\partial \phi^i}. \tag{10}$$

The mean and variance of Y follow from (10) as

$$E(Y) = \mu'_1 = \mu + \sigma\psi(\phi)$$

and

$$\mathrm{Var}(Y) = E^2(X) = \sigma^2\psi(1, \phi) + \sigma^2\psi^2(\phi),$$

where $\psi(\cdot)$ is the digamma function, $\psi(n, \phi)$ is the polygamma function and n is a positive integer.

The central moments (μ_r) and cumulants (κ_r) of Y can be determined from (10) as

$$\mu_r = \sum_{k=0}^{r} (-1)^k \binom{r}{k} \mu_1'^k \mu'_{r-k} \qquad \text{and} \qquad \kappa_r = \mu'_r - \sum_{k=1}^{r-1} \binom{r-1}{k-1} \kappa_k \mu'_{r-k},$$

respectively, where $\kappa_1 = \mu'_1$. Then, $\kappa_2 = \mu'_2 - \mu_1'^2$, $\kappa_3 = \mu'_3 - 3\mu'_2\mu'_1 + 2\mu_1'^3$, $\kappa_4 = \mu'_4 - 4\mu'_3\mu'_1 - 3\mu_2'^2 + 12\mu'_2\mu_1'^2 - 6\mu_1'^4$, etc. The skewness $\gamma_1 = \kappa_3/\kappa_2^{3/2}$ and kurtosis $\gamma_2 = \kappa_4/\kappa_2^2$ follow from the second, third and fourth cumulants. Other kinds of moments such L-moments may also be obtained in closed-form, but we consider only the previous moments for reasons of space.

Figure 2 provides some plots of the skewness (Figure 2a) and kurtosis (Figure 2b) for $\sigma = 1.5$ as a function of μ for some values of ϕ, whereas Figure 3 provides some plots of the skewness (Figure 3a) and kurtosis (Figure 3b) for $\mu = 0$ as a function of ϕ for some values of σ. It can be noted that when the parameter μ increases the LGW distribution is higher and concentrated than the standard normal distribution, i.e., it has heavy tails. Moreover, when the parameter ϕ decreases, the LGW distribution becomes positive asymmetrical (Figure 2). On the other hand, Figure 3 indicates that when the parameter ϕ increases, the shape of the LGW distribution is flatter than that of the normal distribution. For the parameter ϕ greater than three, this distribution has a negative skewness and if the parameter ϕ is less than three, the distribution is positively skewed.

The nth descending factorial moment of Y is

$$\mu'_{(n)} = E(Y^{(r)}) = E\,[Y(Y-1) \times \cdots \times (Y-r+1)] = \sum_{k=0}^{r} s(r,k)\,\mu'_k,$$

where

$$s(r,k) = (k!)^{-1} \left[\frac{d^k x^{(r)}}{dx^k}\right]_{x=0}$$

is the Stirling number of the first kind which counts the number of ways to permute a list of r items into k cycles. So, we can obtain the factorial moments from the ordinary moments given before.

3.2 Mean Deviations

For empirical purposes, the shapes of many distributions can be usefully described by what we call the first incomplete moment, which plays an important role for measuring inequality, for example, income quantiles and Lorenz and Bonferroni curves. The first incomplete moment of Y is given by $m_1(z) = \int_{-\infty}^{z} y\, f(y; \phi, \mu, \sigma) dy$. Changing variable $u = \exp\left(\frac{y-\mu}{\sigma}\right)$ and using a similar approach of Section 3.1, we can write

$$m_1(z) = \frac{1}{\Gamma(\phi)} \left[\mu\,\gamma(\phi, z^\star) + \sigma\, J(\phi, z^\star)\right], \tag{11}$$

where $z^\star = \exp\left(\frac{z-\mu}{\sigma}\right)$, $\gamma(\phi, z^\star) = \int_0^{z^\star} u^{\phi-1}\,\mathrm{e}^{-u}du$ is the incomplete gamma function, and $J(\phi, z^\star) = \int_0^{z^\star} \log(u)\,u^{\phi-1}\,\mathrm{e}^{-u}du$. The integral $J(\phi, z^\star)$ can be expressed in terms of the digamma function $\psi(\phi) = d\log[\Gamma(\phi)]/d\phi$ and the Meijer G-function defined by

$$G^{m,n}_{p,q}\left(x\left|\begin{array}{c}a_1,\ldots,a_p\\ b_1,\ldots,b_q\end{array}\right.\right) = \frac{1}{2\pi\mathrm{i}}\int_L \frac{\prod_{j=1}^{m}\Gamma\left(b_j+t\right)\prod_{j=1}^{n}\Gamma\left(1-a_j-t\right)}{\prod_{j=n+1}^{p}\Gamma\left(a_j+t\right)\prod_{j=m+1}^{q}\Gamma\left(1-b_j-t\right)}\,x^{-t}\,dt,$$

where $\mathrm{i} = \sqrt{-1}$ is the complex unit and L denotes an integration path (see Gradshteyn and Ryzhik, 2000; Section 9.3) for a description of this path. The Meijer G-function contains many integrals with elementary and special functions. Some of these integrals are included in Prudnikov et al. (1986). In the MATHEMATICA software, $\psi(\phi)$ is denoted by PolyGamma$[0, \phi]$ and the Meijer G-function is represented by

$$G^{m,n}_{p,q}\left(x\left|\begin{array}{c}a_1,\ldots,a_p\\ b_1,\ldots,b_q\end{array}\right.\right) = \mathrm{MeijerG}\left[\{\{\mathrm{a}_1,\ldots,\mathrm{a}_\mathrm{n}\},\{\mathrm{a}_{\mathrm{n}+1},\ldots,\mathrm{a}_\mathrm{p}\}\},\{\{\mathrm{b}_1,\ldots,\mathrm{b}_\mathrm{m}\},\{\mathrm{b}_{\mathrm{m}+1},\ldots,\mathrm{b}_\mathrm{q}\}\},\mathrm{x}\right].$$

We can obtain using the MATHEMATICA software

$$J(\phi, z^\star) = -\gamma(\phi, z^\star)\,\log(z^\star) - \mathrm{MeijerG}\left[\{\{\},\{1,1\}\},\{\{0,0,\phi\},\{\}\},\mathrm{z}^\star\right] + \Gamma(\phi)\,\mathrm{PolyGamma}[0,\phi].$$

Applications of equation (11) can be addressed to obtain Bonferroni and Lorenz curves defined for a given probability π by $B(\pi) = m_1(q)/(\pi\mu_1')$ and $L(\pi) = m_1(q)/\mu_1'$, respectively, where $\mu_1' = E(Y)$ and $q = Q(\pi)$ is the qf (7) of Y at π.

The mean deviations about the mean ($\delta_1 = E(|Y - \mu_1'|)$) and about the median ($\delta_2 = E(|Y - M|)$) of Y can be expressed as

$$\delta_1 = 2\mu_1'\,F(\mu_1') - 2m_1(\mu_1') \qquad \text{and} \qquad \delta_2 = \mu_1' - 2m_1(M),$$

respectively, where $\mu_1' = E(Y)$, $M = Median(Y) = Q(0.5)$ is the median computed from (7), $F(\mu_1') = 1 - S(\mu_1')$ is easily calculated from the survival function (6) and $m_1(z)$ is given by (11).

3.3 Generating Function

The moment generating function (mgf) provides the basis of an alternative route to analytical results compared with working directly with the pdf and cdf and it is widely used in the characterization of distributions and the application of the skew-normal test (Meintanis, 2010) and other goodness of fit tests (Ghosh, 2013). Therefore, using a result in Prudnikov et al. (1986, equation 2.6.21.1), we can derive the mgf of Y as

$$M_Y(t) = \frac{\mathrm{e}^{t\mu}}{\Gamma(\phi)}\sum_{i=0}^{\infty}\frac{(t\sigma)^i}{i!}\frac{\partial^i}{\partial\phi^i}\Gamma(\phi).$$

4. Characterizations of LGW Distribution

In this section we present certain characterizations of LGW distribution. The first characterization is based on a simple relationship between two truncated moments. It should be mentioned that for this characterization, the cdf need no have a closed form. We believe, due to the nature of the cdf of LGW, there may not be other possibly interesting characterizations than the ones presented in this section. Our first characterization result borrows from a theorem due to [*Gläzel*, 1987], see Theorem G in the Appendix A. Note that the result holds also when the interval I is not closed. Moreover, as shown in [*Gläzel*, 1990], this characterization is stable in the sense of weak convergence. Here is our first characterization of LGW distribution.

Proposition 1. Let $X : \Omega \to \mathbb{R}$ be a continuous random variable and let $q_1(x) = \exp\left\{(1-\phi)\left(\frac{x-\mu}{\sigma}\right)\right\}$ and $q_2(x) = q_1(x)\exp\left\{-\mathrm{e}^{\left(\frac{x-\mu}{\sigma}\right)}\right\}$ for $x \in \mathbb{R}$. The random variable X belongs to LGW family (5) if and only if the function η defined in Theorem G has the form

$$\eta(x) = \frac{1}{2}\exp\left\{-\mathrm{e}^{\left(\frac{x-\mu}{\sigma}\right)}\right\}, \quad x \in \mathbb{R}.$$

Proof. Let X be a random variable with density (5), then

$$(1-F(x))E\left[q_1(X)\mid X\geq x\right]=\frac{1}{\Gamma(\phi)}\exp\left\{-e^{\left(\frac{x-\mu}{\sigma}\right)}\right\},\quad x\in\mathbb{R},$$

and

$$(1-F(x))E\left[q_2(X)\mid X\geq x\right]=\frac{1}{2\Gamma(\phi)}\exp\left\{-2e^{\left(\frac{x-\mu}{\sigma}\right)}\right\},\quad x\in\mathbb{R},$$

and finally

$$\eta(x)q_1(x)-q_2(x)=-\frac{1}{2}q_1(x)\left\{-\exp\left\{-e^{\left(\frac{x-\mu}{\sigma}\right)}\right\}\right\}<0\quad for\ x\in\mathbb{R}.$$

Conversely, if η is given as above, then

$$s'(x)=\frac{\eta'(x)q_1(x)}{\eta(x)q_1(x)-q_2(x)}=\frac{1}{\sigma}e^{\left(\frac{x-\mu}{\sigma}\right)},\quad x\in\mathbb{R},$$

and hence

$$s(x)=e^{\left(\frac{x-\mu}{\sigma}\right)},\quad x\in\mathbb{R}.$$

Now, in view of Theorem G, X has density (5).

Corollary 1. Let $X:\Omega\to\mathbb{R}$ be a continuous random variable and let $q_1(x)$ be as in Proposition 1. The pdf of X is (5) if and only if there exist functions q_2 and η defined in Theorem G satisfying the differential equation

$$\frac{\eta'(x)q_1(x)}{\eta(x)q_1(x)-q_2(x)}=\frac{1}{\sigma}e^{\left(\frac{x-\mu}{\sigma}\right)},\quad x\in\mathbb{R}.$$

The general solution of the differential equation in Corollary 1 is

$$\eta(x)=\exp\left\{e^{\left(\frac{x-\mu}{\sigma}\right)}\right\}\left[\int\frac{1}{\sigma}\exp\left\{\left(\frac{x-\mu}{\sigma}\right)-e^{\left(\frac{x-\mu}{\sigma}\right)}\right\}(q_1(x))^{-1}q_2(x)\,dx+D\right],$$

where D is a constant. Note that a set of functions satisfying the differential equation in Corollary 1, is given in Proposition 1 with $D=0$. However, it should be also noted that there are other triplets (q_1,q_2,η) satisfying the conditions of Theorem G.

5. The Log-gamma Weibull Regression Model with Censored Data

The last decade is full of works on generalized classes of regression models, which are always precious for applied statisticians. In practical applications, the lifetimes are affected by explanatory variables such as the cholesterol level, blood pressure and many others. Let $\mathbf{x}_i=(x_{i1},\ldots,x_{ip})^\top$ be the explanatory variable vector associated with the ith response variable y_i, for $i=1,\ldots,n$.

We construct a linear regression model for the response variable y_i based on the LGW distribution given by

$$y_i=\mathbf{x}_i^\top\boldsymbol{\beta}+\sigma z_i,\ \ i=1,\ldots,n, \tag{12}$$

where the random error z_i has the pdf (8), $\boldsymbol{\beta}=(\beta_1,\ldots,\beta_p)^\top$, $\sigma>0$ and $\phi>0$ are unknown scalar parameters and $\mathbf{x}_i$ is the vector of explanatory variables modeling the location parameter $\mu_i=\mathbf{x}_i^\top\boldsymbol{\beta}$. Hence, the location parameter vector

$\boldsymbol{\mu} = (\mu_1, \ldots, \mu_n)^\top$ of the LGW regression model has a linear structure $\boldsymbol{\mu} = \mathbf{x}\boldsymbol{\beta}$, where $\mathbf{x} = (\mathbf{x}_1, \ldots, \mathbf{x}_n)^\top$ is a known model matrix.

Consider a sample $(y_1, \mathbf{x}_1), \ldots, (y_n, \mathbf{x}_n)$ of n independent observations, where each random response is defined by $y_i = \min\{\log(t_i), \log(c_i)\}$. We assume non-informative censoring such that the observed lifetimes and censoring times are independent. Let F and C be the sets of individuals for which y_i is the log-lifetime or log-censoring, respectively. We consider non-informative censoring such that the observed lifetimes and censoring times are independent. The log-likelihood function for the vector of parameters $\boldsymbol{\theta} = (\phi, \sigma, \boldsymbol{\beta}^\top)^\top$ from model (12) has the form $l(\boldsymbol{\theta}) = \sum_{i \in F} l_i(\boldsymbol{\theta}) + \sum_{i \in C} l_i^{(c)}(\boldsymbol{\theta})$, where $l_i(\boldsymbol{\theta}) = \log[f(y_i)]$, $l_i^{(c)}(\boldsymbol{\theta}) = \log[S(y_i)]$, $f(y_i)$ is the pdf (5) and $S(y_i)$ is the survival function (6), for $i = 1, \ldots, n$. Therefore, the log-likelihood function for $\boldsymbol{\theta}$ reduces to

$$l(\boldsymbol{\theta}) = -r\log(\sigma) - r\log\left[\Gamma(\phi)\right] + \frac{\phi}{\sigma}\sum_{i \in F}(y_i - \mathbf{x}_i^\top\boldsymbol{\beta}) - \sum_{i \in F}\exp\left(\frac{y_i - \mathbf{x}_i^\top\boldsymbol{\beta}}{\sigma}\right) + \sum_{i \in C}\log\left\{1 - \frac{1}{\Gamma(\phi)}\gamma\left[\exp\left(\frac{y_i - \mathbf{x}_i^\top\boldsymbol{\beta}}{\sigma}\right), \phi\right]\right\}, \tag{13}$$

where $\boldsymbol{\theta} = (\phi, \sigma, \boldsymbol{\beta}^\top)^\top$ is the vector of unknown parameters, r is the observed number of failures and $\gamma(x, \phi)$ is the incomplete gamma function. The components of the score vector $U(\boldsymbol{\theta})$ are given by

$$\frac{\partial l(\boldsymbol{\theta})}{\partial \phi} = -r\psi(\phi) + \frac{1}{\sigma}\sum_{i \in F}(y_i - \mathbf{x}_i^\top\boldsymbol{\beta}) - \sum_{i \in C}\frac{V(\exp(z_i), \phi, 1) - \psi(\phi)\gamma[\exp(z_i), \phi]}{\Gamma(\phi) - \gamma[\exp(z_i), \phi]},$$

$$\frac{\partial l(\boldsymbol{\theta})}{\partial \sigma} = -\frac{r}{\sigma} - \frac{\phi}{\sigma}\sum_{i \in F} z_i + \frac{1}{\sigma}\sum_{i \in F} z_i \exp(z_i) + \frac{1}{\sigma}\sum_{i \in C}\frac{z_i \exp[\phi z_i - \exp(z_i)]}{\Gamma(\phi) - \gamma[\exp(z_i), \phi]}$$

$$\frac{\partial l(\boldsymbol{\theta})}{\partial \beta_j} = -\frac{\phi}{\sigma}\sum_{i \in F} x_{ij} + \frac{1}{\sigma}\sum_{i \in F} x_{ij}\exp(z_i) + \frac{1}{\sigma}\sum_{i \in C}\frac{x_{ij}\exp[\phi z_i - \exp(z_i)]}{\Gamma(\phi) - \gamma[\exp(z_i), \phi]},$$

where $z_i = \frac{y_i - \mathbf{x}_i^\top}{\sigma}$, $\psi(\cdot)$ is the digamma function, $V(x, \phi, 1) = \int_0^x u^{\phi-1}e^{-u}\log(u)du$ and $j = 0, 1, \ldots, p$.

The MLE $\widehat{\boldsymbol{\theta}}$ of $\boldsymbol{\theta}$ can be obtained by maximizing the log-likelihood function (13). We use the matrix programming language Ox (MaxBFGS function) (see Doornik, 2007) to calculate the estimate $\widehat{\boldsymbol{\theta}}$. Initial values for $\boldsymbol{\beta}$ and σ are taken from the fit of the LW regression model with $\phi = 1$.

Under general regularity conditions, the asymptotic distribution of $\sqrt{n}(\widehat{\boldsymbol{\theta}} - \boldsymbol{\theta})$ is multivariate normal $N_{p+3}(0, K(\boldsymbol{\theta})^{-1})$, where $K(\boldsymbol{\theta})$ is the expected information matrix. The asymptotic covariance matrix $K(\boldsymbol{\theta})^{-1}$ of $\widehat{\boldsymbol{\theta}}$ can be approximated by the inverse of the $(p+3) \times (p+3)$ observed information matrix $J(\boldsymbol{\theta})$ and then the inference on the parameter vector $\boldsymbol{\theta}$ can be based on the normal approximation $N_{p+3}(0, J(\boldsymbol{\theta})^{-1})$ for $\widehat{\boldsymbol{\theta}}$. This multivariate normal $N_{p+3}(0, J(\boldsymbol{\theta})^{-1})$ distribution can be used to construct approximate confidence regions for some parameters in $\boldsymbol{\theta}$ and for the hazard and survival functions.

5.1 Bootstrap Re-sampling Method

The bootstrap is a computer-based method for assessing the accuracy of statistical estimates and tests. It was first proposed by Efron (1979). Treat the data as if they were the(true, unknown) population and draw samples (with replacement) from the data as if you were sampling from the population. Repeat the procedure a large number of times(say B) each time computing the quantity of interest. Then, use the B values of the quantity of interest to estimate its unknown distribution.

Let $\mathbf{T}=(T_1, \ldots, T_n)$ be an observed random sample and $\hat{F}$ be the empirical distribution of $\mathbf{T}$. Thus, a bootstrap sample $\mathbf{T}^*$ is constructed by re-sampling with replacement of n elements of the sample $\mathbf{T}$. For the B bootstrap samples generated, $\mathbf{T}_1^*, \ldots, \mathbf{T}_B^*$, the bootstrap replication of the parameter of interest for the bth sample is given by

$$\hat{\boldsymbol{\theta}}_b^* = s(\mathbf{T}_b^*),$$

i.e., the value of $\hat{\boldsymbol{\theta}}$ for sample $\mathbf{T}_b^*$, $b = 1, \ldots, B$.

The bootstrap estimator of the standard error (Efron and Tibshirani, 1993) is the standard deviation of these bootstrap samples, namely

$$\widehat{EP}_B = \left[\frac{1}{(B-1)}\sum_{b=1}^{B}\left(\hat{\theta}_b^* - \bar{\theta}_B\right)^2\right]^{1/2},$$

where $\bar{\theta}_B = \frac{1}{B}\sum_{b=1}^{B}\hat{\theta}_b^*$. Note that B is the number of bootstrap samples generated. According to Efron and Tibshirani (1993), assuming $B \geq 200$, it is generally sufficient to present good results to determine the bootstrap estimates. However, to achieve greater accuracy, a reasonably high B value must be considered. We describe the bias corrected and accelerated (BCa) method for constructing approximated confidence intervals based on the bootstrap re-sampling method (Hashimoto et al., 2013).

6. Sensitivity Analysis

There are basically two approaches to detecting observations that seriously influence the results of a statistical analysis. One approach is the case-deletion approach, and the second approach is one in which the stability of the estimated outputs with respect to the model inputs is studied via various minor model perturbation schemes.

6.1 Global Influence

The assessment of robustness aspects of the parameter estimates in statistical models has been an important concern of various researchers in recent decades. The case deletion measures , which consists of studying the impact on the parameter estimates after dropping individual observations, is probably the most employed technique to detect influential observations (Cook and Weisberg, 1982)

A global influence measure considered by Xie and Wei (2007) is a generalization of the Cook distance defined as a standardized norm $\hat{\boldsymbol{\theta}}_{(i)} - \hat{\boldsymbol{\theta}}$. It is expressed as

$$GD_i(\boldsymbol{\theta}) = (\hat{\boldsymbol{\theta}}_{(i)} - \hat{\boldsymbol{\theta}})^\top [\ddot{\mathbf{L}}(\boldsymbol{\theta})](\hat{\boldsymbol{\theta}}_{(i)} - \hat{\boldsymbol{\theta}}), \quad (14)$$

where $\ddot{\mathbf{L}}(\boldsymbol{\theta})$ is the observed information matrix.

Another measures to evaluate the influence is called of likelihood distance and considers the difference between $\hat{\boldsymbol{\theta}}_{(i)}$ and $\hat{\boldsymbol{\theta}}$.

Thus, the likelihood distance is given by

$$LD_i(\boldsymbol{\theta}) = 2\left[l(\hat{\boldsymbol{\theta}}) - l(\hat{\boldsymbol{\theta}}_{(i)})\right], \quad (15)$$

where $l(\hat{\boldsymbol{\theta}})$ is the value of the logarithm of the likelihood function of the full sample and $l(\hat{\boldsymbol{\theta}}_{(i)})$ is the value of the logarithm of the likelihood function of the sample excluding the i-th observation.

6.2 Local Influence

A second tool for sensitivity analysis is known as local influence. The local influence measures are calculated after a perturbation scheme in the model or data is established. Thus different perturbation schemes can be used according to the purpose of the analysis. Therefore, considering the model defined in (12) and the logarithm of the likelihood function expressed in (13), the following perturbation schemes are used:

a. ***Case perturbation***
Let $0 \leq \omega_i \leq 1$ and $\boldsymbol{\omega}_0 = (1, \ldots, 1)^\top$ be the vector representing no perturbation. The logarithm of the log-likelihood disturbed for each model incorporating different weights for each element of the data is defined by

$$l(\boldsymbol{\theta}) = -\{r\log(\sigma) - r\log[\Gamma(\phi)]\}\sum_{i\in F}\omega_i + \frac{\phi}{\sigma}\sum_{i\in F}\omega_i(y_i - \mathbf{x}_i^\top\boldsymbol{\beta}) - \sum_{i\in F}\left[\omega_i \times \exp\left(\frac{y_i - \mathbf{x}_i^\top\boldsymbol{\beta}}{\sigma}\right)\right] + \sum_{i\in C}\omega_i\log\left\{1 - \frac{1}{\Gamma(\phi)}\gamma\left[\exp\left(\frac{y_i - \mathbf{x}_i^\top\boldsymbol{\beta}}{\sigma}\right), \phi\right]\right\}.$$

b. ***Response variable perturbation***
To verify the sensitivity of the model (12), it is assumed that the response variable $y_i = \log(t_i)(i = 1, \ldots, n)$ is submitted to the additive perturbation scheme such that $y_i^* = y_i + \omega_i \text{sd}(y_i)$, where $\text{sd}(y_i)$ is a scaling factor which can be the

standard deviation of the logarithm of failure times and $\omega_i \in \mathbf{R}$ is a perturbation vector (Silva et al., 2010). In this case, $\omega_0 = (0, \ldots, 0)^\top$ is the vector representing no perturbation and the logarithm likelihood function disturbed is expressed as

$$l(\boldsymbol{\theta}) = -r\log(\sigma) - r\log[\Gamma(\phi)] + \frac{\phi}{\sigma}\sum_{i\in F}(y_i^* - \mathbf{x}_i^\top\boldsymbol{\beta}) - \sum_{i\in F}\exp\left(\frac{y_i^* - \mathbf{x}_i^\top\boldsymbol{\beta}}{\sigma}\right) + \sum_{i\in C}\log\left\{1 - \frac{1}{\Gamma(\phi)}\gamma\left[\exp\left(\frac{y_i^* - \mathbf{x}_i^\top\boldsymbol{\beta}}{\sigma}\right), \phi\right]\right\}.$$

c. ***Explanatory variable perturbation***

Another way of evaluating the sensitivity of the model (12) is to consider small perturbations in a particular continuous explanatory variable, denoted by X_j. In this case, the explanatory variable is submitted to the additive perturbation scheme, such that $x_{ij}^* = x_{ij} + \omega_i \mathrm{sd}(x_{ij})$, wheresd$(x_{ij})$ is scaling factor that can be the standard deviation of the disturbed explanatory variable (Silva et al., 2010). Thus, considering that $\mathbf{x}_i^{*\top}\boldsymbol{\beta} = \beta_0 + \beta_1 x_{i1} + \ldots + \beta_j x_{ij}^* + \ldots + \beta_p x_{ip}$ and $\omega_0 = (0, \ldots, 0)^\top$ is the vector representing no perturbation, the logarithm of the likelihood function is given by

$$l(\boldsymbol{\theta}) = -r\log(\sigma) - r\log[\Gamma(\phi)] + \frac{\phi}{\sigma}\sum_{i\in F}(y_i - \mathbf{x}_i^{*\top}\boldsymbol{\beta}) - \sum_{i\in F}\exp\left(\frac{y_i - \mathbf{x}_i^{*\top}\boldsymbol{\beta}}{\sigma}\right) + \sum_{i\in C}\log\left\{1 - \frac{1}{\Gamma(\phi)}\gamma\left[\exp\left(\frac{y_i - \mathbf{x}_i^{*\top}\boldsymbol{\beta}}{\sigma}\right), \phi\right]\right\}.$$

For all three perturbation schemes, the array of maximum curvature is calculated numerically as

$$\boldsymbol{\Delta} = (\boldsymbol{\Delta}_{vi})_{(p+3)\times n} = \left[\frac{\partial^2 l(\boldsymbol{\theta}|\boldsymbol{\omega})}{\partial\theta_v\partial\omega_i}\right]_{(p+3)\times n},$$

where $v = 1, \ldots, p+3$, $i = 1, \ldots, n$ and $\boldsymbol{\omega} = (\omega_1, \ldots, \omega_n)^\top$ is the vector of weights that penalizes the LGW regression model or the observations.

7. Residual Analysis

In order to study the assumptions of the errors and the presence of outliers, we propose various residuals, for example, Collett (2003), Weisberg (2005) and Colosimo and Giolo (2006). In the context of survival analysis, the deviance residual has been more widely used, because they take into account the information of censored times (Silva et al., 2008). Thus, the deviance residual plot versus the observed times provides a way to verify the adequacy of the adjusted model and help to find atypical observations. The deviance residual is expressed as

$$r_{D_i} = \begin{cases} \mathrm{sign}(\hat{r}_{M_i})\Big\{-2\Big[1 + \log\Big\{1 - \frac{1}{\Gamma(\hat{\phi})}\gamma\Big[\log[1+\exp(\hat{z}_i)], \hat{\phi}\Big]\Big\} + \\ \log\Big\{-\log\Big\{1 - \frac{1}{\Gamma(\hat{\phi})}\gamma\Big[\log[1+\exp(\hat{z}_i)], \hat{\phi}\Big]\Big\}\Big\}\Big]\Big\} & \text{if} \quad i \in F, \\ \mathrm{sign}(\hat{r}_{M_i})\Big\{-2\log\Big\{1 - \frac{1}{\Gamma(\hat{\phi})}\gamma\Big[\log[1+\exp(\hat{z}_i)], \hat{\phi}\Big]\Big\}\Big\}^{1/2} & \text{if} \quad i \in C, \end{cases} \tag{16}$$

where

$$r_{M_i} = \begin{cases} 1 + \log[\hat{S}(y_i; \widehat{\boldsymbol{\theta}})] & \text{if} \quad i \in F, \\ \log[\hat{S}(y_i; \widehat{\boldsymbol{\theta}})] & \text{if} \quad i \in C, \end{cases}$$

is the martingale residuals, sign$(\cdot)$ is a function that leads the values $+1$ if the argument is positive and -1 if the argument is negative and $\hat{z}_i = (y_i - \mathbf{x}_i^\top\hat{\boldsymbol{\beta}})/\hat{\sigma}$ for $i = 1, \ldots, n$.

7.1 Simulation Study

We performed a simulation to assess the MLEs of the LGW regression model with censored data, and also to investigate the behavior of the empirical distribution of martingale and deviance residuals. For the simulation study, the variables

$z_1, \ldots, z_n$ of the LGW distribution (8) were generated by the acceptance-rejection method, as explained in Ross (2006) and Bonat et al. (2012). Therefore, for the sample sizes $n = 100$, $n = 300$ and $n = 500$, the values of the parameters of the distributions are set at $\phi = 0.8$ and 1.5, $\sigma = 1.0$, $\beta_0 = 2.0$ and $\beta_1 = 4.0$. The survival times are generated from the following algorithm (Zeviani, 2012), adapted in this work for censored data.

i. Generate $v \sim \text{uniform}(a_1, b_1)$, where a_1 and b_1 are chosen to represent the support of random variable with pdf (8).

ii. Generate $u \sim \text{uniform}(0, b_2)$, where b_2 was chosen to represent the values of the density (8).

iii. If $u \leq f(v) \Rightarrow z = v$, where $f(\cdot)$ is the function (8).

iv. Generate $x_1 \sim \text{uniform}(0, 1)$.

v. Write $y^* = \beta_0 + \beta_1 x_1 + \sigma z$.

vi. Generate $c \sim \text{uniform}(0, \tau)$, where τ was adjusted to obtain the percentages of right censoring 10% and 30%.

vii. If $y = \min(y^*, c)$ then $y \in F$(Failure), else $y \in C$(Censored).

viii. Otherwise, return to step *i*.

Therefore, 1,000 samples are generated for each combination of n, ϕ, σ and censoring percentages by means of Monte Carlo simulations, and the MLEs of the model parameters are obtained for each of the samples. Then, for each adjusted model, the residuals r_{D_i} (16) are determined. On the other hand, Figures 4-5 display the plots of the residuals versus the expected values of the order statistics of the standard normal distribution. This plot is known as the normal probability plot and serves to assess the departure from the normality assumption of the residuals (Weisberg, 2005). Therefore, the following interpretations are obtained from these plots: the empirical distribution of the deviance residual agrees with the standard normal distribution and as the sample size increases, the empirical distribution of the deviance residual becomes closer to the normal distribution (as illustrated in Figures 4-5).

8. Application

In this study, information from 49 PhD students of a particular program of postgraduate between the periods 1999 to 2012 are used. The interest of the study is to check whether the median completion time is within the maximum period (four years) stipulated by CAPES. Further, we wish to verify whether gender and age in the year of first registration exert some influence on the completion time of the students and if there is interaction between these explanatory variables. Thus, the response variable was defined as the time (in months) of first registration until the end of the PhD (date of defense). However, students who dropped out or who abandoned the course and did not return or who have not completed the course during this period are considered censored times. So, we define the following variables, y: logarithm of time (months), x_1: age (years) and x_2: gender (0= female, 1=male).

In many applications there is qualitative information about the hazard shape, which can help with selecting a particular model. In this context, a device called the total time on test (TTT) plot (Aarset 1987) is useful. The TTT plot is obtained by plotting $\mathsf{G}(r/n) = [(\sum_{i=1}^{r} T_{i:n}) + (n-r)T_{r:n}]/(\sum_{i=1}^{n} T_{i:n})$, where $r = 1, \ldots, n$ and $T_{i:n}$, $i = 1, \ldots, n$ are the order statistics of the sample, against r/n (Mudholkar *et al*, 1996).

First, to verify the behavior of the PhD data, the Kaplan-Meier and the TTT curves are displayed in Figure 6. From these plots, it is noted that the TTT-plot indicates that the time until the end of the PhD of students presents an increasing failure rate (Figure 6a). Moreover, the time not present a level above zero as shown in Figure 6b. There will be evidence that the gender of the students did affect the time until the end of the PhD. Thus, in accordance with what was observed in Figure 6, we can consider the following model:

$$y_i = \beta_0 + \beta_1 x_{1i} + \beta_2 x_{2i} + \beta_3 x_{1i} x_{2i} + \sigma z_i, \quad i = 1, \ldots, 49.$$

To maximize the function (13) and obtain the MLEs of the parameters of the proposed model, we use the `MaxBGFS` subroutine of the matrix programming language `Ox` version `6.20` with initial values $\phi = 1.000$, $\sigma = 0.760$, $\beta_0 = 4.880$, $\beta_1 = -0.153$ and $\beta_2 = 0.026$, obtained from the fit of the Weibull regression model in software `R` (version `2.15.1`). Thus, Table 1 provides the parameter estimates of the models, standard errors and significance of the parameters for the MLEs and non-parametric bootstrap estimates. By examining the figures in this table, we conclude that the estimates by the two methods are very similar. So, there is an evidence that the presence of the interaction between age and gender at the time of completion of a PhD degree.

The next step is to detect possible influential points in the LGW regression model. The measurements of global and local influence are calculated using the matrix programming language `Ox` (version `6.20`). Generalized Cook's distance (14) and likelihood distance (15) are displayed in Figure 7. From this figure, it is noted that the cases ♯8, ♯18 and ♯21 are possible influential observations. For the local influence plots, considering perturbation of cases ($C_{\mathbf{d}_{max}} = 2.1805$), the logarithm of time perturbation ($C_{\mathbf{d}_{max}} = 0.3925$) and explanatory variable perturbation x_1 ($C_{\mathbf{d}_{max}} = 4.4124$), it is noted that the points ♯18 and ♯21 can be considered as possible influential observations as illustrated in Figures 8-9. On the other hand, the plot of the deviance residuals versus the fitted values is displayed in Figure 10. It is observed that there is evidence that the observations ♯7, ♯18 and ♯21 are discrepant (Figure 10a).

Therefore, the sensitivity analysis (global influence and local influence) and residual analysis detect points ♯18 and ♯21 as possible influential observations. These observations identified as potential influential points correspond to the students who have the following descriptions:

i. The observation 18 corresponds to a 31 years old female student, who defended her PhD in a maximum time (60 months).

ii. The observation 21 corresponds to a 31 years old male student, who defended his PhD in a minimum time (35 months).

Thus, to analyze the impact of these observations on the parameter estimates, we adjust the model eliminating individually each observation, and then removing the two observations. In Table 2, we present relative changes (in percentages) of the estimates defined by $\mathbf{RC}_{\theta_j} = \left[(\hat{\theta}_j - \hat{\theta}_{j(i)})/\hat{\theta}_j\right] \times 100$, where $\theta_{j(i)}$ is the MLE without the ith observation.

On the figures of Table 2, we note that the MLEs of the parameters of the LGW regression model are robust to the deletion of influential observations. Moreover, the significance of the estimates of the parameters does not change (at the 5% significance level) after removal of the cases, that is, no changes inferential after removal of observations considered influential in diagnostics plots. Therefore, the observations are kept in the data set.

Finally, we verify the quality of the adjustment range of the LGW regression model by constructing in Figure 11 the normal probability plot for the component of the waste diversion with simulated envelope (Atkinson, 1985). This figure reveals that there is evidence of a good fit of the LGW regression model to the current data. Thus, the fitted model to the data can be expressed as

$$\hat{y}_i = 4.1351 - 0.0045\, x_{i1} - 0,4237\, x_{i2} + 0,0101\, x_{i1}\, x_{i2}, \quad i = 1, \ldots, 49.$$

Thus, according to this model, the time until the end of the PhD for male students varies with age, in this case, older students have the highest degree of time than the younger students.

9. Concluding Remarks

In this paper, we derive explicit expressions for the raw and incomplete moments, quantile and generating functions and mean deviations of the log-gamma Weibull distribution. Based on this distribution, we construct a new log-gamma Weibull regression model to investigate the effect of the age and gender at the time until the end of a PhD student. We also provide diagnostic measures to test the adequacy of the fitted model. In particular, the estimates of the parameters of the new model obtained by two methods are similar. Further, we provide diagnostic analysis for the fitted model to the the times of the students until the end of their PhD. Considering these aspects, we conclude that the fitted model can explain the effect of the variables on the time and that there is an interaction between age and gender of the students at the time until the end of the PhD.

Acknowledgment

This work was supported by FAPESP grant 2010/04496-2, Brazil.

Appendix A

Theorem G. Let $(\Omega, \mathcal{F}, \mathbf{P})$ be a given probability space and let $I = [d, e]$ be an interval for some $d < e$ ($d = -\infty$, $e = \infty$ might as well be a Let $X : \Omega \to I$ be a continuous random variable with the distribution function F and let q_1 and q_2 be two real functions defined on I such that

$$\mathbf{E}\left[q_2(X) \mid X \geq x\right] = \mathbf{E}\left[q_1(X) \mid X \geq x\right] \eta(x), \quad x \in I,$$

is defined with some real function η. Assume that $q_1, q_2 \in C^1(H)$, $\eta \in C^2(H)$ and F is twice continuously differentiable and strictly monotone function on the set I. Finally, assume that the equation $q_1\eta = q_2$ has no real solution in the interior of I. Then F is uniquely determined by the functions q_1 , q_2 and η, particularly

$$F(x) = \int_d^x C\left|\frac{\eta'(u)}{\eta(u)q_1(u) - q_2(u)}\right| \exp(-s(u))\ du\ ,$$

where the function s is a solution of the differential equation $s' = \frac{\eta' q_1}{\eta q_1 - q_2}$ and C is the normalization constant, such that $\int_I dF = 1$.

Appendix B

The Ox 6.21 (object-oriented matrix programming language) code for the simulation study.

B.1. LGLL model

```
#include<oxstd.h>
#include<oxdraw.h>
#include<oxprob.h>
#include<oxfloat.h>
#include<maximize.h>
#include<simula.h>
#pragma link("maximize.oxo")
static decl g_mT;
static decl n;
log_vero(const vP, const adFunc, const avScore, const amHessiam){
decl cont,y,x,xx1,xbeta,z;
decl uns=ones(n,1);
decl vero=zeros(1,n);
decl phi=vP[0];
decl sigma=vP[1];
decl beta0=vP[2];
decl beta1=vP[3];
//
for(cont=0;cont<n;++cont){
y=g_mT[cont][0];
xx1=g_mT[cont][2];
xbeta=beta0+(beta1*xx1);
z=(y-xbeta)/sigma;
x=log(1+exp(z));
//
if(g_mT[cont][1]==1){
vero[0][cont]=-log(sigma)-loggamma(phi)+z-(2*x)+((phi-1)*log(x+1e-10));
}
else{
vero[0][cont]=log(1-probgamma(x,phi,1)+1e-10);
}
}
adFunc[0]=double(vero*uns);
return 1;
}
main(){
ranseed("GM");

// Variaveis locais
decl i,j,theta0,g_mC,g_mZ,g_mY,g_mD,x1,pcens,cont,mu,xbeta,z,x,Sy;

// Numero de simulacoes
```

```
decl r=1000;

// Tamanho da mostra
n=100;

// Porcentagem de censura
pcens=0;

// Parametros da distribuicao
//decl phi1=0.8;
decl phi1=1.5;
decl sig1=1;
decl b0=2;
decl b1=4;
theta0=<1.5;1;2;4>;
//
decl theta=zeros(r,4,0);
decl rmi=zeros(n,r);
decl rm_ord=zeros(n,r);
decl rdi=zeros(n,r);
decl rd_ord=zeros(n,r);
decl nc=pcens*n;
decl testecens=zeros(n,r);
decl testey=zeros(n,r);
decl testex=zeros(n,r);

j=0;
do{
g_mC=zeros(n,1);
g_mZ=zeros(n,1);
g_mD=ones(n,2);
//
i=0;
while(i<n){
// Valores para phi=0.8
//decl xi=-10+(20*ranu(1,1));
//decl yi=0.237*ranu(1,1);
//
// Valores para phi=1.5
decl xi=-10+(20*ranu(1,1));
decl yi=0.265*ranu(1,1);
decl fdp=(1/gammafact(phi1))*exp(xi)*((1+exp(xi))^(-2))*((log(1+exp(xi)+1e-10))^(phi1-1));
if(yi<=fdp){
g_mZ[i]=xi;
i++;
}
}
x1=ranu(n,1);
mu=b0+(b1*x1);
g_mY=mu+(sig1*g_mZ);
//
cont=0;
for(i=0;i<n;i++){
if(cont<nc){
decl cte=5;
g_mC[i]=cte*ranu(1,1);
if(g_mY[i]<g_mC[i]){
```

```
g_mD[i][0]=g_mY[i];
}
else{
g_mD[i][0]=g_mC[i];
g_mD[i][1]=0;
cont=cont+1;
}
}
else{
g_mD[i][0]=g_mY[i];
}
}
print("Contador ", cont);
g_mT=g_mD~x1;
//
//MaxControl(-1,50);
decl vp,dfunc,ir,mhess;
vp=<1.5;1;2;4>;
ir = MaxBFGS(log_vero, &vp, &dfunc, 0, TRUE);
//
// Se convergir
if(ir==MAX_CONV){
print("\n ============================= ", j, " Replica ============================");
theta[j][0]=vp[0];
theta[j][1]=vp[1];
theta[j][2]=vp[2];
theta[j][3]=vp[3];
//
//print("\n Dados simulados ", g_mY);
//
print("\nCONVERGENCE: ",MaxConvergenceMsg(ir) );
print("\nMaximized log-likelihood: ", "%7.3f", dfunc);
print("\nML estimate: ", "%6.3f", vp);
//-------------------- Calculando os residuos -----------------------
for(i=0;i<n;i++){
decl pphi=vp[0];
decl psigma=vp[1];
decl pb0=vp[2];
decl pb1=vp[3];
decl cens=g_mT[i][1];
//
xbeta=pb0+(pb1*g_mT[i][2]);
z=(g_mT[i][0]-xbeta)/psigma;
x=log(1+exp(z));
Sy=1-probgamma(x,pphi,1);
if(cens==1){
rmi[i][j]=1+log(Sy);
if(rmi[i][j]==1){
rmi[i][j]=0.9999999999;
}
rdi[i][j]=(rmi[i][j]/fabs(rmi[i][j]))*((-2*(rmi[i][j]+log(1-rmi[i][j])))^(1/2));
}
else{
rmi[i][j]=log(Sy);
if(rmi[i][j]==0){
rmi[i][j]=-1e-5;
}
```

```
rdi[i][j]=(rmi[i][j]/fabs(rmi[i][j]))*((-(2*rmi[i][j]))^(1/2));
}
//
}
rm_ord[][j]=sortc(rmi[][j]);
rd_ord[][j]=sortc(rdi[][j]);
j++;
}
}while(j<r);
print("\n ------------------------ Resultado final ---------------------------");
//------------------- Residuos ----------------------------
decl rmResult=zeros(n,3);
decl rdResult=zeros(n,3);
for(i=0;i<n;i++){
rmResult[i][0]=min(rm_ord[i][]);
rmResult[i][1]=meanr(rm_ord[i][]);
rmResult[i][2]=max(rm_ord[i][]);
//
rdResult[i][0]=min(rd_ord[i][]);
rdResult[i][1]=meanr(rd_ord[i][]);
rdResult[i][2]=max(rd_ord[i][]);
}
//print("\n Residuo Rd ", rdResult);
//------------------- Media e variancia das r simulacoes ----------------
decl media=meanc(theta);
decl variancia=varc(theta);
decl medidas=media|variancia;
print("\nML estimate : ", media);
//------------------- Erro quadratico medio --------------------------
decl EQM=zeros(1,4,0);
for(j=0;j<4;j++){
decl vicio=media[j]-theta0[j];
EQM[0][j]=variancia[j]+(vicio^2);
}
print("\n EQM: ", meanc(EQM));
```

References

Aarset, M. V. (1987). How to identify a bathtub hazard rate. *IEEE Transactions on Reliability, 36*, 106-108. http://dx.doi.org/10.1109/TR.1987.5222310

Atkinson, A. C. (1985). *Plots, transformations and regression: an introduction to graphical methods of diagnostic regression analysis*. University Press: Oxford.

Bonat, W. H., Krainski, E. T., Ribeiro Jr, P. J., & Zeviani, W. M. (2012). *Mtodos computacional em inferncia estat stica*. Região Brasileira da Sociedade Internacional de Biometria: Piracicaba.

Carrasco, J. M. F., Ortega, E. M. M., & Paula, G. A. (2008). Log-Modified Weibull Regression Models with Censored Data: Sensitivity and Residual Analysis. *Computational Statistics and Data Analysis, 52*, 4021-4029. http://dx.doi.org/10.1016/j.csda.2008.01.027

Collett, D. (2003). *Modeling survival data in medical research*. Chapman & Hall: London.

Colosimo, E. A., & Giolo, S. R. (2006). *Anlise de sobrevivncia aplicada*. Edgard Blcher: São Paulo.

Cook, R. D., & Wweisberg, S. (1982). *Residuals and influence in regression*. Chapman and Hall: New York.

Cox, D. R., & Oakes, D. (1984). *Analysis of survival data*. Chapman and Hall: New York.

Doornik, J. A. (2007). *An Object-Oriented Matrix Language Ox 5*. Timberlake Consultants Press: London.

Efron, B. (1979). Bootstrap methods: another look at the jackknife. *The Annals of Statistics, 7*, 1-26. http://dx.doi.org/10.1214/aos/1176344552

Efron, B., & Tibshirani, R. J. (1993). *An introduction to the bootstrap*. Chapman & Hall: New York. http://dx.doi.org/10.1007/978-1-4899-4541-9

Ghosh, S. (2013). Normality testing for a long-memory sequence using the empirical moment generating function. *Journal of Statistical Planning and Inference, 143*, 944-954. http://dx.doi.org/10.1016/j.jspi.2012.10.016

Glänzel, W. (1987). A characterization theorem based on truncated moments and its application to some distribution families. *Mathematical Statistics and Probability Theory (Bad Tatzmannsdorf, 1986), Vol. B*. Reidel, Dordrecht, 75-84. http://dx.doi.org/10.1007/978-94-009-3965-3_8

Glänzel, W. (1990). Some consequences of a characterization theorem based on truncated moments, *Statistics: A Journal of Theoretical and Applied Statistics, 21*, 613-618.

Gradshteyn, I. S., & Ryzhik, I. M. (2000). *Table of Integrals, Series and Products*. Academic Press: San Diego.

Kalbfleisch, J. D., & Prentice, R. L. (2002). *The statistical analysis of failure time data*. John Wiley: New York. http://dx.doi.org/10.1002/9781118032985

Hashimoto, E. M., Cordeiro, G. M., & Ortega, E. M. M. (2013). The new Neyman type A beta Weibull model w ith long-term survivors.*Computational Statistics, 28*, 933-954. http://dx.doi.org/10.1007/s00180-012-0338-9

Horta, J. S. B., & Moraes, M. C. M. (2005). O sistema CAPES de avaliaão da ps-graduaão: da rea de educaão ?grande rea de cincias humanas. *Revista Brasileira de Educaão, 30*, 95-181.

Lawless, J. F. (2003). *Statistical models and methods for lifetime data*. John Wiley & Sons: New Jersey.

Leiva, V., Barros, M., Paula, G. A., & Galea, M. (2007). Influence Diagnostics in Log-Birnbaum-Saunders Regression Models with Censored Data. *Computational Statistics and Data Analysis, 51*, 5694-5707. http://dx.doi.org/10.1016/j.csda.2006.09.020

Meintanis, S. G. (2010). Testing skew normality via the moment generating function. *Mathematical Methods of Statistics, 19*, 64-72. http://dx.doi.org/10.3103/S1066530710010047

Moreira, N. P., Silveira, S. F. R., Ferreira, M. A. M., & Cunha, N. R. S. (2010). Eficincia e qualidade dos programas de ps-gradua ão das instituiões federais de ensino superior usurias do programa de fomento ?ps-graduaão. *Ensaio: Avaliaão e Polticas Pblicas em Educaão, 18*, 365-388.

Mudholkar, G. S., Srivastava, D. k., & Kollia, G. D. (1996). A generalization of the Weibull distribution with application to the analysis of survival data. *Journal of the American Statistical Association, 91*, 1575-1583. http://dx.doi.org/10.1080/01621459.1996.10476725

Nadarajah, S, Cordeiro, G. M., & Ortega, E. M. M. (2015). The Zografos-Balakrishnan-G Family of Distributions: Mathematical Properties and Applications. *Communications in Statistics - Theory and Methods, 44*, 186-215.

Ortega, E. M. M., Cordeiro, G. M., & Hashimoto, E. M. (2011). A log-linear regression model for the beta-Weibull distribution. *Communication in Statistics: Simulation and Computation, 40*, 1206-1235. http://dx.doi.org/10.1080/03610918.2011.568150

Prudnikov, A. P., Brychkov, Y. A., & Marichev, O. I. (1986). *Integrals and series*. Vol. I, Taylor & Francis: London.

Ristic, M. M., & Balakrishnan, N. (2012). The gamma-exponentiated exponential distribution. *Journal of Statistical Computation and Simulation, 82*, 1191-1206. http://dx.doi.org/10.1080/00949655.2011.574633

Ross, S. M. (2006). *Simulation*. Elsevier Academic Press: Boston.

Silva, G. O., Ortega, E. M. M., Garibay, V. C., & Barreto, M. L. (2008). Log-Burr XII regression models with censored Data. *Computational Statistics and Data Analysis, 52*, 3820-3842. http://dx.doi.org/10.1016/j.csda.2008.01.003

Silva, G. O., Ortega, E. M. M., & Cordeiro, G. M. (2010). The beta modified Weibull distribution. *Lifetime Data Analysis, 16*, 409-430. http://dx.doi.org/10.1007/s10985-010-9161-1

Silva, G. O., Ortega, E. M. M., & Paula, G. A. (2011). Residuals for log-Burr XII regression models in survival analysis. *Journal of Applied Statistics, 38*, 1435-1445. http://dx.doi.org/10.1080/02664763.2010.505950

Stacy, E. W. (1962). A generalization of the gamma distribution. *The Annals of Mathematical Statistics, 33*, 1187-1192. http://dx.doi.org/10.1214/aoms/1177704481

Weisberg, S. (2005). *Applied linear regression*. John Wiley & Sons: New York. http://dx.doi.org/10.1002/0471704091

Xie, F. C., & Wei, B. C. (2007). Diagnostics analysis in censored generalized Poisson regression model. *Journal of Statistical Computation and Simulation, 77*, 695-708. http://dx.doi.org/10.1080/10629360600581316

Zeviani, W. M. (2012). *Estatstica computacional*. Available in:<http://www.leg.ufpr.br/ ~walmes /ensino/ce083-2012-01/ >. Access in: 30 set. 2012

Zografos, K., & Balakrishnan, N. (2009). On families of beta-and generalized gama-generated distributions and associated inference. *Statistical Methodology, 6*, 344-362. http://dx.doi.org/10.1016/j.stamet.2008.12.003

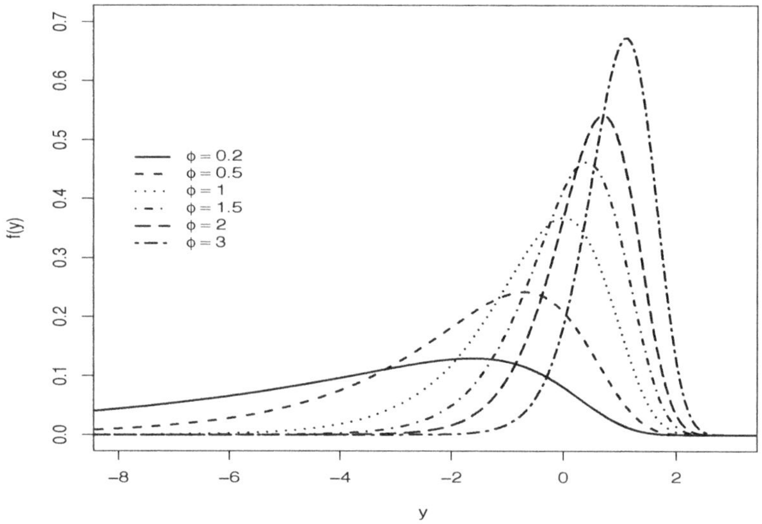

Figure 1. The log-gama-Weibull density function

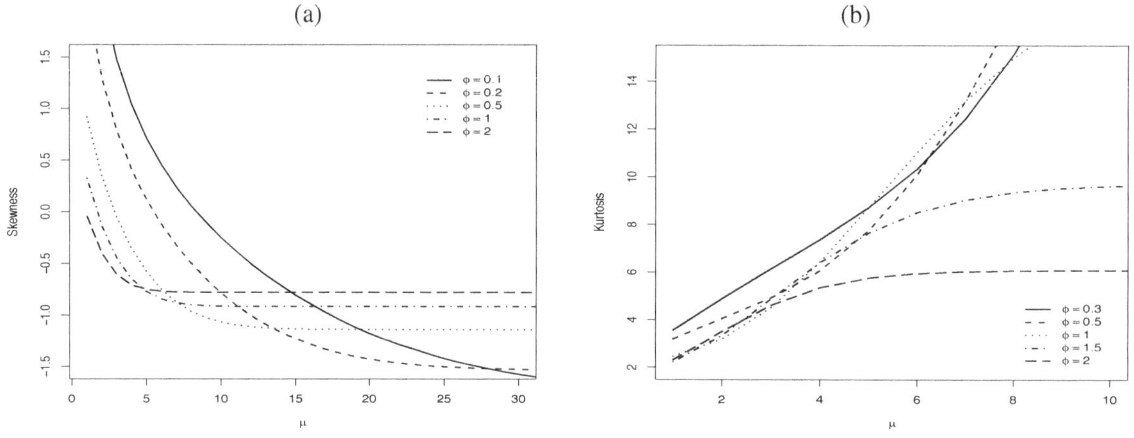

Figure 2. Skewness and kurtosis of the LGW distribution as functions of μ with $\sigma = 1.5$ and different values of ϕ.

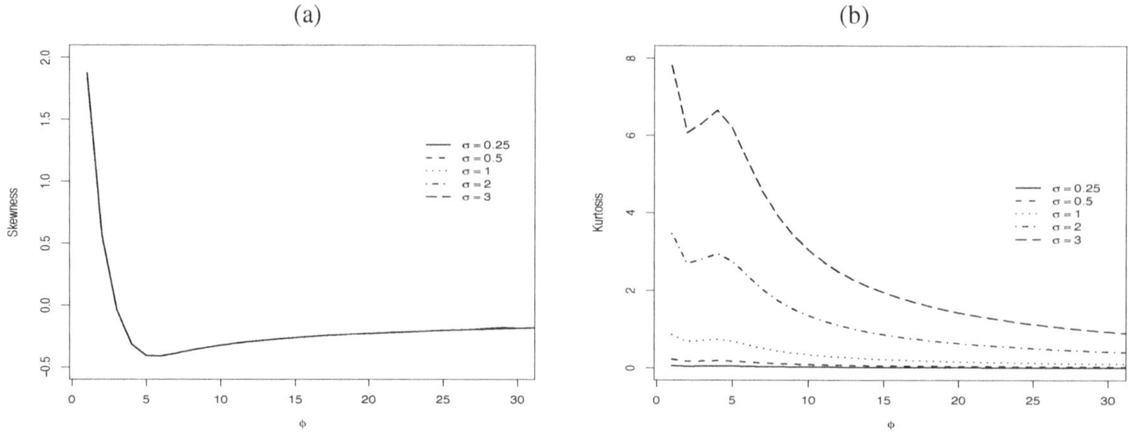

Figure 3. Skewness and kurtosis of the LGW distribution as functions of ϕ with $\mu = 0$ and different values of σ.

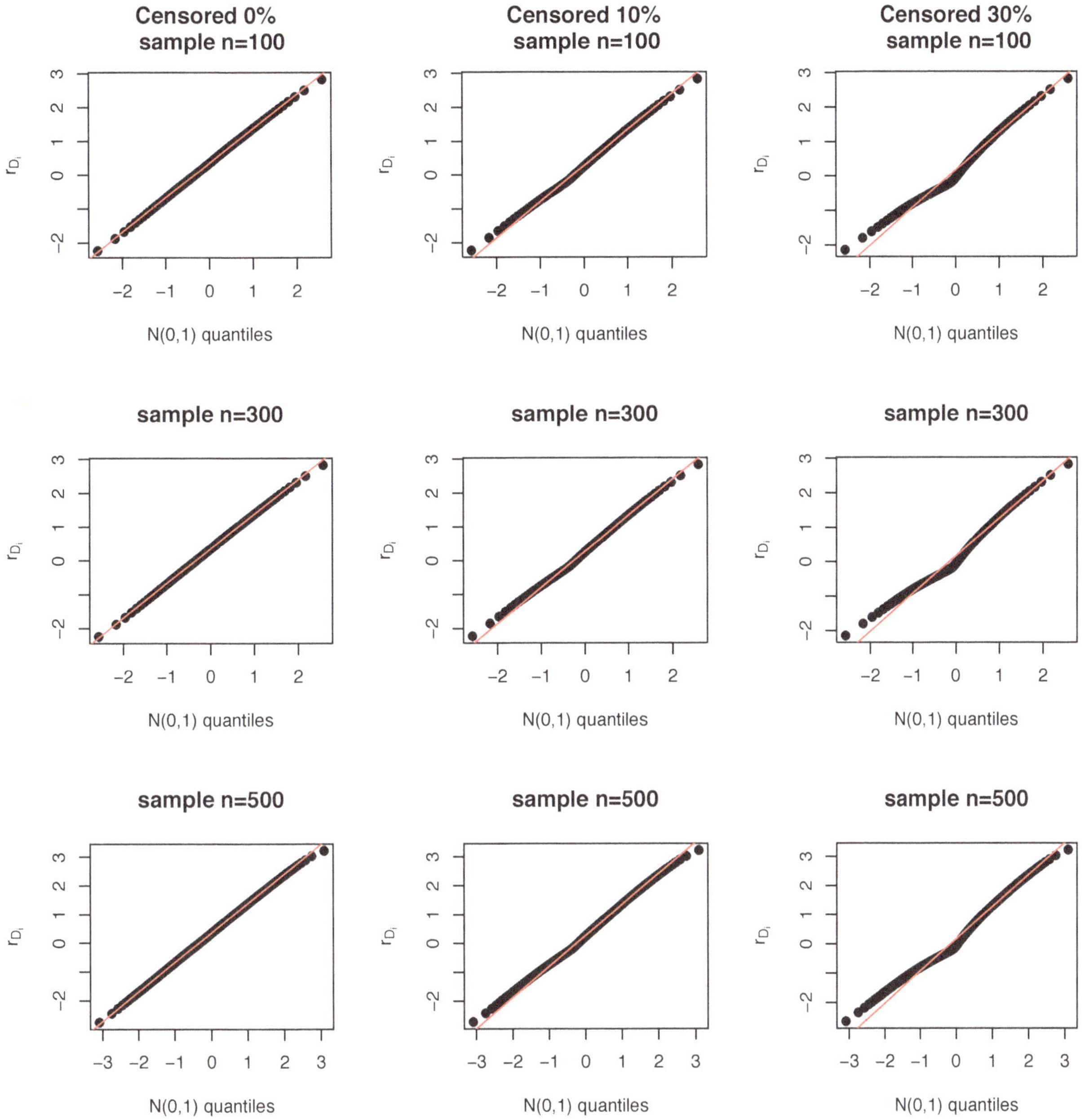

Figure 4. Normal probability plots for r_{D_i} in the LGW regression model with $\phi = 0.8$

Table 1. MLEs and non-parametric bootstrap estimates for the parameters of the regression model for the current data.

	MLEs			Non-parametric bootstrap		
Parameter	Estimate	S.E.	p-value	Estimate	S.E.	95% C.I. BCa
ϕ	0.9520	1.0377	–	1.5895	0.8656	(0.2163, 1.6857)
σ	0.0630	0.0445	–	0.0801	0.0285	(0.0172, 0.0942)
β_0	4.1351	0.1590	0.0000	4.0647	0.0919	(4.0517, 4.3508)
β_1	-0.0045	0.0026	0.0884	-0.0040	0.0018	(-0.0088, -0.0023)
β_2	-0.4237	0.1112	0.0001	-0.4124	0.1061	(-0.6320, -0.2734)
β_3	0.0101	0.0030	0.0006	0.0098	0.0026	(0.0066, 0.0155)

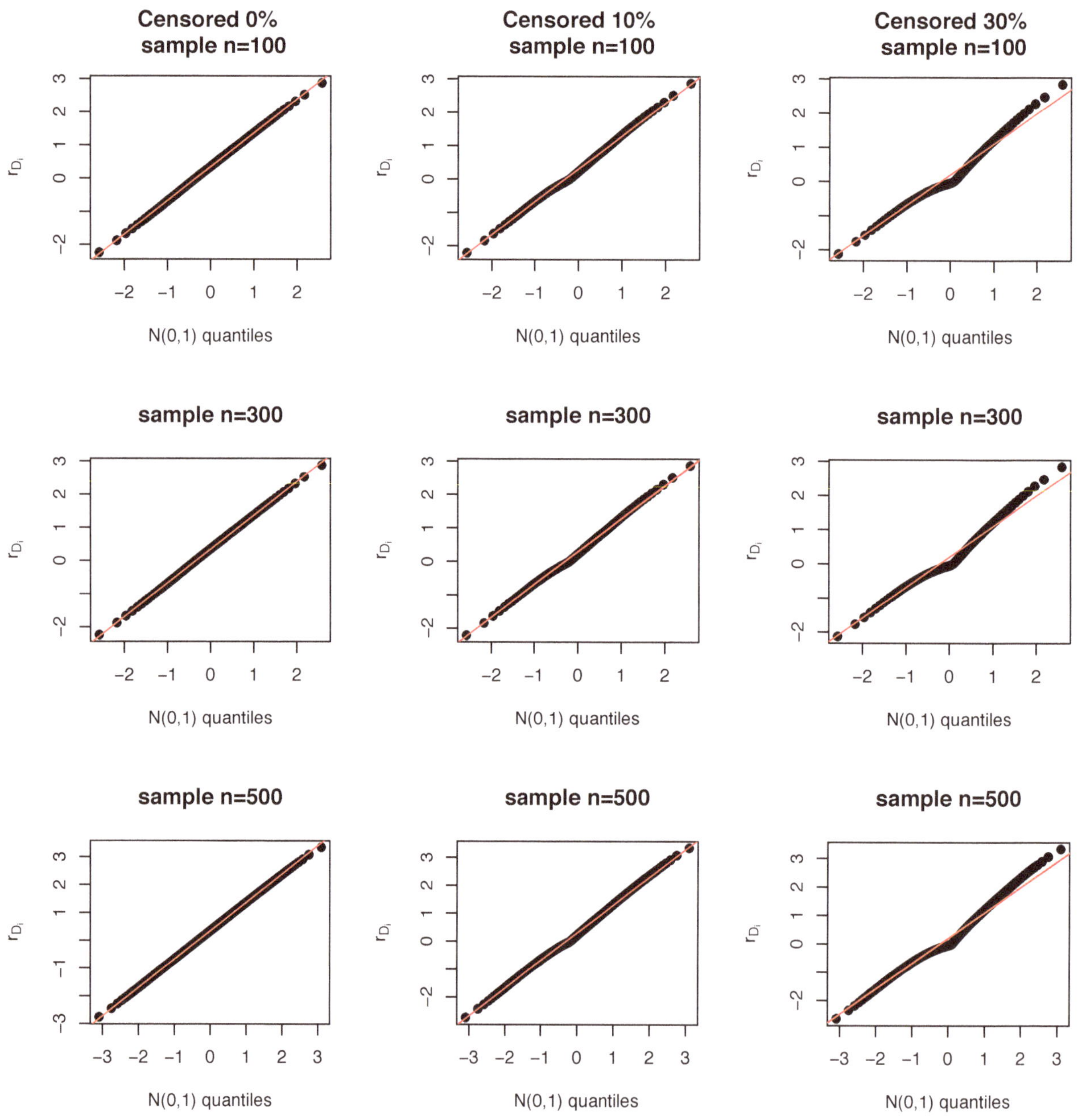

Figure 5. Normal probability plots for r_{D_i} in the LGW regression model with $\phi = 1.5$

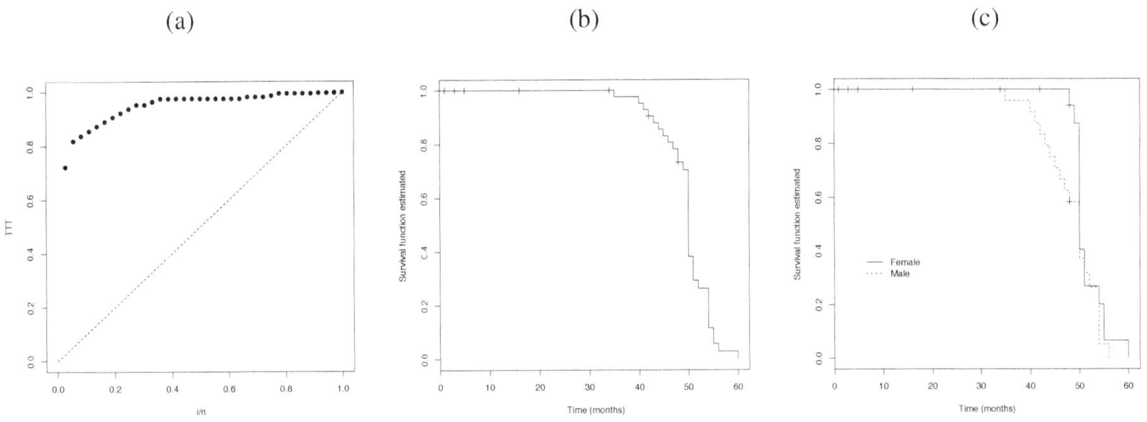

Figure 6. (a) TTT-plot. Survival curve estimate by Kaplan-Meier method for: (b) time. (c) Explanatory variable (gender)

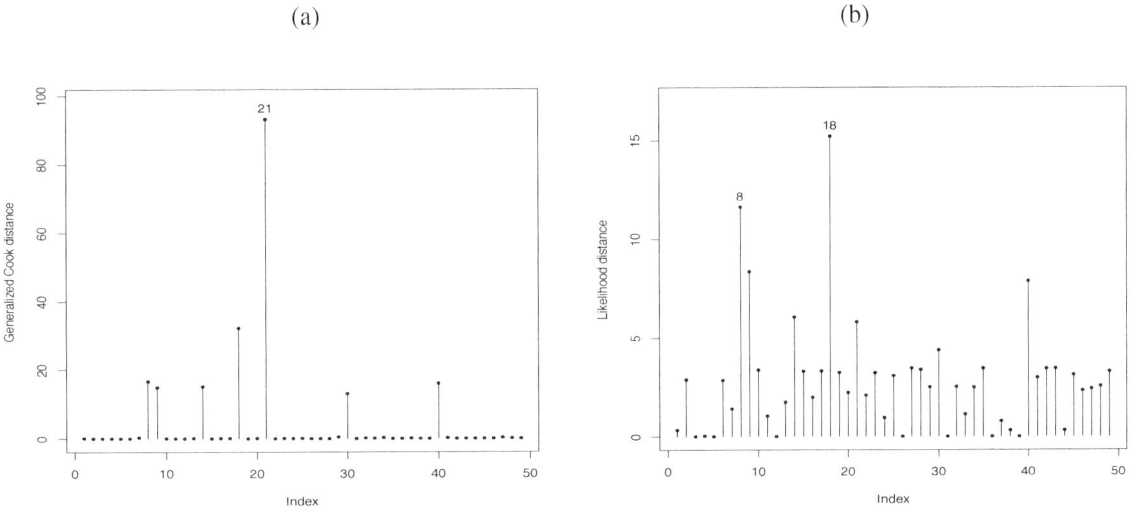

Figure 7. Index plot of global influence from the regression model fitted to the current data. (a) Generalized Cook distance. (b) Likelihood distance

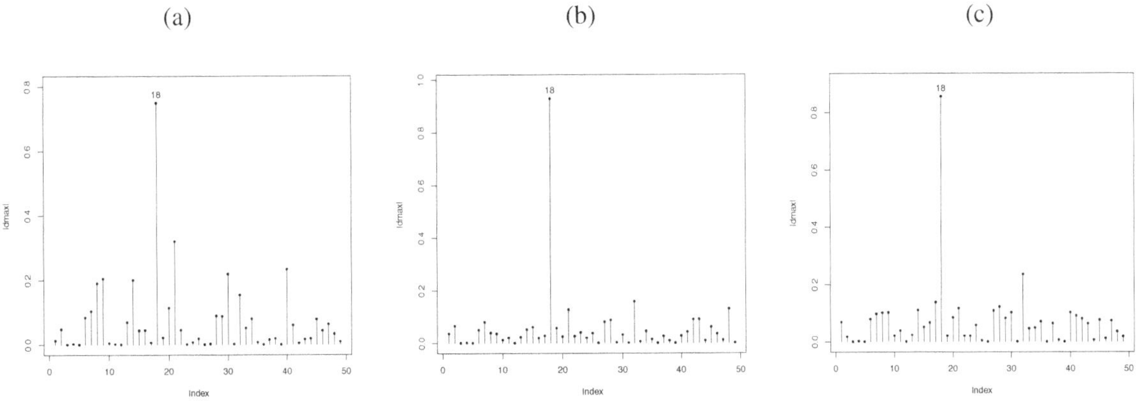

Figure 8. Index plot of $\mathbf{d}_{max}$ from the regression model fitted to the current data. (a) Case-weight perturbation. (b) Response variable perturbation. (c) Explanatory variable perturbation, x_1

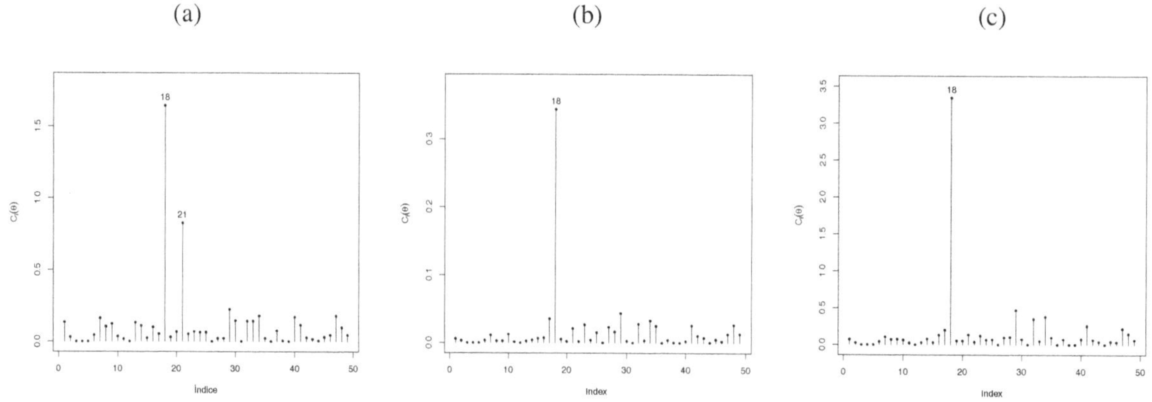

Figure 9. Index plot of local total C_i from the regression model fitted to the current data. (a) Case-weight perturbation. (b) Response variable perturbation. (c) Explanatory variable perturbation, x_1

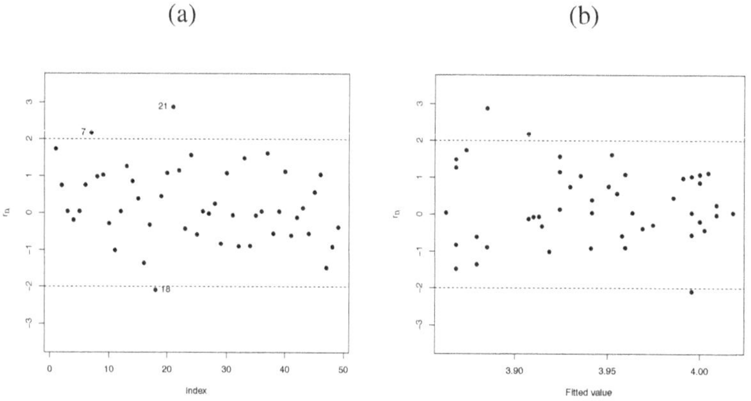

Figure 10. Residual analysis of the LGW regression model fitted to the current data. (a and b) Deviance residual

Table 2. Relative changes [-**RC**-in %], estimates and the corresponding p-values in parentheses for the regression coefficients to explain the logarithm of time.

Set	$\hat{\phi}$	$\hat{\sigma}$	$\hat{\beta}_0$	$\hat{\beta}_1$	$\hat{\beta}_2$	$\hat{\beta}_3$
A	-	-	-	-	-	-
	0.9520	0.0630	4.1351	-0.0045	-0.4237	0.0101
	(-)	(-)	(0.0000)	(0.0884)	(0.0001)	(0.0006)
A-{♯18}	[4]	[15]	[3]	[53]	[27]	[25]
	0.9257	0.0536	4.0282	-0.0021	-0.3096	0.0076
	(-)	(-)	(0.0000)	(0.2810)	(0.0007)	(0.0019)
A-{♯21}	[-351]	[-120]	[6]	[22]	[6]	[6]
	4.3450	0.1388	3.8747	-0.0035	-0.3981	0.0095
	(-)	(-)	(0.0000)	(0.1625)	(0.0004)	(0.0016)
A-{♯18, ♯21}	[49]	[51]	[2]	[51]	[31]	[29]
	0.4919	0.0310	4.0603	-0.0022	-0.2913	0.0072
	(-)	(-)	(0.0000)	(0.1892)	(0.0003)	(0.0009)

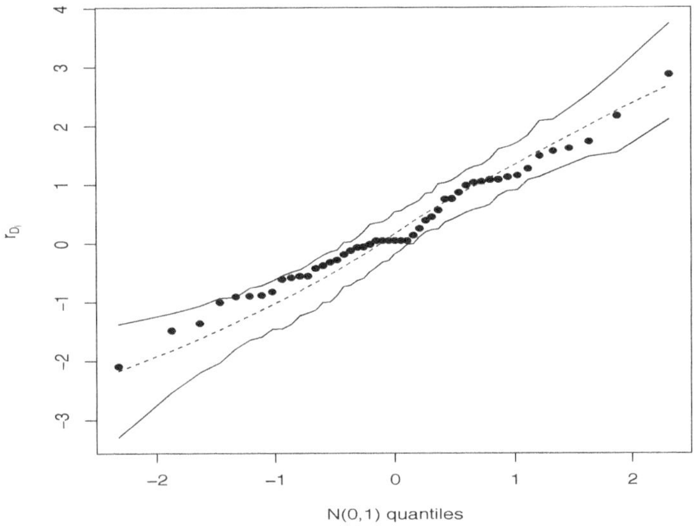

Figure 11. Normal probability plot for the deviance residual with envelopes

12

Extreme Value Theory: a New Characterization of the Distribution Function for the Mixed Method

Kané Ladji[1], Diawara Daouda[2], & Diallo Moumouni[3]

[1,3] Faculty of Economics and Management (F. S. E. G) Bamako-Mali

[2] Zhongnan University of Economics and Law Wuhan China

Correspondence: Kané Ladji, Assistant Master Professor, Faculty of Economics and Management, Bamako-Mali. E-mail: fsegmath@gmail.com

Abstract

Consider the sample $X_1, X_2, \dots, X_N$ of N independent and identically distributed (iid) random variables with common cumulative distribution function (cdf) F, and let F_u be their conditional excess distribution function. We define the ordered sample by $X_1 \le X_2 \le \dots \le X_N$. Pickands (1975), Balkema and de Haan (1974) posed that for a large class of underlying distribution functions F, and large u, F_u is well approximated by the Generalized Pareto Distribution. The mixed method is a method for determining thresholds. This method consists in minimizing the variance of a convex combination of other thresholds.

The objective of the mixed method is to determine by which probability distribution one can approach this conditional distribution. In this article, we propose a theorem which specifies the conditional distribution of excesses when the deterministic threshold tends to the end point.

Keywords: distribution function, Generalized Pareto Distribution (GPD), Mixed Method (MM).

2000 Mathematics Subject Classifications: 60G52, 60G70, 62G20, 62G32, 60E07, 62E20

1. Introduction

Pareto distribution is traditionally used by reinsurer's excess of loss mainly because of its good mathematical properties, particularly from the simplicity of the formulas resulting from its application. The new mixed method (MM) was proposed in [1, 2, 3, 4] to determine a threshold $U = \sum_{k=1}^{p} \alpha_k U_k + \alpha_3 U_3$ with $1 \le p \le 2$, at which a unit is declared atypical minimizing the variance of a convex combination of thresholds obtained by the mean excess function and generalized Pareto distribution (extreme quantile were estimated with a probability of 99.9% being an extreme value for the distribution of amounts of sinister with a confidence level of 95%). This method allows a compromi between the GPD method and FME method, between a minimum strategy GPD and maximum strategy FME (Mean Excess Function). It is more correlated with the GPD method and relatively smooth.

2. Method

This article focuses on two major paragraphs. The first paragraph (see paragraph 3.1) is based on determining a threshold $U = \sum_{k=1}^{p} \alpha_k U_k + \alpha_3 U_3$ with $1 \le p \le 2$ by the mixed method (MM) and last paragraph (see paragraph 3.2) is to determine a distribution function of the laws of the mixed method. Let U_3: the threshold beyond which a unit is declared as extreme, obtained by the GPD function and U: the threshold beyond which a unit is declared as extreme, obtained by the mixed method (MM).Let $X_1, X_2, \dots, X_N$ N random variables (iid) common distribution function F. We are looking from the distribution F of X to define a conditional distribution F_{U_3} compared to U_3 threshold for random variables exceeding this threshold. It defines the excess over the threshold U_3 as the set of random variables y_j defined by: $y_j = X_j - U_3$ for $j \in E(U\_3) = \{j \in \{1,2,\dots,N\}/Xj > U3\}$. The function of distribution of the excess over the threshold U_3 is defined by:

$$F_{U_3}(y) = P(X - U_3 \leq y/X > U_3) = \begin{cases} \dfrac{F(U_3 + y) - F(U_3)}{1 - F(U_3)} & if \quad y \geq 0, \\ 0 & if \quad y < 0, \end{cases}$$

Thus, for large threshold U_3, the law of excess is approximated by a generalized Pareto law:

$$F_{U_3}(y) \approx F^{GPD}_{\xi,\sigma(U_3)}(y).$$

In this article, we will show that:

$$F_U(y) \approx F^{MM}_{\xi,\sigma(\mathrm{U})}(y),$$

where $F^{MM}_{\xi,\sigma(\mathrm{U})}(y)$ is the distribution function of the law of the mixed method and U is the threshold beyond which a unit is declared as extreme, obtained by the mixed method (MM).

Theorem Pickands (1975), Balkema and de Haan (1974) assures us that the law of the excess may be approaching a generalized Pareto law. In this article, we will use the theorem Pickands (1975), Balkema and de Haan (1974) to show that the law of the excess can be approached by a law of the mixed method.

3. Results

3.1. Determination of Threshold U By the Mixed Method (MM)

The new mixed method (MM) was proposed in [1, 2, 3, 4] to determine a threshold $U = \sum_{k=1}^{p} \alpha_k U_k + \alpha_3 U_3$ with $1 \leq p \leq 2$, at which a unit is declared atypical minimizing the variance of a convex combination of thresholds obtained by the mean excess function and generalized Pareto distribution (extreme quantile were estimated with a probability of 99.9% being an extreme value for the distribution of amounts of sinister with a confidence level of 95%).

Let U_1 be the threshold beyond which a unit is declared as extreme, obtained by the record values, U_2 be the threshold beyond which a unit is declared as extreme, obtained by the mean excess function and U_3 the threshold beyond which a unit is declared as extreme, obtained by the GPD function with $U_1 < U_2 < U_3$. Let $U = \alpha U_p + (1 - \alpha)U_q$ with $0 < \alpha < 1$, minimizes the variance U, p, $q = 1, 2, 3$ and $p < q$. We get.

$$\alpha = \frac{V(X_{U_q}) - Cov(X_{U_p}, X_{U_q})}{V(X_{U_p}) + V(X_{U_q}) - 2Cov(X_{U_p}, X_{U_q})}$$

For $U = \alpha U_1 + (1 + \alpha)U_2 - 2\alpha U_3$ with $\alpha \in \mathbb{R}$. We get :

$$\alpha = \frac{-V(X_{U_2}) - Cov(X_{U_1}, X_{U_2}) + 2Cov(X_{U_2}, X_{U_3})}{V(X_{U_1}) + V(X_{U_2}) + 4V(X_{U_3}) + 2Cov(X_{U_1}, X_{U_2}) - 4Cov(X_{U_1}, X_{U_3}) - 4Cov(X_{U_2}, X_{U_3})}$$

Consider the sample $X_1, X_2, \dots, X_N$ of N independent and identically distributed (iid) random variables. We define the ordered sample by $X_1 \leq X_2 \leq \cdots \leq X_N$. Let $X_{U_j}, j = 1,2,3$ thresholds obtained by different methods. We consider a statistical series to a variable X_{U_j} ,taking the amount $X_1, X_2, \dots, X_N$ and X_{U_j} ,which have been sorted in ascending order: $X_1 \leq X_2 \leq \cdots \leq X_k \leq U_j \leq \cdots \leq X_N$. We consider a statistical series 2 variables X and Y, taking the amount $X_1, X_2, \dots, X_N$ and $Y_1, Y_2, \dots, Y_N$. Which have been sorted in ascending order: $X_1 \leq X_2 \leq \cdots \leq X_N$ and $Y_1 \leq Y_2 \leq \cdots \leq Y_N$. We write:

- The means of X and Y are : $\bar{X} = \dfrac{1}{N}\sum_{i=1}^{N} X_i$ et $\bar{Y} = \dfrac{1}{N}\sum_{i=1}^{N} Y_i$

- The variances of X and Y are: $V(X)=\frac{1}{N}\sum_{i=1}^{N}(X_i-\bar{X})^2$ et $V(Y)=\frac{1}{N}\sum_{i=1}^{N}(Y_i-\bar{Y})^2$

- The covariance of X and Y is: $Cov(X,Y)=\frac{1}{N}\sum_{i=1}^{N}(X_i-\bar{X})(Y_i-\bar{Y})$

Example 1. Threshold Calculation

The data base provides a sample of 2020 observations for 4 wheel vehicle for personal use during the year 2013. The data come from a Malian insurance company and concern the amounts of claims caused by the insured of a risk class. This file contains only the amounts of claims during the insurance year. U_1 be the threshold beyond which a unit is declared as extreme, obtained by the record values. U_2 be the threshold beyond which a unit is declared as extreme, obtained by the mean excess function. U_3 the threshold beyond which a unit is declared as extreme, obtained by the GPD function and U the threshold beyond which a unit is declared as extreme, obtained by the method MM.

Let N be the number of claims and $X_1, X_2, \dots, X_N$ the realizations of X, which is the random variable representing the amounts of loss. As usual we assume mutual independence of random variables.

Table 1. Determination of threshold U by the mixed method (MM)

Record values	Mean excess function	GPD function	MM method
U_1	U_2	U_3	$U=\sum_{k=1}^{p}\alpha_k U_k+\alpha_3 U_3$ with $1\le p\le 2$
$U_1=11{,}5$		$U_3=12{,}5$	$U=\alpha U_1+(1-\alpha)U_3=11{,}88$ with $\alpha=0{,}69$
	$U_2=12$	$U_3=12{,}5$	$U=\alpha U_2+(1-\alpha)U_3=12{,}19$ with $\alpha=0{,}63$
$U_1=11{,}5$	$U_2=12$	$U_3=12{,}5$	$U=\alpha U_1+(1+\alpha)U_2-2\alpha U_3=12{,}10$ with $\alpha=-0{,}06$

3.2 Law (distribution) of The Mixed Method

In this section, we will give the main result of this paper is to write a new law of the mixed method (MM). Let U_3: the threshold beyond which a unit is declared as extreme, obtained by the GPD function and U: the threshold beyond which a unit is declared as extreme, obtained by the mixed method (MM). Let $X_1, X_2, \dots, X_N$ N random variables (iid) common distribution function F. We are looking from the distribution F of X to define a conditional distribution F_{U_3} compared to U_3 threshold for random variables exceeding this threshold. It defines the excess over the threshold U_3 as the set of random variables y_j defined by: $y_j=X_j-U_3$ for $j\in E_{U_3}=\{j\in\{1,2,\dots,N\}/X_j>U_3\}$. It defines the excess over the threshold U_3 as the set of random variables y_j defined by:

$$F_{U_3}(y) = P(X - U_3 \leq y/X > U_3) = \begin{cases} \dfrac{F(U_3 + y) - F(U_3)}{1 - F(U_3)} & if \quad y \geq 0, \\ 0 & if \quad y < 0, \end{cases}$$

The objective of the mixed method is to determine by which probability distribution one can approach this conditional distribution. In this article, we propose the following theorem (**Theorem 2**) which specifies the conditional distribution of excesses when the deterministic threshold tends to the end point X_F.

Theorem 1 (Pickands (1975), Balkema and de Haan (1974)): Let F_{U_3} be the conditional distribution of the excess over a threshold U_3, combined with unknown distribution function F. This function F belongs to the domain of attraction of G_ξ if and only if there exist a positive function σ such

$$\lim_{U_3 \to X_F} \text{Sup}_{0<y<X_F - U_3} \left| F_{U_3}(y) - F^{GPD}_{\xi,\sigma(U_3)}(y) \right| = 0,$$

Where $F^{GPD}_{\xi,\sigma(U_3)}$ is the distribution function of GPD, define by:

$$F^{GPD}_{\xi,\sigma(U_3)}(y) = \begin{cases} 1 - \left(1 + \dfrac{y\xi}{\sigma}\right)^{\frac{-1}{\xi}} & if\ \xi \neq 0 \\ 1 - e^{\frac{-y}{\sigma}} & if\ \xi = 0 \end{cases}$$

for $y \in [0, (X_F - U_3)]$ if $\xi \geq 0$ and $y \in \left[0, Min\left(\frac{-\xi}{\rho}, X_F - U_3\right)\right]$ if $\xi < 0$ with $X_F = sup\{X \in \mathbb{R}, F(X) < 1\}$

Theorem 2: Let F_U be the conditional distribution of the excess over a threshold $U = \sum_{k=1}^{p} \alpha_k U_k + \alpha_3 U_3$ with $1 \leq p \leq 2$, combined with unknown distribution function F. This function F belongs to the domain of attraction of G_ξ if and only if there exists a positive function σ such

$$\lim_{U \to X_F} \text{Sup}_{0<y<X_F - U} \left| F_U(y) - F^{MM}_{\xi,\sigma(U)}(y) \right| = 0,$$

Where $F^{MM}_{\xi,\sigma(U)}$ is the distribution function of mixed method (MM), define by:

$$F^{MM}_{\xi,\sigma(U)}(y) = \begin{cases} 1 - \left(1 + \dfrac{y\xi}{\sigma}\right)^{\frac{-1}{\xi}} & if\ \xi \neq 0 \\ 1 - e^{\frac{-y}{\sigma}}\ if\ \xi = 0 \end{cases}$$

for $y \in [0, (X_F - U)]$ if $\xi \geq 0$ and $y \in \left[0, Min\left(\frac{-\xi}{\rho}, X_F - U\right)\right]$ if $\xi < 0$ with $X_F = sup\{X \in \mathbb{R}, F(X) < 1\}$.

Proof: The conditional distribution F_U of the excesses above the threshold U with is defined by:

$F_U(y) = P(X - U \leq y/X > U) = \frac{F(U+y) - F(U)}{1 - F(U)}$ for $0 \leq y \leq X_F - U$.

This is equivalent to:

$F_U(x) = P(X < x/X > U) = \frac{F(x) - F(U)}{1 - F(U)}$ for $x \geq U$.

The proof of $F_U(y) \approx F^{MM}_{\xi,\sigma(U)}(y)$ results directly from the evidence of $F_{U_3}(y) \approx F^{GPD}_{\xi,\sigma(U_3)}(y)$ (theorem 1) and $U = \sum_{k=1}^{p} \alpha_k U_k + \alpha_3 U_3$ with $1 \leq p \leq 2$.

Example 2: Winting $F^{MM}_{\xi,\sigma(U)}(y)$.

U_1: be the threshold beyond which a unit is declared as extreme, obtained by the record values, *U_2*: be the threshold beyond which a unit is declared as extreme, obtained by the mean excess function, *U_3*: the threshold beyond which a unit

is declared as extreme, obtained by the GPD function and U: the threshold beyond which a unit is declared as extreme, obtained by the MM function. Let N be the number of claims and $X_1, X_2, \dots, X_N$ the realizations of X, which is the random variable representing the amounts of sinister.

Table 2. Knowing the parameters (σ, ξ) and the thresholds, we can write the distribution functions of GPD and MM.

Parameters GPD ξ, σ	Threshold GPD U_3	Threshold MM U	distribution function of GPD	distribution function of MM
			$F^{GPD}_{\xi,\sigma(U_3)}(y) = 1 - \left(1 + \frac{y\xi}{\sigma}\right)^{\frac{-1}{\xi}}$ With $y = X - U_3$	$F^{MM}_{\xi,\sigma(U)}(y) = 1 - \left(1 + \frac{y\xi}{\sigma}\right)^{\frac{-1}{\xi}}$ With $y = X - U$
$\xi = -0{,}3293$ $\sigma = 1{,}5576$	$U_3 = 12{,}5$	$U = 11{,}88$	$1 - \left(1 - \frac{0{,}3293y}{1{,}5576}\right)^{\frac{1}{0{,}3293}}$ $y = X - 12{,}5$	$1 - \left(1 - \frac{0{,}3293y}{1{,}5576}\right)^{\frac{1}{0{,}3293}}$ $y = X - 11{,}88$
$\xi = -0{,}3293$ $\sigma = 1{,}5576$	$U_3 = 12{,}5$	$U = 12{,}19$	$1 - \left(1 - \frac{0{,}3293y}{1{,}5576}\right)^{\frac{1}{0{,}3293}}$ $y = X - 12{,}5$	$1 - \left(1 - \frac{0{,}3293y}{1{,}5576}\right)^{\frac{1}{0{,}3293}}$ $y = X - 12{,}19$
$\xi = -0{,}3293$ $\sigma = 1{,}5576$	$U_3 = 12{,}5$	$U = 12{,}10$	$1 - \left(1 - \frac{0{,}3293y}{1{,}5576}\right)^{\frac{1}{0{,}3293}}$ $y = X - 12{,}5$	$1 - \left(1 - \frac{0{,}3293y}{1{,}5576}\right)^{\frac{1}{0{,}3293}}$ $y = X - 12{,}10$

Example 3: Threshold Calculation By the Graphical Method.

Knowledge of parameters (σ, ξ) allows to determine graphically the threshold U_3 by the GPD method and U by MM method (mixed method). To do this, we will write a program on the MAPLE software to determine these thresholds.

Distribution Function $F^{MM}(\xi, \sigma, u)$:

$$Distribution := \mathbf{proc}(\zeta, \sigma, u)\,\mathbf{local}\ i, j;\ plot\left(1 - \left(1 + \frac{\zeta.(U-u)}{\sigma}\right)^{\frac{-1}{\zeta}}, U = 0..20, thickness = 2, color = black\right)\mathbf{end};$$

$$Density := \mathbf{proc}(\zeta, \sigma, u)\,\mathbf{local}\ i, j;\ plot\left(\frac{1}{\sigma}\cdot\left(1 + \frac{\zeta.(x-u)}{\sigma}\right)^{\frac{-1-\zeta}{\zeta}}, ,\ = 0..20, thickness = 2, color = black\right)\mathbf{end};$$

Table 3. Knowing the parameters (σ, ξ) and the thresholds, we can graphically read the thresholds

<table>
<tr>
<td>
GPD method:

$\xi = -0{,}3293,\ \sigma = 1{,}5576$ with $U_3 = 12{,}5$

$$F^{GPD}_{\xi,\sigma(U_3)}(y) = 1 - \left(1 - \frac{0{,}3293y}{1{,}5576}\right)^{\frac{1}{0{,}3293}}$$

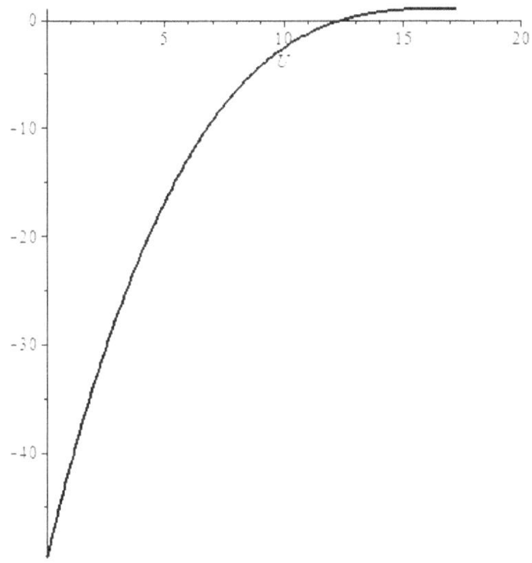

The threshold can be read graphically $U_3 = 12{,}5$
</td>
<td>
MM method:

$\xi = -0{,}3293,\ \sigma = 1{,}5576$ with $U = 11{,}88$

$$F^{MM}_{\xi,\sigma(U)}(y) = 1 - \left(1 - \frac{0{,}3293y}{1{,}5576}\right)^{\frac{1}{0{,}3293}}$$

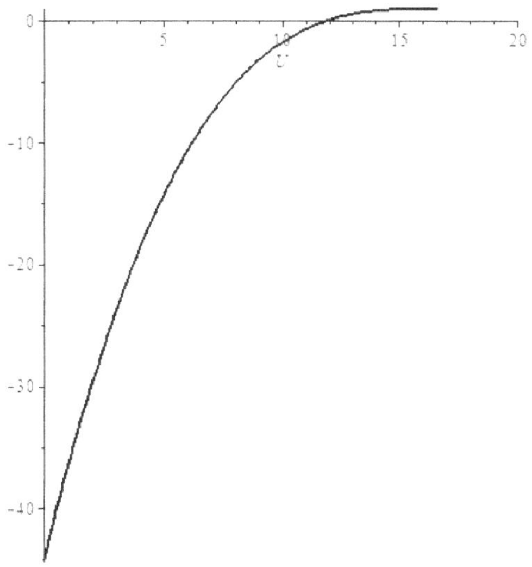

The threshold can be read graphically $U = 11{,}88$
</td>
</tr>
<tr>
<td>
MM method:

$\xi = -0{,}3293,\ \sigma = 1{,}5576$ with $U = 12{,}10$

$$F^{MM}_{\xi,\sigma(U)}(y) = 1 - \left(1 - \frac{0{,}3293y}{1{,}5576}\right)^{\frac{1}{0{,}3293}}$$

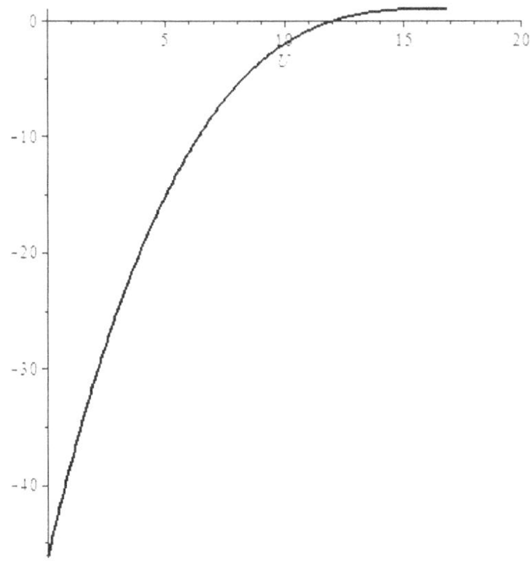

The threshold can be read graphically $U = 12{,}10$
</td>
<td>
MM method:

$\xi = -0{,}3293,\ \sigma = 1{,}5576$ with $U = 12{,}19$

$$F^{MM}_{\xi,\sigma(U)}(y) = 1 - \left(1 - \frac{0{,}3293y}{1{,}5576}\right)^{\frac{1}{0{,}3293}}$$

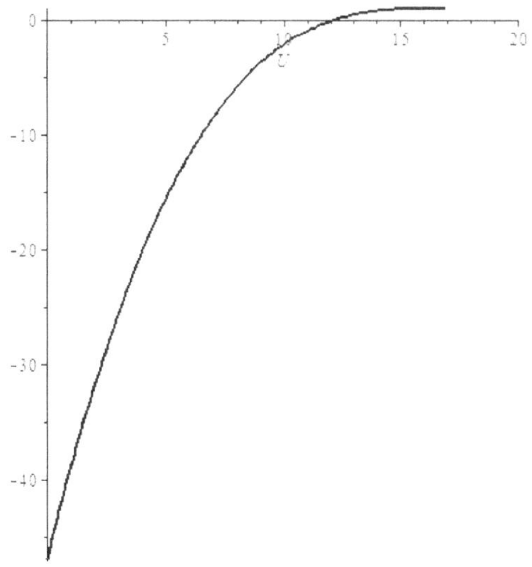

The threshold can be read graphically $U = 12{,}19$
</td>
</tr>
</table>

4. Conclusions

In the literature, various methods have been proposed to estimate the parameters (σ, ξ) of the GDP law: the method of maximum likelihood, method of moments, the Bayesian method, the estimator Pickands (1975) and the Hill estimator (1975). Note that these last two estimators can only be used for the index of the tail of the distribution ξ. Knowledge of parameters (σ, ξ) by the first two methods allows to determine graphically the threshold U_3 by the GPD method and U by MM method (mixed method). Therefore, we must carefully determine the parameters by different methods.

References

Alexander, J., & McNeil, S. T. (1997). The peaks over thersholds for estimating high quantiles of loss distribution, International ASTIN Colloquium, 70-94.

Allowen, A. (2008). Variations autour des Boxplots. Departement Systématique et Evolution.

Balkema, A., & Dehaan, L. (1974). Residual life time at great age. *The Annals of probability, 2*(5), 792-804. http://dx.doi.org/10.1214/aop/1176996548

Charles, S. (2007). Assurances et probabilités. http://math.univ-lille1.fr /~suquet

De Haan, L., & Ferreira, A. (1984). Extreme value theory. Springer-Verlag.

Diawara Daouda. (2015). Detection of a serious sinister in a rate box: An application of the extreme value theory in car insurance. *International Journal of Current Research, 7*(6), 17342–17346.

Embrechts, P., Kluppelberg, C., & Mikosch, T. (1997). *Modelingextremal events for insurance and finance.* Springer, Berlin. http://dx.doi.org/10.1007/978-3-642-33483-2

Gnedenko, B. V. (1943). Sur la distribution limite du terme maximum d'une série aléatoire. *Ann. Math., 44,* 423-453. http://dx.doi.org/10.2307/1968974

Mathieu, R. (2006). A User's Guide to the POT Package (Version 1.4)

Noureddine, B., Michel, G., & Olga, V. (2009). Les sinistres graves en assurance automobile: Une nouvelle approche par la théorie des valeurs extrêmes. *Revue MODULAD, 39.*

Pickands, J. (1975), Statistical inference using extreme order statistics. *Ann. Statist., 3,* 119-131. http://dx.doi.org/10.1214/aos/1176343003

Tukey, J. W. (1977). Exploratory Data Analysis. Ed.Addison-Wesley.

Xu, Y. M., Daouda, D., & Ladji, K. (2016). An application of extreme value theory in automobile insurance: A new approach to the convex combination of two variables thresholds minimizing the variance of this combination. *Advances and Applications in Statistics, 48*(2), 91–108.

Xu, Y. M., Ladji, K., & Daouda, D. (2016). Mixed method of extreme value theory, with application to the calculation of the portion of each claim payable by the reinsurance of excess of sinister. *International Journal of Statistics and Probability, 5*(2), 1927–7032.

13

The Topp-Leone Generated Weibull Distribution: Regression Model, Characterizations and Applications

Gokarna R. Aryal[1], Edwin M. Ortega[2], G. G. Hamedani[3], & Haitham M. Yousof[4]

[1] Department of Mathematics, Statistics and CS, Purdue University Northwest, USA

[2] Departamento de Ciências Exatas, Universidade de São Paulo, Piracicaba, SP, Brazil

[3] Department of Mathematics, Statistics and Computer Science, Marquette University, USA

[4] Department of Statistics, Mathematics and Insurance, Benha University, Egypt

Correspondence: Gokarna R. Aryal, Department of Mathematics, Statistics and Computer Science, Purdue University Northwest, Hammond, IN, USA. E-mail: aryalg@pnw.edu

Abstract

This paper introduces a new four-parameter lifetime model called the Topp Leone Generated Weibull (TLGW) distribution. This distribution is a generalization of the two parameter Weibull distribution using the genesis of Topp-Leone distribution. We derive many of its structural properties including ordinary and incomplete moments, quantile and generating functions and order statistics. Parameter estimation using maximum likelihood method and simulation results to assess effectiveness of the distribution are discussed. Also, for the first time, we introduce a regression model based on the new distribution. We prove empirically the importance and flexibility of the new model in modeling various types of real data sets.

Keywords: Topp-Leone distribution, Weibull distribution, order statistics, parameter estimation, regression model, simulation

1. Introduction

The Weibull distribution has an undeniable popularity in probability and statistics due to its versatility of modeling real world data. Yet there are many cases where the classical Weibull distribution is unable to capture true phenomenon under study. Therefore, several of its generalizations have been proposed and studied. A generalized form of Weibull distribution is obtained by inducting one or more parameter(s) to the 2-parameter Weibull distribution. It has been proven that several of these generalized distribution are more flexible and are capable of modeling real world data better than the classical Weibull distribution. A state-of-the-art survey on the class of such generalized Weibull distributions can be found in Lai, Xie, & Murthy (2001) and Nadarajah (2009).

Some generalization of the Weibull distribution studied in the literature includes, but are not limited to, exponentiated Weibull (Mudholkar & Srivastava, 1993; Mudholkar, Srivastava, & Freimer, 1995; Mudholkar, Srivastava, & Kollia, 1996), additive Weibull (Xie & Lai, 1995), Marshall-Olkin extended Weibull (Ghitany, Al-Hussaini, & Al-Jarallah, 2005), beta Weibull (Famoye, Lee, & Olumolade, 2005), modified Weibull (Sarhan & Zaindin, 2009), beta modified Weibull (Silva, Ortega, & Cordeiro, 2010), transmuted Weibull (Aryal & Tsokos, 2011), extended Weibull (Xie, Tang, & Goh, 2002), modified Weibull (Lai, Xie, & Murthy, 2003), Kumaraswamy Weibull (Cordeiro, Ortega, & Nadarajah, 2010), Kumaraswamy modified Weibull (Cordeiro, Ortega, & Silva, 2012), Kumaraswamy inverse Weibull (Shahbaz, Shazbaz, & Butt, 2012), exponentiated generalized Weibull (Cordeiro, Ortega, & Cunha 2013), McDonald modified Weibull (Merovci & Elbatal, 2013), beta inverse Weibull (Hanook, Shahbaz, Mohsin,& Kibria, 2013), transmuted exponentiated generalized Weibull (Yousof et al., 2015), McDonald Weibull (Cordeiro, Hashimoto, & Ortega, 2014), gamma Weibull (Provost, Saboor, & Ahmad, 2011), transmuted modified Weibull (Khan & King, 2013), beta Weibull (Lee, Famoye, & Olumolade, 2007), generalized transmuted Weibull (Nofal, Afify, Yousof, & Cordeiro, 2015), transmuted additive Weibull (Elbatal & Aryal, 2013), exponentiated generalized modified Weibull (Aryal & Elbatl, 2015), transmuted exponentiated additive Weibull (Nofal, Afify, Yousof, & Louzada, 2016), Marshall Olkin additive Weibull (Afify, Cordeiro, Yousof, Saboor, & Ortega, 2016) and Kumaraswamy transmuted exponentiated additive Weibull (Nofal, Afify, Yousof, Granzotto,& Louzada, 2016) distributions.

In this paper we introduce a new generalization of the Weibull distribution using the genesis of the Topp-Leone distribution and is named as Topp-Leone Generated Weibull (TLGW) distribution. Consider the Topp-Leone generated family of distributions proposed by Rezaei, Sadr, Alizadeh & Nadarajah (2016) with its probability density function (pdf) and cumulative distribution function (cdf) given by,

$$f_{TLG}(x) = 2\alpha\theta g(x)[G(x)]^{\theta\alpha-1}\left[1 - G(x)^{\theta}\right]\left[2 - G(x)^{\theta}\right]^{\alpha-1}, \quad x \geq 0 \tag{1}$$

$$F_{TLG}(x) = \left\{G(x)^{\theta}\left[2 - G(x)^{\theta}\right]\right\}^{\alpha}, \quad x \geq 0 \tag{2}$$

respectively. In this paper we will use this generalization to the Weibull (W) distribution whose pdf and cdf are given, respectively, by

$$g(x) = \beta\eta^{\beta}x^{\beta-1}\,\mathrm{e}^{-(\eta x)^{\beta}} \tag{3}$$

and

$$G(x) = 1 - \mathrm{e}^{-(\eta x)^{\beta}}. \tag{4}$$

By inserting (3) and (4) into (1), we can write the pdf of the TLGW distribution as

$$f_{TLGW}(x) = 2\alpha\theta\beta\eta^{\beta}x^{\beta-1}\,\mathrm{e}^{-(\eta x)^{\beta}}\left(1 - \mathrm{e}^{-(\eta x)^{\beta}}\right)^{\theta\alpha-1}\left[1 - \left(1 - \mathrm{e}^{-(\eta x)^{\beta}}\right)^{\theta}\right]\left[2 - \left(1 - \mathrm{e}^{-(\eta x)^{\beta}}\right)^{\theta}\right]^{\alpha-1} \tag{5}$$

The corresponding cdf of TLGW distribution is given by

$$F_{TLGW}(x) = \left(1 - \mathrm{e}^{-(\eta x)^{\beta}}\right)^{\alpha\theta}\left[2 - \left(1 - \mathrm{e}^{-(\eta x)^{\beta}}\right)^{\theta}\right]^{\alpha} \tag{6}$$

Figure 1 illustrates the graphical behavior of the pdf of TLGW distribution for selected parameter values. As we shall see in the sequel, this is a rather flexible family compared to the Weibull distribution.

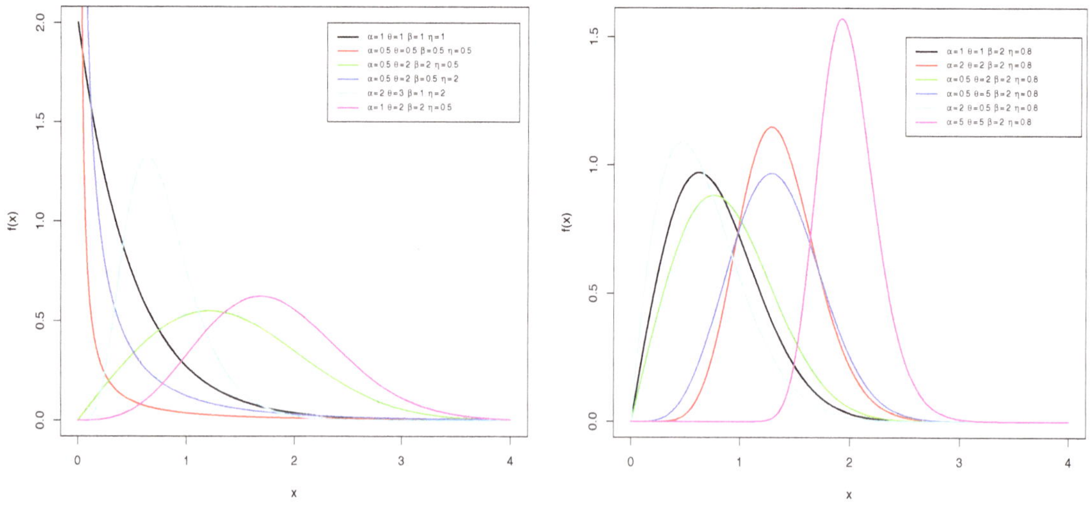

Figure 1. Probability density function of TLGW distribution

In order to derive mathematical and statistical properties of the TLGW distribution, the series expansion of its pdf and cdf will be useful. The cdf in (6) can be expressed as

$$F(x) = \sum_{k=0}^{\infty} \Upsilon_k \Pi_{[(\alpha+k)\theta]}(x) \tag{7}$$

where $\boldsymbol{\Upsilon}_k = (-1)^k\, 2^{\alpha-k}\binom{\alpha}{k}$ and $\boldsymbol{\Pi}_{(\alpha+k)\theta}(x) = \left(1 - e^{-(\eta x)^\beta}\right)^{(\alpha+k)\theta}$. Observe that $\boldsymbol{\Pi}_\gamma(x) = \left(1 - e^{-(\eta x)^\beta}\right)^\gamma$ is the cdf of exponentiated-Weibull (exp-W) distribution with power parameter γ. This means the TLGW distribution can be expressed as a linear mixture of the exp-W distribution. Similarly, pdf (5) can be expressed as

$$f(x) = \sum_{k=0}^{\infty} \boldsymbol{\Upsilon}_k \boldsymbol{\pi}_{[(\alpha+k)\theta]}(x) \qquad (8)$$

with

$$\boldsymbol{\pi}_\gamma(x) = \underbrace{\gamma\beta\eta^\beta x^{\beta-1}\, e^{-(\eta x)^\beta}}_{g(x;\beta,\eta)} \underbrace{\left[1 - e^{-(\eta x)^\beta}\right]^{\gamma-1}}_{G(x;\beta,\eta)^{\gamma-1}}.$$

Equation (8) reveals that the density of X can be expressed as a linear mixture of exp-W densities. So, several mathematical properties of the new family can be obtained from those of the exp-W distribution.

The paper is unfolded as follows. In Section 2, we obtain some mathematical properties including moments, cumulants, generating function, residual, reversed residual life functions, stress-strength model and order of the proposed distribution. In Section 3, we provide a useful characterization of the new distribution. In Section 4, the model parameters are estimated by maximum likelihood and a simulation study is performed. In Section 5, we present a regression model based on the TLGW distribution with censored data. In Section 6, the usefulness of the new distribution is illustrated by means of four real data sets, where we prove empirically that it outperforms some well-known lifetime distributions. Finally, Section 7 offers some concluding remarks.

2. Mathematical Properties

In this section we will provide some mathematical properties of the TLGW distribution including the moments, incomplete moments, mean deviations, order statistic etc.

2.1 Moments, Cumulants and Generating Function

The rth ordinary moment of X is given by $\mu_r' = E(X^r) = \int_{-\infty}^{\infty} x^r f_{TLGW}(x)\,dx$. For any $r > -\beta$, we have

$$\mu_r' = \sum_{k,h=0}^{\infty} \boldsymbol{\Psi}_{k,h}\, \Gamma\left(1 + \frac{r}{\beta}\right), \qquad (9)$$

where

$$\boldsymbol{\Psi}_{k,h} = \boldsymbol{\Upsilon}_k \frac{(-1)^h\, \Gamma((\alpha+k)\theta + 1)}{h!\,\eta^r\,\Gamma((\alpha+k)\theta + 1 - h)\,(h+1)^{\frac{r+\beta}{\beta}}}.$$

Henceforth, $Y_{[(\alpha+k)\theta]}$ denotes the exp-W distribution with power parameter $[(\alpha+k)\theta]$. Setting $r = 1$ in (9), we have the mean of X. The rth central moment of X, say M_r, follows as $M_r = E(X-\mu)^r = \sum_{h=0}^{r} (-1)^h \binom{r}{h} (\mu_1')^r \mu_{r-h}'$. The skewness and kurtosis measures also can be calculated from the ordinary moments using well-known relationships. The cumulants (κ_n) of X follow recursively from $\kappa_n = \mu_n' - \sum_{r=0}^{n-1}\binom{n-1}{r-1}\kappa_r \mu_{n-r}'$, where $\kappa_1 = \mu_1'$, $\kappa_2 = \mu_2' - \mu_1'^2$, $\kappa_3 = \mu_3' - 3\mu_2'\mu_1' + \mu_1'^3$, etc. The moment generating function (mgf) of X is given by $M_X(t) = E\left(e^{tX}\right)$. Clearly, It can be derived using equation (9), for $r > -\beta$, as

$$M_X(t) = \sum_{k=0}^{\infty} \boldsymbol{\Upsilon}_k\, M_{[(\alpha+k)\theta]}(t) = \sum_{k=0}^{\infty} \boldsymbol{\Upsilon}_k \sum_{r=0}^{\infty} \frac{t^r}{r!}\mu_r' = \sum_{r,k,h=0}^{\infty} \boldsymbol{\Psi}_{k,h} \frac{t^r}{r!}\, \Gamma\left(1 + \frac{r}{\beta}\right).$$

The effect of the parameters α and θ on mean, variance, skewness and kurtosis, for given values of $\beta = 2$ and $\eta = 2$, are displayed in Figures 2 and 3.

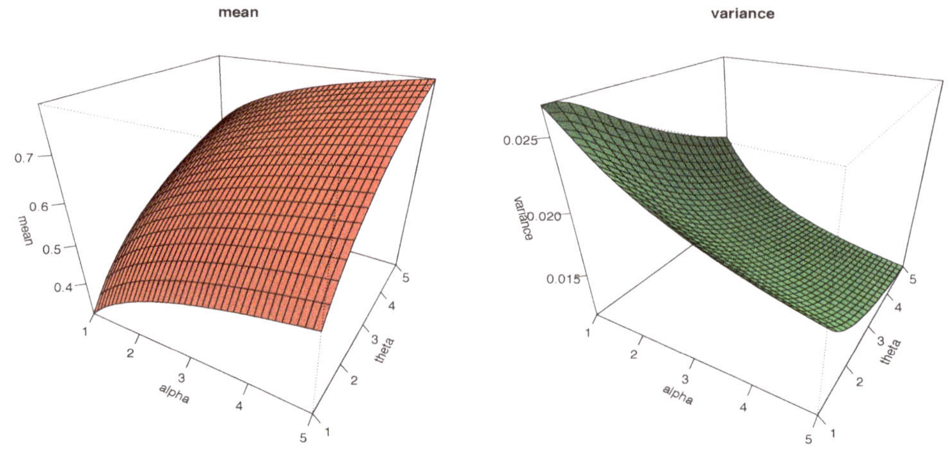

Figure 2. Plots of mean and variance of TLGW distribution

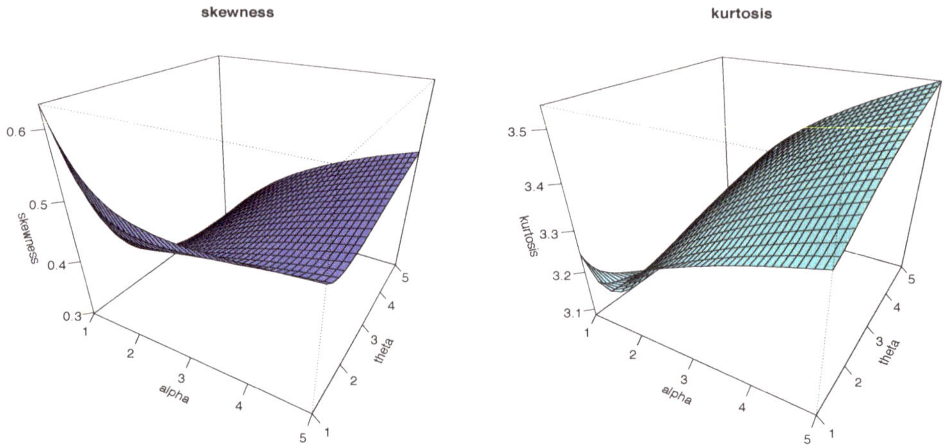

Figure 3. Plots of skewness and kurtosis of TLGW distribution

2.2 Incomplete Moments and Mean Deviations

The main applications of the first incomplete moment refer to the mean deviations and the Bonferroni and Lorenz curves. These curves are very useful in economics, reliability, demography, insurance and medicine. The sth incomplete moment, say $\varphi_s(t)$, of X for $s > -\beta$ can be expressed from (8) as

$$\begin{aligned}\varphi_s(t) &= \sum_{k=0}^{\infty} \Upsilon_k \int_{-\infty}^{t} x^s \pi_{[(\alpha+k)\theta]}(x)\,dx \\ &= \sum_{k,h=0}^{\infty} \Upsilon_k \frac{(-1)^h \Gamma((\alpha+k)\theta+1)}{h!\eta^s \Gamma((\alpha+k)\theta+1-h)(h+1)^{\frac{s+\beta}{\beta}}} \Gamma\left(1+\frac{s}{\beta}, \left(\frac{\eta}{y}\right)^{\beta}\right). \qquad (10)\end{aligned}$$

The mean deviations about the mean [$\delta_1 = E(|X - \mu_1'|)$] and about the median [$\delta_2 = E(|X - M|)$] of X are given by $\delta_1 = 2\mu_1' F(\mu_1') - 2\varphi_1(\mu_1')$ and $\delta_2 = \mu_1' - 2\varphi_1(M)$, respectively, where $\mu_1' = E(X)$, $M = \text{Median}(X) = Q(0.5)$ is the median, $F(\mu_1')$ is easily calculated from (4) and $\varphi_1(t)$ is the first incomplete moment given by (10) with $s = 1$. A general equation for $\varphi_1(t)$ can be derived from (10) as

$$\varphi_1(t) = \sum_{k,h=0}^{\infty} \Upsilon_k \frac{(-1)^h \Gamma((\alpha+k)\theta+1)}{h!\eta\Gamma((\alpha+k)\theta+1-h)(h+1)^{\frac{1+\beta}{\beta}}} \Gamma\left(1+\frac{1}{\beta},\left(\frac{\eta}{y}\right)^{\beta}\right).$$

2.3 Residual and Reversed Residual Life Functions

The nth moment of the residual life, say $m_n(t) = E[(X-t)^n \mid X > t]$, $n = 1, 2, \ldots$, uniquely determines $F(x)$. The nth moment of the residual life of X is given by $m_n(t) = \frac{1}{R(t)} \int_t^{\infty} (x-t)^n dF(x)$. Therefore,

$$m_n(t) = \frac{1}{R(t)} \sum_{k,h=0}^{\infty} \Upsilon_k^{\star} \frac{(-1)^h \Gamma((\alpha+k)\theta+1)}{h!\eta^n\Gamma((\alpha+k)\theta+1-h)(h+1)^{\frac{r+\beta}{\beta}}} \Gamma\left(1+\frac{n}{\beta},\left(\frac{\eta}{t}\right)^{\beta}\right),$$

where $\Upsilon_k^{\star} = \Upsilon_k \sum_{r=0}^{n} \binom{n}{r} (-t)^{n-r}$. Another interesting function is the mean residual life (MRL) function or the life expectation at age t defined by $m_1(t) = E[(X-t) \mid X > t]$, which represents the expected additional life length for a unit which is alive at age t. The MRL of X can be obtained by setting $n = 1$ in the last equation. The nth moment of the reversed residual life, say $M_n(t) = E[(t-X)^n \mid X \le t]$ for $t > 0$ and $n = 1, 2, \ldots$, uniquely determines $F(x)$. We obtain $M_n(t) = \frac{1}{F(t)} \int_0^t (t-x)^n dF(x)$. Then, the nth moment of the reversed residual life of X becomes

$$M_n(t) = \frac{1}{F(t)} \sum_{k,h=0}^{\infty} \Upsilon_k^{\star\star} \frac{(-1)^h \Gamma((\alpha+k)\theta+1)}{h!\eta^n\Gamma((\alpha+k)\theta+1-h)(h+1)^{\frac{n+\beta}{\beta}}} \Gamma\left(1+\frac{n}{\beta},\left(\frac{\eta}{t}\right)^{\beta}\right),$$

where $\Upsilon_k^{\star\star} = \Upsilon_k \sum_{r=0}^{n} (-1)^r \binom{n}{r} t^{n-r}$. The mean inactivity time (MIT) or mean waiting time (MWT), also called the mean reversed residual life function, is given by $M_1(t) = E[(t-X) \mid X \le t]$ and it represents the waiting time elapsed since the failure of an item on condition that this failure had occurred in $(0, t)$. The MIT of the TLGW distributions can be obtained easily by setting $n = 1$ in the above equation.

2.4 A Stress-Strength Model

Stress-strength model is the most widely approach used for reliability estimation. This model is used in many applications of physics and engineering such as strength failure and system collapse. In stress-strength modeling, $\mathbf{R} = \Pr(X_2 < X_1)$ is a measure of reliability of the system when it is subjected to random stress X_2 and has strength X_1. The system fails if and only if the applied stress is greater than its strength and the component will function satisfactorily whenever $X_1 > X_2$. R can be considered as a measure of system performance and naturally arise in electrical and electronic systems. Other interpretation can be that, the reliability, say $\mathbf{R}$, of the system is the probability that the system is strong enough to overcome the stress imposed on it. Let X_1 and X_2 be two independent random variables have TLGW$(\alpha_1, \theta_1, \eta, \beta)$ and TLGW$(\alpha_2, \theta_2, \eta, \beta)$ distributions . Then, we can write

$$\mathbf{R} = \int_0^{\infty} f_1(x; \alpha_1, \theta_1, \eta, \beta) F_2(x; \alpha_2, \theta_2, \eta, \beta) dx = \sum_{k,j=0}^{\infty} \mathbf{\Omega}_{k,j} \int_0^{\infty} \boldsymbol{\pi}_{[(\alpha_1+k)\theta_1]+[(\alpha_2+j)\theta_2]}(x) dx,$$

where

$$\mathbf{\Omega}_{k,j} = \frac{(-1)^{k+j} 2^{\alpha_1+\alpha_2-k-j} \binom{\alpha_1}{k}\binom{\alpha_2}{j} ((\alpha_1+k)\theta_1)}{[(\alpha_1+k)\theta_1 + (\alpha_2+j)\theta_2]}$$

and

$$\boldsymbol{\pi}_{[(\alpha_1+k)\theta_1+(\alpha_2+j)\theta_2]}(x) = [(\alpha_1+k)\theta_1 + (\alpha_2+j)\theta_2] \beta\eta^{\beta} x^{\beta-1} e^{-(\eta x)^{\beta}} \left[1 - e^{-(\eta x)^{\beta}}\right]^{[(\alpha_1+k)\theta_1+(\alpha_2+j)\theta_2]-1}.$$

Thus, $\mathbf{R}$ can be simply expressed as

$$\mathbf{R} = \sum_{k,j=0}^{\infty} \mathbf{\Omega}_{k,j}.$$

2.5 Order Statistics

Let $X_1, \ldots, X_n$ be a random sample from the TLGW distribution and let $X_{(1)}, \ldots, X_{(n)}$ be the corresponding order statistics. The pdf of ith order statistic can be written as

$$f_{i:n}(x) = \frac{f(x)}{\mathrm{B}(i, n-i+1)} \sum_{j=0}^{n-i} (-1)^j \binom{n-i}{j} F^{j+i-1}(x), \tag{11}$$

where $B(\cdot,\cdot)$ is the beta function. Substituting (5) and (6) in (11) the pdf of $X_{i:n}$ can be expressed as the pdf of $X_{i:n}$ can be expressed as

$$f_{i:n}(x) = \sum_{j=0}^{n-i} \sum_{r,k=0}^{\infty} \mathbf{m}_{j,r,k}\, \boldsymbol{\pi}_{r+k}(x).$$

where

$$\mathbf{m}_{j,r,k} = \frac{r(-1)^j \Upsilon_r \mathbf{f}_{j+i-1,k}}{(r+k)\mathrm{B}(i, n-i+1)}$$

and $\mathbf{f}_{j+i-1,k}$ can be obtained recursively from

$$\mathbf{f}_{j+i-1,k} = \frac{1}{(kb_0)} \sum_{m=0}^{k} [m(j+i)-k] \Upsilon_m \mathbf{f}_{j+i-1,k-m},$$

for $k \geq 1$ with $\mathbf{f}_{j+i-1,0} = b_0^{j+i-1}$. Then, the density function of the TLGW order statistics is a mixture of exp-W density. Based on the last equation, we note that the properties of $X_{i:n}$ follow from those of Y_{r+k}. For example, the moments of $X_{i:n}$ can be expressed as, for $q > -\beta$,

$$E\left(X_{i:n}^q\right) = \sum_{j=0}^{n-i} \sum_{r,k=0}^{\infty} \mathbf{m}_{j,r,k} E\left(Y_{r+k}^q\right) = \sum_{j=0}^{n-i} \sum_{r,k,h=0}^{\infty} \mathbf{m}^{\star}_{j,r,k,h} \Gamma\left(1+\frac{q}{\beta}\right), \tag{12}$$

where

$$\mathbf{m}^{\star}_{j,r,k,h} = \mathbf{m}_{j,r,k} \frac{(-1)^h \Gamma(r+k+1)}{h! \eta^q \Gamma(r+k+1-h)(h+1)^{\frac{q+\beta}{\beta}}}.$$

The L-moments are analogous to the ordinary moments but can be estimated by linear combinations of order statistics. They exist whenever the mean of the distribution exists, even though some higher moments may not exist and are relatively robust to the effects of outliers. Based upon the moments in equation (12), we can derive explicit expressions for the L-moments of X as infinite weighted linear combinations of the means of suitable TLGW order statistics. They are linear functions of expected order statistics defined by

$$\lambda_r = \frac{1}{r} \sum_{d=0}^{r-1} (-1)^d \binom{r-1}{d} E(X_{r-d:r}),\ r \geq 1.$$

3. Characterizations

In this section we present certain characterizations of TLGW distribution. These characterizations are based on the ratio of two truncated moments. Due to the nature of this distribution, we believe that these may be the only possible characterizations of TLGW distribution. Our first characterization result employs a theorem due to Glänzel (1987) see Theorem 1 in Appendix A. Note that the result holds also when the interval H is not closed. Moreover, it could also be applied when the cdf F does not have a closed form. As shown in Glänzel (1990), this characterization is stable in the sense of weak convergence.

Proposition 3.1. Let $X : \Omega \to (0, \infty)$ be a continuous random variable and let $q_1(x) = \left(1 - e^{-(\eta x)^\beta}\right)^{\theta(1-\alpha)} \left[2 - \left(1 - e^{-(\eta x)^\beta}\right)^\theta\right]^{1-\alpha}$ and $q_2(x) = q_1(x)\left[1 - \left(1 - e^{-(\eta x)^\beta}\right)^\theta\right]$ for $x > 0$. The random variable X belongs to the family (5) if and only if the function ξ defined in Theorem 1 has the form

$$\xi(x) = \frac{2}{3}\left[1 - \left(1 - e^{-(\eta x)^\beta}\right)^\theta\right], \quad x > 0.$$

Proof. Let X be a random variable with pdf (5), then

$$(1 - F(x)) E[q_1(x) \mid X \geq x] = \alpha\left[1 - \left(1 - e^{-(\eta x)^\beta}\right)^\theta\right]^2, \quad x > 0,$$

and

$$[1-F(x)]\,E\left[q_2(x)\mid X\geq x\right]=\frac{2\alpha}{3}\left[1-\left(1-e^{-(\eta x)^\beta}\right)^\theta\right]^3,\quad x>0,$$

and finally

$$\xi(x)\,q_1(x)-q_2(x)=-\frac{1}{3}q_1(x)\left[1-\left(1-e^{-(\eta x)^\beta}\right)^\theta\right]<0\quad for\ x>0.$$

Conversely, if ξ is given as above, then

$$s'(x)=\frac{\xi'(x)\,q_1(x)}{\xi(x)\,q_1(x)-q_2(x)}=\frac{2\theta\beta\eta^\beta x^{\beta-1}e^{-(\eta x)^\beta}\left(1-e^{-(\eta x)^\beta}\right)^{\theta-1}}{1-\left(1-e^{-(\eta x)^\beta}\right)^\theta},\quad x>0,$$

and hence

$$s(x)=\ln\left\{\left[1-\left(1-e^{-(\eta x)^\beta}\right)^\theta\right]^{-2}\right\},\quad x>0.$$

Now, in view of Theorem 1, X has density (5).

Corollary 3.1. Let $X:\Omega\rightarrow(0,\infty)$ be a continuous random variable and let $q_1(x)$ be as in Proposition 3.1. Then X has pdf (5) if and only if there exist functions q_2 and ξ defined in Theorem 1 satisfying the differential equation

$$\frac{\xi'(x)\,q_1(x)}{\xi(x)\,q_1(x)-q_2(x)}=\frac{2\theta\beta\eta^\beta x^{\beta-1}e^{-(\eta x)^\beta}\left(1-e^{-(\eta x)^\beta}\right)^{\theta-1}}{1-\left(1-e^{-(\eta x)^\beta}\right)^\theta},\quad x>0.$$

Corollary 3.2. The general solution of the differential equation in Corollary 3.1 is

$$\xi(x)=\left[1-\left(1-e^{-(\eta x)^\beta}\right)^\theta\right]^{-1}\left[-\int 2\theta\beta\eta^\beta x^{\beta-1}e^{-(\eta x)^\beta}\left(1-e^{-(\eta x)^\beta}\right)^{\theta-1}(q_1(x))^{-1}q_2(x)\,dx+D\right],$$

where D is a constant. Note that a set of functions satisfying the differential equation in Corollary 3.1, is given in Proposition 3.1 with $D=0$. However, it should be also noted that there are other triplets (q_1,q_2,ξ) satisfying the conditions of Theorem 1.

4. Parameter Estimation

Subsection 4.1 provides procedures for maximum likelihood estimation of the TLGW distribution. Subsection 4.2 assesses the performance of the maximum likelihood estimators (MLEs) in terms of biases and mean squared errors by means of a simulation study.

4.1 Parameter Estimation

Several methods for parameter estimation have been proposed in the literature but the maximum likelihood method is the most commonly employed one. The maximum likelihood estimators enjoy desirable properties and can be used for constructing confidence intervals and regions and also in test statistics. The normal approximation for these estimators, in large samples, can be easily handled either analytically or numerically. So, we consider the estimation of the unknown parameters of this family from complete samples only by maximum likelihood. Let $x_1,\ldots,x_n$ be a random sample from the TLGW distribution. Let $\boldsymbol{\tau}=(\alpha,\theta,\beta,\eta)^T$ be the 4×1 parameter vector. To determine the MLE of $\boldsymbol{\tau}$, we use the log-likelihood function (ℓ) of TLGW distribution given by

$$\begin{aligned}\ell(\boldsymbol{\tau}) &= n\ln 2+n\ln\alpha+n\ln\theta+n\ln\beta+n\beta\ln\eta+(\beta-1)\sum_{i=1}^{n}\ln(x_i)-\sum_{i=1}^{n}(\eta x_i)^\beta\\ &\quad+(\theta\alpha-1)\sum_{i=1}^{n}\ln s_i+\sum_{i=1}^{n}\ln\left(1-s_i^\theta\right)+(\alpha-1)\sum_{i=1}^{n}\ln\left(2-s_i^\theta\right),\end{aligned}$$

where $s_i = \left(1 - e^{-(\eta x_i)^\beta}\right)$. Let $z_i = \frac{\partial s_i}{\partial \beta} = (\eta x_i)^\beta e^{-(\eta x_i)^\beta} \ln(\eta x_i)$ and $q_i = \beta \eta^{\beta-1} x^\beta e^{-(\eta x)^\beta}$. Then the components of the score vector are given by

$$\begin{aligned}
\frac{\partial \ell(\boldsymbol{\tau})}{\partial \alpha} &= \frac{n}{\alpha} + \theta \sum_{i=1}^{n} \ln s_i + \sum_{i=1}^{n} \ln\left(2 - s_i^\theta\right), \\
\frac{\partial \ell(\boldsymbol{\tau})}{\partial \theta} &= \frac{n}{\theta} + \alpha \sum_{i=1}^{n} \ln s_i - \sum_{i=1}^{n} \frac{s_i^\theta \ln s_i}{\left(1 - s_i^\theta\right)} - (\alpha - 1) \sum_{i=1}^{n} \frac{s_i^\theta \ln s_i}{\left(2 - s_i^\theta\right)}, \\
\frac{\partial \ell(\boldsymbol{\tau})}{\partial \beta} &= \frac{n}{\beta} + n \ln \eta + \sum_{i=1}^{n} \ln(x_i) - \sum_{i=1}^{n} \frac{\ln(\eta x)}{(\eta x)^{-\beta}} + (\theta\alpha - 1) \sum_{i=1}^{n} \frac{z_i}{s_i} - \sum_{i=1}^{n} \frac{\theta z_i s_i^{\theta-1}}{1 - s_i^\theta} - (\alpha - 1) \sum_{i=1}^{n} \frac{\theta z_i s_i^{\theta-1}}{2 - s_i^\theta}
\end{aligned}$$

and

$$\frac{\partial \ell(\boldsymbol{\tau})}{\partial \eta} = \frac{n\beta}{\eta} - \beta \eta^{\beta-1} \sum_{i=1}^{n} x_i^\beta + (\theta\alpha - 1) \sum_{i=1}^{n} \frac{q_i}{s_i} - \theta \sum_{i=1}^{n} \frac{q_i s_i^{\theta-1}}{\left(1 - s_i^\theta\right)} - (\alpha - 1) \sum_{i=1}^{n} \frac{q_i s_i^{\theta-1}}{\left(2 - s_i^\theta\right)}.$$

Now, setting the nonlinear system of equations $\frac{\partial \ell(\boldsymbol{\tau})}{\partial \alpha} = 0$, $\frac{\partial \ell(\boldsymbol{\tau})}{\partial \theta} = 0$, $\frac{\partial \ell(\boldsymbol{\tau})}{\partial \beta} = 0$ and $\frac{\partial \ell(\boldsymbol{\tau})}{\partial \eta} = 0$ and solving them simultaneously yields the MLE $\widehat{\boldsymbol{\tau}} = (\widehat{\alpha}, \widehat{\theta}, \widehat{\beta}, \widehat{\eta})^T$. To solve these equations, it is usually more convenient to use nonlinear optimization methods such as the quasi-Newton algorithm to numerically maximize ℓ. For interval estimation of the parameters, we obtain the 4×4 observed information matrix $J(\boldsymbol{\tau}) = \{\frac{\partial^2 \ell}{\partial r \partial s}\}$ (for $r, s = \alpha, \theta, \beta, \eta$), whose elements can be computed numerically. Under standard regularity conditions when $n \to \infty$, the distribution of $\widehat{\boldsymbol{\tau}}$ can be approximated by a multivariate normal $N_4(0, J(\widehat{\boldsymbol{\tau}})^{-1})$ distribution to construct approximate confidence intervals for the parameters. Here, $J(\widehat{\boldsymbol{\tau}})$ is the total observed information matrix evaluated at $\widehat{\boldsymbol{\tau}}$.

4.2 Simulation Study

In this section, we present some simulations for different sample sizes to assess the accuracy of the MLEs. Simulating random variables from well defined probability distributions has been discussed in the computational statistics literature, e.g; the inverse transformation method, the rejection and acceptance sampling technique, etc. An ideal technique for simulating from the TLGW distribution is the inversion method. We can simulate random variable X by

$$X = \frac{1}{\eta}\left[-\ln\left\{1 - \left(1 - \sqrt{1 - U^{\frac{1}{\alpha}}}\right)^{\frac{1}{\theta}}\right\}\right]^{\frac{1}{\beta}},$$

where U is a uniform random number in $(0, 1)$. For selected combinations of θ, α, β and η we generate samples of sizes $n = 50, 100, 200, 300, 500$ and 1000 from the TLGW distribution. We repeat the simulations $N = 1,000$ times and evaluate the mean estimates and the root mean squared errors (RMSEs). The empirical results obtained using the statistical computing software R are given in Tables 1 and 2. It can be observed that as sample size increases the mean squared error decreases. Therefore, the maximum likelihood method works very well to estimate the parameters of TLGW distribution.

Table 1. Empirical means and the RMSEs of TLGW distribution

	$\theta = 0.5$	$\alpha = 1.5$	$\beta = 2.0$	$\eta = 2.0$
n	$\widehat{\theta}$	$\widehat{\alpha}$	$\widehat{\beta}$	$\widehat{\eta}$
50	0.6557	3.5869	3.6344	2.4805
	(1.7665)	(4.6465)	(3.0606)	(6.6074)
100	0.6102	2.9638	2.9702	2.0982
	(0.7534)	(3.0763)	(2.0948)	(0.8570)
200	0.5622	2.4859	2.5936	1.9948
	(0.5062)	(2.3428)	(1.3669)	(0.4574)
300	0.5528	2.3607	2.4361	1.9857
	(0.4576)	(2.0398)	(1.0502)	(0.3922)
500	0.5865	2.1773	2.2545	2.0171
	(0.4487)	(1.7359)	(0.7393)	(0.3690)
1000	0.5599	1.9159	2.1635	2.0049
	(0.3763)	(1.1891)	(0.5083)	(0.2988)

Table 2. Empirical means and the RMSEs of TLGW distribution

	$\theta = 2.0$	$\alpha = 0.5$	$\beta = 1.5$	$\eta = 2.5$
n	$\widehat{\theta}$	$\widehat{\alpha}$	$\widehat{\beta}$	$\widehat{\eta}$
50	3.0078	1.3099	2.1327	2.9353
	(3.4059)	(10.4155)	(1.8721)	(2.3342)
100	2.9044	0.6648	1.7716	2.6559
	(2.8306)	(0.6846)	(0.9444)	(0.9228)
200	2.8347	0.5538	1.5785	2.6469
	(2.2698)	(0.4076)	(0.4318)	(0.5841)
300	2.7806	0.5285	1.5396	2.6439
	(2.0799)	(0.3081)	(0.3188)	(0.5508)
500	2.6320	0.5084	1.5199	2.6043
	(1.7402)	(0.2567)	(0.2367)	(0.4239)
1000	2.4849	0.5019	1.5100	2.5806
	(1.4294)	(0.2098)	(0.1781)	(0.3650)

5. The Log-Topp Leone Generated Weibull Regression Model With Censored Data

Let X be a random variable having the density function (5). The random variable $Y = \log(X)$ has a log-Topp Leone Generated Weibul (LTLGW) distribution, whose density function (parameterized in terms of $\beta = 1/\sigma$ and $\eta = \exp(-\mu)$) can be expressed as

$$\begin{aligned} f(y) &= \frac{2\alpha\theta}{\sigma}\exp\left[\left(\frac{y-\mu}{\sigma}\right)-\exp\left(\frac{y-\mu}{\sigma}\right)\right]\left[1-\left\{1-\exp\left[-\exp\left(\frac{y-\mu}{\sigma}\right)\right]\right\}^{\theta}\right] \\ &\times\left\{1-\exp\left[-\exp\left(\frac{y-\mu}{\sigma}\right)\right]\right\}^{\alpha\theta-1}\left[2-\left\{1-\exp\left[-\exp\left(\frac{y-\mu}{\sigma}\right)\right]\right\}^{\theta}\right]^{\alpha-1}, \end{aligned} \quad (13)$$

where $-\infty < y < \infty$, $\sigma > 0$ and $-\infty < \mu < \infty$.

We refer to equation (13) as the (new) LTLGW distribution, say $X \sim \text{TLGW}(\alpha, \theta, \sigma, \mu)$, where μ is a location parameter,σ is a dispersion parameter and α and θ are shape parameters. Thus,

$$\text{if} \quad X \sim \text{LTLGW}(\alpha, \theta, \eta, \beta) \quad \text{then} \quad Y = \log(X) \sim \text{LTLGW}(\alpha, \theta, \sigma, \mu).$$

The plots of (13) in Figure 5 for selected parameter values show great flexibility of the density function in terms of the parameters α and θ.

The survival function corresponding to (13) becomes

$$S(y) = 1-\left\{1-\exp\left[-\exp\left(\frac{y-\mu}{\sigma}\right)\right]\right\}^{\alpha\theta}\left[2-\left\{1-\exp\left[-\exp\left(\frac{y-\mu}{\sigma}\right)\right]\right\}^{\theta}\right]^{\alpha}. \quad (14)$$

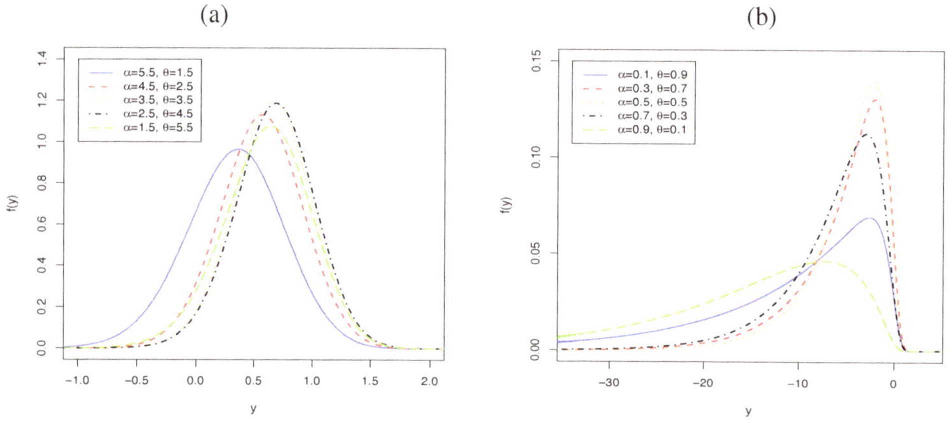

Figure 4. Plots of the LTLGW density for some parameter values. (a) For different values of $\alpha \in (1.0, 5.0)$ and $\beta \in (1.0, 5.0)$ for $\mu = 0$ and $\sigma = 0$. (b) For different values of $\alpha \in (0, 1.0)$ and $\beta \in (0, 1.0)$ for $\mu = 0$ and $\sigma = 0$.

We define the standardized random variable $Z = (Y - \mu)/\sigma$ with density function

$$\pi(z;\alpha,\theta) = 2\,\alpha\,\theta \exp\left[z - \exp(z)\right]\{1 - \exp\left[-\exp(z)\right]\}^{\alpha\theta-1}\left[1 - \{1 - \exp\left[-\exp(z)\right]\}^{\theta}\right]\left[2 - \{1 - \exp\left[-\exp(z)\right]\}^{\theta}\right]^{\alpha-1}. \quad (15)$$

In many practical applications, the lifetimes are affected by explanatory variables such as the cholesterol level, blood pressure, weight and many others. Parametric regression models to estimate univariate survival functions for censored data are widely used. A parametric model that provides a good fit to lifetime data tends to yield more precise estimates of the quantities of interest. Based on the LTLGW density function, we propose a linear location-scale regression model for censored data linking the response variable y_i and the explanatory vector $\mathbf{v}_i^T = (v_{i1}, \ldots, v_{ip})$ as follows

$$y_i = \mathbf{v}_i^T \boldsymbol{\gamma} + \sigma z_i, \quad i = 1, \ldots, n, \quad (16)$$

where the random error z_i has density function (15), $\boldsymbol{\gamma} = (\gamma_1, \ldots, \gamma_p)^T$, $\sigma > 0$, $a > 0$ and $b > 0$ are unknown parameters. The parameter $\mu_i = \mathbf{v}_i^T \boldsymbol{\gamma}$ is the location of y_i. The location parameter vector $\boldsymbol{\mu} = (\mu_1, \ldots, \mu_n)^T$ is given by a linear model $\boldsymbol{\mu} = \mathbf{V}\boldsymbol{\gamma}$, where $\mathbf{V} = (\mathbf{v}_1, \ldots, \mathbf{v}_n)^T$ is a known model matrix. The LTLGW regression model (16) opens new possibilities for fitting many different types of censored data. It is an extension of an accelerated failure time model using the TLGW distribution for censored data.

Consider a sample $(y_1, \mathbf{v}_1), \ldots, (y_n, \mathbf{v}_n)$ of n independent observations, where each random response is defined by $y_i = \min\{\log(x_i), \log(c_i)\}$. We assume non-informative censoring such that the observed lifetimes and censoring times are independent. Let F and C be the sets of individuals for which y_i is the log-lifetime or log-censoring, respectively. Conventional likelihood estimation techniques can be applied here. The log-likelihood function for the vector of parameters $\boldsymbol{\tau} = (\alpha, \theta, \sigma, \boldsymbol{\gamma}^T)^T$ from model (16) has the form

$$l(\boldsymbol{\tau}) = \sum_{i\in F} l_i(\boldsymbol{\tau}) + \sum_{i\in C} l_i^{(c)}(\boldsymbol{\tau}),$$

where $l_i(\boldsymbol{\tau}) = \log[f(y_i|\mathbf{v}_i)]$, $l_i^{(c)}(\boldsymbol{\tau}) = \log[S(y_i|\mathbf{v}_i)]$, $f(y_i|\mathbf{v}_i)$ is the density (13) and $S(y_i|\mathbf{v}_i)$ is the survival function (14) of Y_i. The total log-likelihood function for $\boldsymbol{\tau}$ reduces to

$$\begin{aligned} l(\boldsymbol{\tau}) \;=\; & r\log\left(\frac{2\,\alpha\,\theta}{\sigma}\right) + \sum_{i\in F} z_i - \sum_{i\in F}\exp(z_i) + (\alpha\,\theta - 1)\sum_{i\in F}\log\{1 - \exp[-\exp(z_i)]\} + \sum_{i\in F}\log\left[1 - \{1 - \exp[-\exp(z_i)]\}^{\theta}\right] \\ & +(\alpha - 1)\sum_{i\in F}\log\left[2 - \{1 - \exp[-\exp(z_i)]\}^{\theta}\right] + \sum_{i\in C}\log\left\{1 - \left\{1 - \exp[-\exp(z_i)]\right\}^{\alpha\theta}\left[2 - \{1 - \exp[-\exp(z_i)]\}^{\theta}\right]^{\alpha}\right\}, \quad (17) \end{aligned}$$

where $z_i = (y_i - \mathbf{v}_i^T\boldsymbol{\gamma})/\sigma$ and r is the number of uncensored observations (failures). The score functions for the parameters a, b, β, σ and $\boldsymbol{\gamma}$ are given by $U_\alpha(\boldsymbol{\tau})$, $U_\theta(\boldsymbol{\tau})$, $U_\sigma(\boldsymbol{\tau})$ and $U_a(\gamma_j)$ for $j = 1, \ldots, p$.

The MLE $\widehat{\boldsymbol{\tau}}$ of $\boldsymbol{\tau}$ is obtained by solving the nonlinear equations $U_\alpha(\boldsymbol{\tau}) = 0$, $U_\theta(\boldsymbol{\tau}) = 0$, $U_\sigma(\boldsymbol{\tau}) = 0$ and $U_{\gamma_j}(\boldsymbol{\tau}) = 0$ simultaneously. These equations cannot be solved analytically and statistical software can be used to solve them numerically.

We can use iterative techniques such as a Newton-Raphson type algorithm to calculate the estimate $\widehat{\boldsymbol{\tau}}$.

We use the subroutine NLMixed in SAS to compute $\widehat{\boldsymbol{\tau}}$. Initial values for $\boldsymbol{\gamma}$ and σ are taken from the fit of the log-Weibull regression model. The fit of the LTLGW model produces the estimated survival function for y_i given by

$$S(y_i;\hat{\alpha},\hat{\theta},\hat{\sigma},\widehat{\boldsymbol{\gamma}}^T) = 1-\{1-\exp[-\exp(\hat{z}_i)]\}^{\hat{\alpha}\hat{\theta}}\left[2-\{1-\exp[-\exp(\hat{z}_i)]\}^{\hat{\theta}}\right]^{\hat{\alpha}}, \tag{18}$$

where $\hat{z}_i = (y_i - \mathbf{v}_i^T\hat{\boldsymbol{\gamma}})/\hat{\sigma}$.

Under standard regularity conditions, the approximate multivariate normal distribution $N_{p+4}(0, J(\boldsymbol{\tau})^{-1})$ for $\widehat{\boldsymbol{\tau}}$ can be used in the classical way to construct confidence intervals for the parameters in $\boldsymbol{\tau}$, where $J(\boldsymbol{\tau})$ is the observed information matrix. Further, we can use LR statistics for comparing some special sub-models with the LTLGW model. We consider the partition $\boldsymbol{\tau} = (\boldsymbol{\tau}_1^T, \boldsymbol{\tau}_2^T)^T$, where $\boldsymbol{\tau}_1$ is a subset of parameters of interest and $\boldsymbol{\tau}_2$ is a subset of the remaining parameters. The LR statistic for testing the null hypothesis $H_0 : \boldsymbol{\tau}_1 = \boldsymbol{\tau}_1^{(0)}$ versus the alternative hypothesis $H_1 : \boldsymbol{\tau}_1 \neq \boldsymbol{\tau}_1^{(0)}$ is $w = 2\{\ell(\widehat{\boldsymbol{\tau}}) - \ell(\widetilde{\boldsymbol{\tau}})\}$, where $\widetilde{\boldsymbol{\tau}}$ and $\widehat{\boldsymbol{\tau}}$ are the estimates under the null and alternative hypotheses, respectively. The statistic w is asymptotically (as $n \to \infty$) distributed as χ_k^2, where k is the dimension of the subset of parameters $\boldsymbol{\tau}_1$ of interest.

6. Applications

In this section, we compare the fit of the TLGW distribution to two real uncensored data sets. Further, we fit the new regression model based on the LTLGW distribution to the censored data set studied by Efron (1988). All the computations were performed using the procedure NLMixed in SAS and the R statistical software.

6.1 Application 1: Uncensored Data Sets

In this subsection we will provide some applications of TLGW distribution to model real data set. We will compare the fit of the model with another generalization proposed and studied by Alzaatreh, Lee, & Famoye (2013) (we call it ALF generator). The corresponding cdf of ALF generator is given by $F(x) = 1 - e^{-\alpha\{-\log[1-G(x)]\}^{\theta}}$. Therefore the cdf of ALFW distribution is $F_{ALFW}(x) = 1 - e^{-\alpha[(\eta x)^{\beta}]^{\theta}}$.

The measures of goodness-of-fit including the log-likelihood function evaluated at the MLEs ($\hat{\ell}$), Akaike information criterion (AIC), Anderson-Darling (A^*), Cramer-von Mises (W^*) and Kolmogorov-Smirnov (K-S), are calculated to compare the fitted models. In general, the smaller the values of these statistics, the better the fit to the data.

Data set 1: This data set consists of the waiting times(in seconds), between 65 successive eruptions of the Kiama Blowhole. These values were recorded with the aid of digital watch on July 12, 1998 by Jim Irish and has been referenced by several authors including da Silva, Thiago, Maciel, Campos, & Cordeiro(2013) and Pinho, Cordeiro & Nobre(2015). The actual data are:
83, 51, 87, 60, 28, 95, 8, 27, 15, 10, 18, 16, 29, 54, 91, 8, 17, 55, 10, 35, 47, 77, 36, 17, 21, 36, 18, 40, 10, 7, 34, 27, 28, 56, 8, 25, 68, 146, 89, 18, 73, 69, 9, 37, 10, 82, 29, 8, 60, 61, 61, 18, 169, 25, 8, 26, 11, 83, 11, 42, 17, 14, 9, 12.
In Table 3 we provide the MLEs and their corresponding standard errors (in parentheses) of the model parameters. The values of $\widehat{\ell}$, AIC, A^* , W^* and K-S statistic are provided in Table 4.

Table 3. Estimated parameters- Kiama Blowhole data

model	α	β	θ	η
TLGW	4.7964	0.5371	1.3044	0.0503
	(2.9824)	(0.2172)	(0.8547)	(0.0585)
ALFW	0.6500	0.9525	1.3378	0.0324
	(1.7765)	(3.7659)	(5.2894)	(0.0695)

Table 4. The $\hat{\ell}$, AIC, A^*, W^* and K-S Statistics-Kiama Blowhole data

model	$-\hat{\ell}$	AIC	A^*	W^*	K-S	p-value(K-S)
TLGW	294.3426	600.0873	0.8042	0.1102	0.0929	0.6373
ALFW	296.9001	601.8003	1.0079	0.1471	0.1113	0.4060

Data set 2: This data set includes an active repair time (in hours) for an airborne communication transceiver reported by Balakrishnan, Leiva, Sanhuzea, & Cabrera (2009), which was originally given by Chhikara & Folks (1989). The actual observations are listed below.
0.2, 0.3, 0.5, 0.5, 0.5, 0.5, 0.6, 0.6, 0.7, 0.7, 0.7, 0.8, 0.8, 1.0, 1.0, 1.0, 1.0, 1.1, 1.3, 1.5, 1.5, 1.5, 1.5, 2.0, 2.0, 2.2, 2.5, 2.7, 3.0, 3.0, 3.3, 3.3, 4.0, 4.0, 4.5, 4.7, 5.0, 5.4, 5.4, 7.0, 7.5, 8.8, 9.0, 10.3, 22.0, 24.5.

In Table 5 we provide the MLEs and the corresponding standard errors (in parentheses) of the model parameters. The values of $\widehat{\ell}$, AIC, W^* A^* and K-S statistic are provided in Table 6.

Table 5. Estimated parameters- repair time data

model	α	β	θ	η
TLGW	4.0709	0.3368	2.3032	2.1797
	(3.0202)	(0.1359)	(1.9328)	(4.5138)
ALFW	0.4004	0.2324	3.8658	0.8163
	(2.591)	(1.4755)	(24.5386)	(5.8762)

Table 6. The $\hat{\ell}$, AIC, A^*, W^* and K-S Statistic-repair time data

model	$-\hat{\ell}$	AIC	A^*	W^*	K-S	p-value(K-S)
TLGW	100.3413	208.6826	0.3699	0.0578	0.0919	0.8313
ALFW	104.4697	216.9394	0.9009	0.1298	0.1204	0.5174

From Tables 4 and 6 it is noted that the values for $-\widehat{\ell}$, AIC, A^*, W^* and K-S test statistic are lower for TLGW distribution than for ALFW distribution indicating that the TLGW distribution fits both the Kiama Blowhole data and repair time data better than ALFW distribution. Graphical comparisons of TLGW and ALFW models for the Kiama Blowhole data are provided in Figure 5. Similarly, the graphical comparisons of repair time data are provided in Figure 6. It is evident that the TLGW distribution fits both of these data better than ALFW distribution.

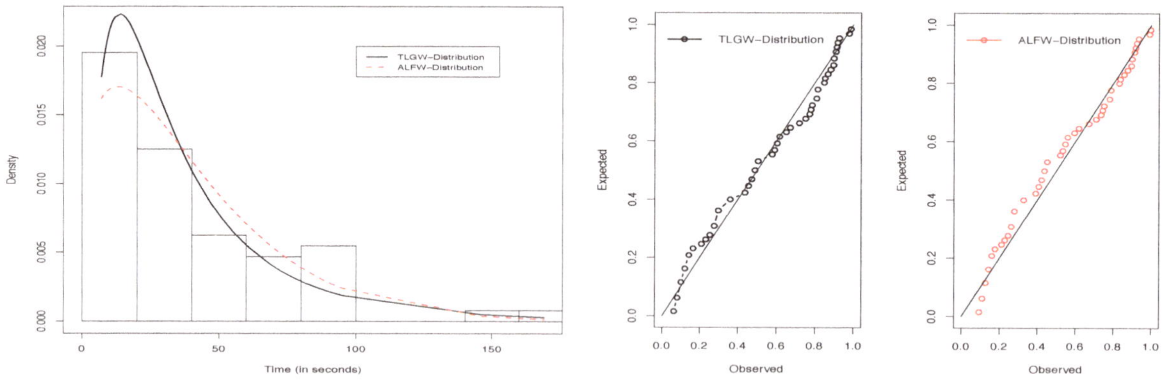

Figure 5. Fitted pdf and QQ plots of TLGW and ALFW distributions for the Kiama Blowhole data

6.2 Application 2: Regression Model With Censored Data

For an application of the LTLGW regression model, we consider the data from a two-arm clinical trial discussed earlier by Efron (1988). Efron (1988) and Mudholkar et al. (1996) observed that the empirical hazard functions for both samples start near zero, suggesting an initial high-risk period in the beginning, a decline for a while and then stabilization after about one year. Specifically, Efron's data from a head and neck cancer clinical trial consist of survival times of 51 patients in arm A (17.6% censored data) who were given radiation therapy and 45 patients in arm B (31.1% censored data) who were given radiation plus chemotherapy. Mudholkar et al.(1996) analyzed these data separately using the exponentiated Weibull distribution. Here, we use the LTLGW regression model.

Let x_i be the survival time (in days) for the ith observation and v_{i1} be the two-Arm clinical trial (0=Arm A, 1=Arm B)

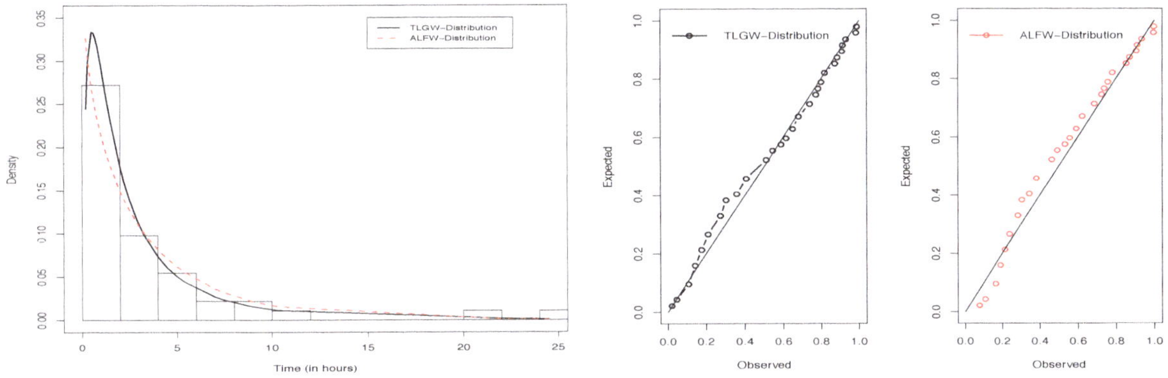

Figure 6. Fitted pdf and QQ plots of TLGW and ALFW distributions for the repair time data

(for other details, see Mudholkar et al. (1996). We propose the model

$$y_i = \gamma_0 + \gamma_1 v_{i1} + \sigma z_i,$$

where the random variable $y_i = \log(x_i)$ follows the LTLGW distribution (13) for $i = 1, 2, \ldots, 96$.

An alternative approach for modeling these data can be provided by the log-Weibull (LW) distribution. There are various extensions of this lifetime distribution; see, for example, the log-beta Weibull (LBW) (Ortega, et al. (2011)) distribution. The LBW density function is given by

$$f(y; a, b, \sigma, \mu) \quad = \quad \frac{1}{\sigma B(a,b)} \exp\Big\{\Big(\frac{y-\mu}{\sigma}\Big) - b\exp\Big(\frac{y-\mu}{\sigma}\Big)\Big\}\Big\{1 - \exp\Big[-\exp\Big(\frac{y-\mu}{\sigma}\Big)\Big]\Big\}^{a-1}, \tag{19}$$

where $-\infty < y < \infty$, $\sigma > 0$ and $-\infty < \mu = \gamma_0 + \gamma_1 v_1 < \infty$. The special case $a = b = 1$ corresponds to the LW distribution.

The MLEs of the model parameters are computed using the procedure NLMixed in SAS. As initial values for $\boldsymbol{\gamma}$ and σ in the iterative algorithm for maximizing the log-likelihood function (17), we adopt the fitted values obtained by fitting the LW regression model. The MLEs of the parameters and the AIC, CAIC and BIC statistics for some models are listed in Table 6.2.

Table 7. MLEs of the parameters from the LTLGW regression model fitted to the Efron data, the corresponding SEs (given in parentheses), p-values in [.] and the basic statistics.

Model	α	θ	σ	β_0	β_1	AIC	CAIC	BIC
LTLGW	25.2819	45.4687	11.9181	-14.9062	0.5202	295.3	296.0	308.2
	(5.6059)	(8.0643)	(2.6627)	(3.5488)	(0.2646)			
				[0.5282]	[0.0522]			
	a	b	σ	β_0	β_1			
LBW	244.20	162.98	19.4117	7.2415	0.5905	300.0	300.7	312.8
	(0.7614)	(0.4759)	(2.2409)	(0.2525)	(0.4340)			
				[<0.0001]	[0.1768]			
LW	1	1	1.1800	6.0387	0.7486	312.6	312.8	320.3
			(0.1081)	(0.1828)	(0.2772)			
				[<0.0001]	[0.0082]			

These results from Table 6.2 indicate that the LTLGW model has the lowest AIC, CAIC and BIC values among the fitted models, so the LTLGW model provides an appropriate fit for these data.

Further, the fitted LTLGW regression model suggests that x_1 is significant at 6% and that there is a significant difference between the two-Arm clinical trial. We plot in Figure 6.2 the empirical survival function and the estimated survival functions for the LTLGW, LBW and LW models. These plots suggest that the LTLGW model provides a suitable fit.

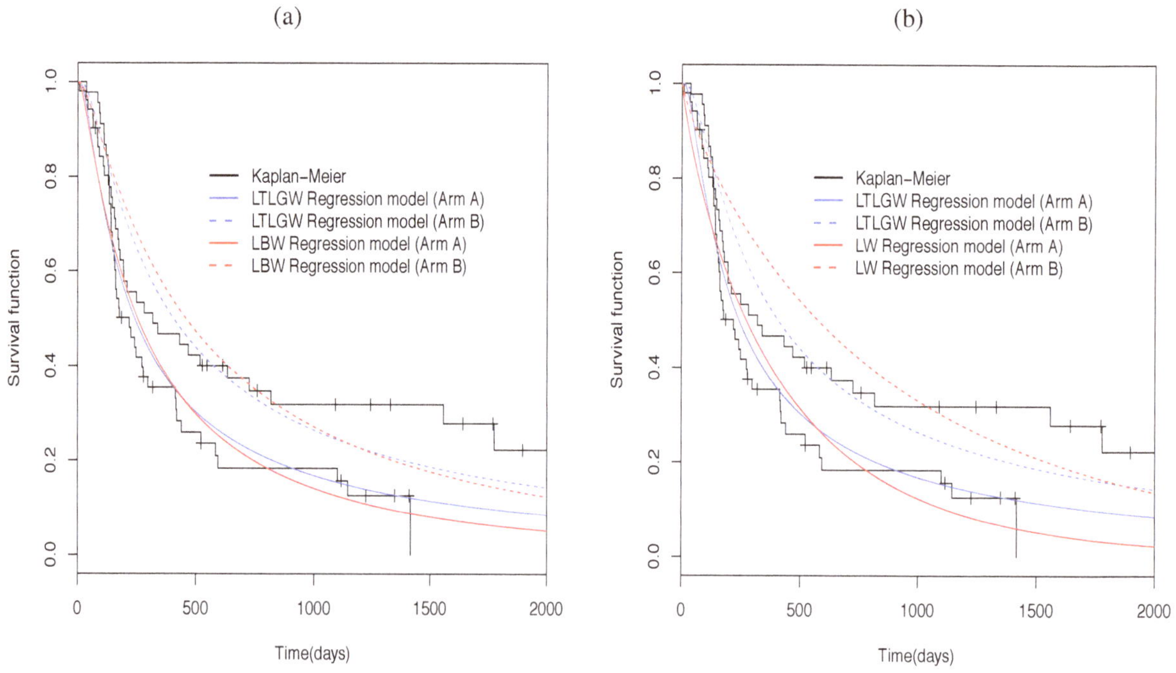

Figure 7. Estimated survival functions and the empirical survival for the Efron data. (a) LTLGW versus LBW regression model. (b) LTLGW versus LW regression model.

7. Concluding Remarks

In this paper, we propose a new four parameter model, called the *Topp-Leone Generated Weibull* (TLGW) distribution, which extends the Weibull distribution. In fact, the TLGW distribution is motivated by the extensive use of the Weibull distribution in many applied areas and also for the fact that the generalization provides more flexibility to analyze real data. The TLGW density function can be expressed as a mixture of epx-W densities. We derive explicit expressions for the ordinary and incomplete moments, moment generating function, entropies, mean residual life and mean inactivity time. We also provide some useful characterizations of the new model. We discuss the maximum likelihood estimation of the model parameters. We introduce the so-called *log-Topp-Leone Generated Weibull* (LTLGW) distribution, based on this new distribution, we propose a LTLGW regression model very suitable for modeling censored and uncensored lifetime data. Applications to real data modeling has been discussed and it has been shown that the proposed model provides consistently better fit than the other competitive models.

References

Afify, A. Z., Cordeiro, G. M., Yousof, H. M., Saboor, A., & Ortega, E. M. M. (2016). The Marshall-Olkin additive Weibull distribution with variable shapes for the hazard rate. *Hacettepe Journal of Mathematics and Statistics*, Forthcoming.

Alzaatreh, A., Lee, C., & Famoye, F. A. (2013). A new method for generating families of continuous distributions, *Metron 71*, 63-79. http://dx.doi.org/10.1007/s40300-013-0007-y

Aryal, G. R., & Tsokos, C. P. (2011). Transmuted Weibull distribution: a generalization of the Weibull probability distribution. *European Journal of Pure and Applied Mathematics*, *4*, 89-102.

Aryal, G., & Elbatal, I. (2015). On the Exponentiated Generalized Modified Weibull Distribution. *Communications for Statistical Applications and Methods*, *22*, 333-348. http://dx.doi.org/10.5351/csam.2015.22.4.333

Balakrishanan, N., Leiva, V., Sanhuzea, A., & Cabrera, E. (2009). Mixture inverse Gaussian distributions and its transformations, moments and applications. *Statistics*, *43*, 91-104. http://dx.doi.org/10.1080/02331880701829948

Chhikara, R. S., & Folks, J. L. (1989). *The inverse Gaussian distribution*, Marcel Dekker, New York.

Cordeiro, G. M., Hashimoto, E. M., & Ortega, E. M. M. (2014). The McDonald Weibull model. *Statistics: A Journal of Theoretical and Applied Statistics*, *48*, 256-278.http://dx.doi.org/10.1080/02331888.2012.748769

Cordeiro, G. M., Ortega, E. M., & da Cunha, D. C. C. (2013). The exponentiated generalized class of distributions. *Journal of Data Science*, *11*, 1-27.

Cordeiro, G. M., Ortega, E. M., & Nadarajah, S. (2010). The Kumaraswamy Weibull distribution with application to failure data. *Journal of the Franklin Institute*, *347*, 1399-1429. http://dx.doi.org/10.1016/j.jfranklin.2010.06.010

Cordeiro, G. M., Ortega, E. M., & Silva, G. O. (2012). The Kumaraswamy modified Weibull distribution: theory and applications. *Journal of Statistical Computation and Simulation*, *84*, 1387-1411. http://dx.doi.org/10.1080/00949655.2012.745125

da Silva, R., Thiago, A., Maciel, D., Campos, R., & Cordeiro, G. (2013). A new lifetime model: the gamma extended Fréchet distribution. *Journal of Statistical Theory and Applications*, *12*, 39-54. http://dx.doi.org/10.2991/jsta.2013.12.1.4

Efron, B. (1988). Logistic regression, survival analysis and the Kaplan-Meier curve. *Journal of the American Statistical Association*, *83*, 414-425. http://dx.doi.org/10.2307/2288857

Elbatal, I., & Aryal, G. (2013). On the transmuted additive Weibull distribution. *Austrian Journal of Statistics*, *42*, 117–132.

Famoye, F., Lee, C., & Olumolade, O. (2005). The Beta-Weibull Distribution, *Journal of Statistical Theory and Applications*, *4*, 121-136.

Ghitany, M. E., Al-Hussaini, E. K., & Al-Jarallah, R. A. (2005). Marshall-Olkin extended Weibull distribution and its application to censored data. *Journal of Applied Statistics*, *32*, 1025-1034. http://dx.doi.org/10.1080/02664760500165008

Glänzel, W. (1987). A characterization theorem based on truncated moments and its application to some distribution families, *Mathematical Statistics and Probability Theory (Bad Tatzmannsdorf, 1986), Vol. B, Reidel, Dordrecht*, 75-84. http://dx.doi.org/10.1007/978-94-009-3965-3_8

Glänzel, W. (1990). Some consequences of a characterization theorem based on truncated moments, *Statistics: A Journal of Theoretical and Applied Statistics*, *21*, 613-618. http://dx.doi.org/10.1080/02331889008802273

Hanook, S., Shahbaz, M. Q., Mohsin, M., & Kibria, G. (2013). A Note on Beta Inverse Weibull Distribution, *Communications in Statistics - Theory and Methods*, *42*, 320-335. http://dx.doi.org/10.1080/03610926.2011.581788

Khan, M. S., & King, R. (2013). Transmuted modified Weibull distribution: a generalization of the modified Weibull probability distribution. *European Journal of Pure and Applied Mathematics*, *6*, 66-88.

Lai, C. D., Xie, M., & Murthy, D. N. P. (2001). Bathtub-shaped failure rate life distributions, *Chapter 3, in Advances in Reliability, vol. 20 of Handbook of Statistics*, pp. 69?04. http://dx.doi.org/10.1016/s0169-7161(01)20005-4

Lai, C. D., Xie, M., & Murthy, D. N. P. (2003). A modified Weibull distribution. *IEEE Transactions on Reliability*, *52*, 33-37.

Lee, C., Famoye, F., & Olumolade, O. (2007). Beta-Weibull distribution: some properties and applications to censored data. *Journal of modern applied statistical methods*, *6*, 17.

Merovci, F. and Elbatal, I. (2013). The McDonald modified Weibull distribution: properties and applications. arXiv preprint arXiv:1309.2961.

Mudholkar, G. S., & Srivastava, D. K. (1993). Exponentiated Weibull family for analyzing bathtub failure-real data. *IEEE Transactions on Reliability*, *42*, 299-302. http://dx.doi.org/10.1109/24.229504

Mudholkar, G. S., Srivastava, D. K., & Freimer, M. (1995). The exponentiated Weibull family: a reanalysis of the bus-motor-failure data. *Technometrics*, *37*, 436-445. http://dx.doi.org/10.1080/00401706.1995.10484376

Mudholkar, G. S., Srivastava, D. K., & Kollia, G. D. (1996). A generalization of the Weibull distribution with application to the analysis of survival data. *Journal of the American Statistical Association*, *91*, 1575-1583. http://dx.doi.org/10.1080/01621459.1996.10476725

Nadarajah, S. (2009). Bathtub-shaped failure rate functions. *Quality and Quantity*, *43*, 855- 863. http://dx.doi.org/10.1007/s11135-007-9152-9

Nofal, Z. M., Afify, A. Z., Yousof, H. M., & Cordeiro, G. M. (2015). The generalized transmuted-G family of distributions. Communications in Statistics-Theory and Methods, Forthcoming. http://dx.doi.org/10.1080/03610926.2015.1078478

Nofal, Z. M., Afify, A. Z., Yousof, H. M., & Louzada, F.(2016). Transmuted exponentiated additive Weibull distribution: properties and applications, Forthcoming.

Nofal, Z. M., Afify, A. Z., Yousof, H. M., Granzotto, D. C. T., & Louzada, F. (2016). Kumaraswamy transmuted exponentiated additive Weibull distribution. *International Journal of Statistics and Probability, 5*, 78-99. http://dx.doi.org/10.5539/ijsp.v5n2p78

Ortega, E. M. M., Cordeiro, G. M., & Hashimoto, E. M. (2011). A log-linear regression model for the Beta-Weibull distribution. *Communications in Statistics- Simulation and Computations*, *40*, 1206-1235. http://dx.doi.org/10.1080/03610918.2011.568150

Pinho, L. G., Cordeiro, G. M., & Nobre, J. S. (2015). The Harris extended exponential distribution. *Communications in Statistics- Theory and methods*, *44*, 3486-3502. http://dx.doi.org/10.1080/03610926.2013.851221

Provost, S. B., Saboor, A., & Ahmad, M. (2011). The gamma-Weibull distribution, *Pak. Journal Stat.*, *27*, 111–131.

Rezaei, S., Sadr, B. B., Alizadeh, M., & Nadarajah, S. (2016). Topp-Leone generated family of distributions: Properties and applications. *Communications in Statistics- Theory and Methods*, Forthcoming. http://dx.doi.org/10.1080/03610926.2015.1053935

Sarhan, A. M., & Zaindin, M. (2009). Modified Weibull distribution, *Applied Sciences*,*11*, 123-136.

Shahbaz, M. G., Shahbaz, S., & Butt, N. M. (2102). The Kumaraswamy-inverse Weibull distribution. *Pak. Journal Stat. Oper. Res.*, *8*, 479-489. http://dx.doi.org/10.18187/pjsor.v8i3.520

Silva, G. O., Ortega, E. M. M., & Cordeiro, G. M. (2010). The beta modified Weibull distribution. *Lifetime Data Analysis*, *16*, 409-430. http://dx.doi.org/10.1007/s10985-010-9161-1

Xie, M., Tang, Y., & Goh, T. N. (2002). A modified Weibull extension with bathtub failure rate function. *Reliability Engineering and System Safety*, *76*, 279-285.http://dx.doi.org/10.1016/s0951-8320(02)00022-4

Xie, M. & Lai, C. D. (1995). Reliability analysis using an additive Weibull model with bathtub-shaped failure rate function. *Reliability Engineering and System Safety*, *52*, 87-93. http://dx.doi.org/10.1016/0951-8320(95)00149-2

Yousof, H. M., Afify, A. Z., Alizadeh, M., Butt, N. S., Hamedani, G. G., & Ali, M. M. (2015). The transmuted exponentiated generalized-G family of distributions. *Pak. J. Stat. Oper. Res.*, *11*, 441-464. http://dx.doi.org/10.18187/pjsor.v11i4.1164

Appendix A

Theorem 1. Let $(\Omega, \mathcal{F}, \mathbf{P})$ be a given probability space and let $H = [d, e]$ be an interval for some $d < e$ ($d=-\infty$, $e = \infty$ might as well be allowed). Let $X : \Omega \to H$ be a continuous random variable with the distribution function F and let q_1 and q_2 be two real functions defined on H such that

$$\mathbf{E}\left[q_2(X) \mid X \geq x\right] = \mathbf{E}\left[q_1(X) \mid X \geq x\right] \xi(x), \quad x \in H,$$

is defined with some real function ξ. Assume that $q_1, q_2 \in C^1(H)$, $\xi \in C^2(H)$ and F is twice continuously differentiable and strictly monotone function on the set H. Finally, assume that the equation $\xi q_1 = q_2$ has no real solution in the interior of H. Then F is uniquely determined by the functions q_1, q_2 and ξ , particularly

$$F(x) = \int_a^x C \left| \frac{\xi'(u)}{\xi(u) q_1(u) - q_2(u)} \right| \exp(-s(u)) \, du ,$$

where the function s is a solution of the differential equation $s' = \frac{\xi' q_1}{\xi q_1 - q_2}$ and C is the normalization constant, such that $\int_H dF = 1$.

14

Statistical Study of Monthly Rainfall Trends by Using the Transmuted Power Lindley Distribution

Daniele C. T. Granzotto[1], Josmar Mazucheli[1] & Francisco Louzada[2]

[1] Universidade Estadual de Maringá, DEs, PR, Brazil

[2] Universidade de São Paulo, ICMC, SP, Brazil

Correspondence: Daniele C. T. Granzotto, Universidade Estadual de Maringá, DEs, PR, Brazil. E-mail: dctgranzotto@uem.br

Abstract

In this article, we generalize the power Lindley distribution using a quadratic rank transmutation map to develop a transmuted power Lindley distribution. The new distribution exhibits, in addition to decreasing, increasing and bathtub hazard rate, depending on its parameters also unimodal hazard rate. A comprehensive mathematical properties of this distribution is provided. Some expressions for the moments, order statistics, quantiles function are derived. The model parameters are estimated by the maximum likelihood method. A Monte Carlo experiment on the finite sample behavior of the MLEs is performed. A real climatological data set was used in order to show the applicability of the new model and different statistics of fit were used as selection criteria.

Keywords: power Lindley distribution, transmutation map, reliability, rainfall data

1. Introduction

The Lindley distribution was proposed by Lindley (1958) and have been widely used in survival analysis and reliability fields. This distribution uses a mixture of exponential and length biased exponential distributions to illustrate the different between fiducial and posterior distributions. Late years, different applications and modifications have been proposed for this model such as: Ghitany et al. (2008) that argue that the Lindley distribution could be a better lifetime model than the exponential distribution through a numerical example; Nadarajah et al. (2011) and Zakerzadeh and Dolati (2009) in the proposition of a generalization; Merovci and Elbatal (2014) introduced a new lifetime distribution;Warahena-Liyanage and Pararai (2014) proposed an exponentiated power Lindley distribution with applications.

Those cited papers translate the concern with the proposition of new survival probability models based on the Lindley distribution. Also, various are the papers extending standard survival distributions in general, designed to serve as statistical survival models for a wide range of real lifetime phenomena. The challenge is the derivation of statistical survival probability models or simply survival distributions of real world lifetime phenomena that can represent more consistently the random behavior of experimental observations.

A convenient way to construct new distributions, in particular survival ones, are transmutation maps proposed by Shaw and Buckley (2007). The transmutation maps comprise the functional composition of a cumulative distribution function on a distribution with the inverse cumulative distribution (quantile) function of a non-Gaussian distributions, see for example Aryal and Tsokos (2009, 2011) that transmuted some models of Gamma distribution family and Granzotto and Louzada (2014); Louzada and Granzotto (2015) that proposed the transmuted log-logistic distribution and the regression extension of this model.

In this paper, we introduce a new lifetime distribution by transmuted and compounding power Lindley distribution named Transmuted Power Lindley (TPL) distribution. Briefly, it is the functional composition of a cumulative distribution function on a distribution with the inverse cumulative distribution (quantile) function of a non-Gaussian distribution, see Shaw and Buckley (2007). In this case, it incorporates a new third parameter (in our case ęË), what introduces a skewnwess and preserve the moments of the distribution base, see for example Shaw and Buckley (2007) and Granzotto and Louzada (2014). Although the TPL model is a positive distribution that can be applied for modeling on several areas such as reliability analysis, reliability along with engineer, hydrology, economics (income inequality) datasets; in this paper we proposed to analyse a real climatological dataset.

The paper is organized as follows. A background with the Lindley and its generalization are presented in Section 2 beyond the genesis of the transmutation map and the distributions Lindley and power Lindley. The derivation of the transmuted generalized Lindley distribution is presented in Section 3. Various important properties such as moments, moment generating function, quantiles, residual life, etc, for the transmuted Lindley distribution, as well as the minimum, maximum and median order statistics are presented in Section 4. Section 5 presents the maximum likelihood estimates and the asymptotic confidence intervals of the unknown parameters. In Section 6, the results of a simulation study is provided as well as the new distribution is illustrated in a climatological real data set, where we presented seven different statistics of fit were used as selection criteria. Final remarks are presented in Section 7.

2. Background

In this section we present a review of the Lindley and the Power Lindley distributions along with the transmutation map method, that are necessary to introduce the new model, TPL.

2.1 The Lindley Distribution

Proposed by Lindley (1958), the Lindley distribution is a exponential mixture that is important for studying stress-strength reliability modeling. Let X be a nonnegative random variable denoting the lifetime of an individual in some population. The random variable X is said to be Lindley distributed if the cumulative distribution function (c.d.f.) is given by

$$F_L(x,\theta) = 1 - (1 + \frac{\theta x}{\theta+1})e^{-\theta x}, \qquad x > 0, \theta > 0, \tag{1}$$

and the corresponding probability density function (p.d.f.) is given by

$$f_L(x,\theta) = \frac{\theta^2}{\theta+1}(1+x)e^{-\theta x}; \qquad x > 0, \theta > 0. \tag{2}$$

Ghitany et al. (2008) argue that the Lindley distribution could be a better lifetime model than the exponential distribution through a numerical example. In addition, they show that the hazard function of the Lindley distribution does not exhibit a constant hazard rate, indicating the flexibility of the Lindley distribution over the exponential distribution.

2.2 The Power Lindley Distribution

Ghitany et al. (2013) proposed new distribution, so called Power Lindley (GL) distribution, for modeling lifetime data. As the authors showed in their paper, they aim to discuss some properties of the power Lindley distribution which was formulated by using a power transformation $X = T^{1/\alpha}$. The paper included the shapes of the density and hazard rate functions, the moments and some associated measures, the quantile function, and the limiting distributions of order statistics. Also, the maximum likelihood estimation of the model parameters and their asymptotic standard errors are derived.

Let X be a nonnegative random variable denoting the lifetime of an individual in some population. The random variable X is said to be power Lindley distributed with parameters θ and α if its cumulative density function (c.d.f.) is given by

$$F_{PL}(x,\theta,\alpha) = \left[1 - (1 + \frac{\theta x^\alpha}{\theta+1})e^{-\theta x^\alpha}\right], \tag{3}$$

where $\theta > 0$ and $\alpha > 0$. The corresponding probability density function (p.d.f.) and the hazard (failure) rate function are given, respectively, by

$$f_{PL}(x,\theta,\alpha) = \frac{\alpha\theta^2}{\theta+1}(1+x^\alpha)x^{\alpha-1}e^{-\theta x^\alpha}; \quad x > 0, \theta, \alpha > 0 \tag{4}$$

and

$$h_{PL}(x,\theta,\alpha) = \frac{\alpha\theta^2}{\theta+1+\theta x^\alpha}(1+x^\alpha)x^{\alpha-1}. \tag{5}$$

Note that equation (3) has two parameters, θ and α, just like the gamma, lognormal, Weibull and exponentiated exponential distributions. Note also that equation (5) has the attractive feature of allowing for monotonically decreasing, monotonically increasing and bathtub shaped hazard rate functions while not allowing for constant hazard rate functions.

2.3 Transmutation Map

Let F_1 and F_2 be the cumulative distribution functions, of two distributions with a common sample space. The general rank transmutation as given in **?** is defined as

$$G_{R12}(u) = F_2(F_1^{-1}(u)) \text{ and } G_{R21}(u) = F_1(F_2^{-1}(u)).$$

Note that the inverse cumulative distribution function also known as quantile function is defined as

$$F^{-1}(y) = \inf_{x\in R}\{F(x) \geq y\} \text{ for } y \in [0,1].$$

The functions $G_{R12}(u)$ and $G_{R21}(u)$ both map the unit interval $I = [0, 1]$ into itself, and under suitable assumptions are mutual inverses and they satisfy $G_{Rij}(0) = 0$ and $G_{Rij}(1) = 1$. A Quadratic Rank Transmutation Map (QRTM) is defined as

$$G_{R12}(u) = u + \lambda u(1-u), |\lambda| \leq 1, \tag{6}$$

from which it follows that the cdf's satisfy the relationship

$$F_2(x) = (1+\lambda)F_1(x) - \lambda F_1(x)^2 \tag{7}$$

which on differentiation yields,

$$f_2(x) = f_1(x)\left[(1+\lambda) - 2\lambda F_1(x)\right], \tag{8}$$

where $f_1(x)$ and $f_2(x)$ are the corresponding pdfs associated with cdf $F_1(x)$ and $F_2(x)$ respectively. An extensive information about the quadratic rank transmutation map is given in Shaw and Buckley (2007). Observe that at $\lambda = 0$ we have the distribution of the base random variable. The following Lemma proved that the function $f_2(x)$ in given (8) satisfies the property of probability density function. Note that $f_2(x)$ given in (8) is a well defined probability density function. Rewriting $f_2(x)$ as $f_2(x) = f_1(x)\left[(1 - \lambda(2F_1(x) - 1)\right]$ we observe that $f_2(x)$ is nonnegative. We need to show that the integration over the support of the random variable is equal one. Consider the case when the support of $f_1(x)$ is $(-\infty, \infty)$. In this case we have

$$\begin{aligned}\int_{-\infty}^{\infty} f_2(x)dx &= \int_{-\infty}^{\infty} f_1(x)\left[(1+\lambda) - 2\lambda F_1(x)\right]dx \\ &= (1+\lambda)\int_{-\infty}^{\infty} f_1(x)dx - \lambda\int_{-\infty}^{\infty} 2f_1(x)F_1(x)dx \\ &= (1+\lambda) - \lambda = 1.\end{aligned}$$

Similarly, other cases where the support of the random variable is a part of real line follows. Hence $f_2(x)$ is a well defined probability density function. We call $f_2(x)$ the transmuted probability density of a random variable with base density $f_1(x)$. Also note that when $\lambda = 0$ then $f_2(x) = f_1(x)$.

3. The Transmuted Power Lindley Distribution

Let X be a nonnegative random variable denoting the lifetime of an individual in some population. The random variable X is said to be Transmuted Power Lindley (TPL) with parameters θ, α and λ if its cumulative density function (c.d.f.) is given by

$$\begin{aligned}F_{TPL}(x,\theta,\alpha,\lambda) &= G(x)\left[(1+\lambda) - \lambda G(x)\right] \quad (9)\\ &= \left[1 - \left(1 + \frac{\theta x^\alpha}{\theta+1}\right)e^{-\theta x^\alpha}\right]\left[1 + \lambda\left(1 + \frac{\theta x^\alpha}{\theta+1}\right)e^{-\theta x^\alpha}\right]\end{aligned}$$

where $\theta, \alpha > 0$ and $\lambda \in (-1, 1)$. The corresponding probability density function (p.d.f.) of the transmuted power Lindley is given by

$$\begin{aligned}f_{TPL}(x,\theta,\alpha,\lambda) &= \frac{\alpha\theta^2}{\theta+1}(1+x^\alpha)x^{\alpha-1}e^{-\theta x^\alpha} \quad (10)\\ &\quad \times\left[1 + \lambda - 2\lambda\left(1 - \left(1 + \frac{\theta x^\alpha}{\theta+1}\right)e^{-\theta x^\alpha}\right)\right].\end{aligned}$$

The transmuted power Lindley distribution is an extended model to analyse data from complex situations and it generalizes some of the widely used distributions in reliability analysis. The power Lindley distribution is clearly a special case for $\lambda = 0$ (see, Ghitany et al. (2013)). Also, the density and cumulative density curves of transmuted power model, for different parameters values can be seen in Figures 1, upper left and right panels, respectively.

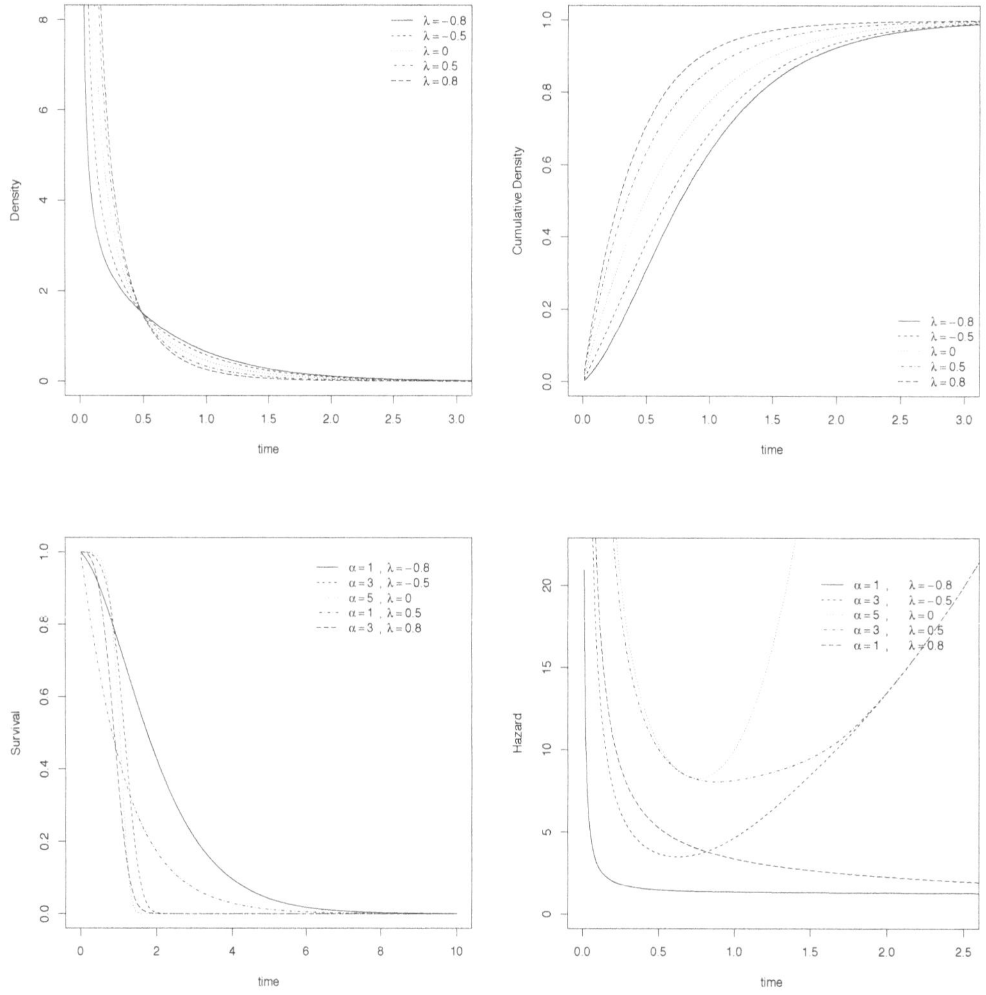

Figure 1. Upper panels: Density and Cumulative Density for fixed $\alpha = \theta = 1$; Lower panels: Survival and Hazard curves for fixed $\theta = 1$.

The reliability function of the transmuted power Lindley model is denoted by $R_{TPL}(t)$ and is defined as

$$R_{TPL}(t, \theta, \alpha, \lambda) = 1 - F_{TPL}(t) \tag{11}$$

$$= 1 - \left[1 - (1 + \frac{\theta x^{\alpha}}{\theta + 1})e^{-\theta x^{\alpha}}\right]\left[1 + \lambda(1 + \frac{\theta x^{\alpha}}{\theta + 1})e^{-\theta x^{\alpha}}\right].$$

For different parameters values the estimated curves can be seen in Figures 1, left lower panels. One of the characteristic in reliability analysis is the hazard rate function defined by

$$h_{TPL}(t, \theta, \alpha, \lambda) = \frac{f_{TPL}(t)}{1 - F_{TPL}(t)} \tag{12}$$

It is important to note that the units for $h_{TPL}(t)$ is the probability of failure per unit of time, distance or cycles. These failure rates are defined with different choices of parameters, see Figures 1, right lower panels.

The cumulative hazard function of the model is defined as

$$H_{TPL}(t,\theta,\alpha,\lambda) = -\ln\left|1-\left[1-(1+\frac{\theta x^{\alpha}}{\theta+1})e^{-\theta x^{\alpha}}\right]\right. \tag{13}$$

$$\left.\times\left[1+\lambda(1+\frac{\theta x^{\alpha}}{\theta+1})e^{-\theta x^{\alpha}}\right]\right|. \tag{14}$$

It is important to note that the units for $H_{TPL}(t)$ is the cumulative probability of failure per unit of time, distance or cycles. For all choice of parameters the distribution has the decreasing patterns of cumulative instantaneous failure rates.

4. Statistical Properties of TPL

4.1 Quantiles and Random Number Generation

The quantile x_q of the TPL is obtained from the following equation

$$F(x_q)=\left[1-(1+\frac{\theta x_q^{\alpha}}{\theta+1})e^{-\theta x_q^{\alpha}}\right]\left[1+\lambda(1+\frac{\theta x_q^{\alpha}}{\theta+1})e^{-\theta x_q^{\alpha}}\right]=q$$

setting $\phi=(1+\frac{\theta x_q^{\alpha}}{\theta+1})e^{-\theta x_q^{\alpha}}$ then we have

$$(1-\phi)(1+\lambda\phi)=q$$

by solving the above equation with respect to ϕ we get

$$\phi=\left[\frac{1-q}{\lambda}+\frac{(\lambda+1)^2}{4\lambda^2}\right]^{1/2}+\frac{\lambda-1}{2\lambda}.$$

Hence we can obtain the quantile x_q of the transmuted generalized Lindley as follows

$$(1+\frac{\theta x_q^{\alpha}}{\theta+1})e^{-\theta x_q^{\alpha}}=\left[\frac{1-q}{\lambda}+\frac{(\lambda+1)^2}{4\lambda^2}\right]^{1/2}+\frac{\lambda-1}{2\lambda}. \tag{15}$$

The above equation has no closed form solution in x_q, so we have to use a numerical technique to get the quantiles. In particular, put $q=0.5$ in equation (15) one gets the median of transmuted power Lindley $(\alpha,\theta,\lambda,x)$.

Thus, random number generation as x of the transmuted power Lindley $(\alpha,\theta,\lambda,x)$ is defined by the following relation

$$\left[1-(1+\frac{\theta x_q^{\alpha}}{\theta+1})e^{-\theta x_q^{\alpha}}\right]\left[1+\lambda(1+\frac{\theta x_q^{\alpha}}{\theta+1})e^{-\theta x_q^{\alpha}}\right]=u$$

where $u\sim U(0,1)$. This yields,

$$(1+\frac{\theta x_q^{\alpha}}{\theta+1})e^{-\theta x_q^{\alpha}}=\left[\frac{1-u}{\lambda}+\frac{(\lambda+1)^2}{4\lambda^2}\right]^{1/2}+\frac{\lambda-1}{2\lambda}. \tag{16}$$

Equation (8) above does not have a closed form solution so we generate u as uniform random variables from U(0, 1) and solve for x in order to generate random numbers from transmuted power Lindley distribution.

4.2 Moments

In this subsection we discuss the r_{th} moment for transmuted power Lindley distribution. Moments are necessary and important in any statistical analysis, especially in applications. It can be used to study the most important features and characteristics of a distribution (e.g., tendency, dispersion, skewness and kurtosis).

Then, the r_{th} moment is given by

$$\begin{aligned} E(X^r) &= \int_0^{+\infty} x^r \frac{\alpha\theta^2}{\theta+1}(1+x^\alpha)x^{\alpha-1}e^{-\theta x^\alpha} \\ &\times \left[1+\lambda-2\lambda\left(1-(1+\frac{\theta x^\alpha}{\theta+1})e^{-\theta x^\alpha}\right)\right]dx \\ &= \frac{\alpha\theta^2}{\theta+1}\int_0^{+\infty}(1+x^\alpha)x^{r+\alpha-1}e^{-\theta x^\alpha}\left[1-\lambda+2\lambda e^{-\theta x^\alpha}+\frac{2\lambda\theta}{\theta+1}x^\alpha e^{-\theta x^\alpha}\right]. \end{aligned} \tag{17}$$

By using Gamma function, in order to solve the equation above, we have

$$E(X^r) = \frac{\alpha\theta^{\frac{r-1}{\alpha}}}{\theta+1}\sum_{k=1}^{2}\theta^{1-k}\Gamma\left(\frac{r-1}{\alpha}+1+k\right)\left[1+\lambda 2^{-(\frac{r-1}{\alpha}+k)}\right]. \tag{18}$$

The mean of the model and the variance are given, respectively by

$$E(X) = \frac{\alpha}{\theta^3(\theta+1)}\left[\theta(2+\lambda)+(4+\lambda)\right] \tag{19}$$

and

$$\begin{aligned} V(X) &= \frac{\alpha\theta^{\frac{1+2\alpha}{\alpha}}}{\theta+1}\left\{\Gamma\left(\frac{1+2\alpha}{\alpha}\right)\left(1+\lambda 2^{-(1+\alpha)/\alpha}\right)+\right. \\ &\left.\Gamma\left(\frac{1+3\alpha}{\alpha}\right)\left(1+\lambda 2^{-(1+2\alpha)/\alpha}\right)\right\} - \frac{\alpha^2}{\theta^6(\theta+1)^2}\left[\theta(2+\lambda)+(4+\lambda)\right]^2. \end{aligned} \tag{20}$$

4.3 Distribution of the Order Statistics

According to Aryal and Tsokos (2011), suppose we have a system containing two components with each of them having independent and identical "base" distribution, for example power Lindley. If the components are connected in series then the overall system will have transmuted baseline distribution with $\lambda = 1$ whereas if the components are parallel then the overall system will have a transmuted baseline.

It has been observed that a transmuted power Lindley distribution with $\lambda = 1$ is the distribution of $\min(X_1, X_2)$ and a transmuted power Lindley distribution with $\lambda = -1$ is the distribution of the $\max(X_1, X_2)$ where X_1 and X_2 are independent and identically distributed 2-parameter power Lindley random variables.

In fact, the order statistics have many applications in reliability and life testing. The order statistics arise in the study of reliability of a system. Let $X_1, X_2, \ldots, X_n$ be a simple random sample from TPL$(\alpha, \theta, \lambda, x)$ with cumulative distribution function and probability density function as in (9) and (11), respectively. Let $X_{(1:n)} \leq X_{(2:n)} \leq \ldots \leq X_{(n:n)}$ denote the order statistics obtained from this sample. In reliability literature, $X_{(i:n)}$ denote the lifetime of an $(n-i+1)-$ out- of- n system which consists of n independent and identically components. Then the pdf of $X_{(i:n)}$, $1 \leq i \leq n$ is given by

$$f_{i::n}(x) = \frac{1}{\beta(i, n-i+1)}\left[F(x_{(i)})\right]^{i-1}\left[1-F(x_{(i)})\right]^{n-i} f(x_{(i)}) \tag{21}$$

also, the joint pdf of $X_{(i:n)}$, $X_{(j:n)}$ and $1 \leq i \leq j \leq n$ is

$$f_{i::j:n}(x_i, x_j) = C\ \left[F(x_i)\right]^{i-1}\left[F(x_j)-F(x_i)\right]^{j-i-1}\left[1-F(x_j)\right]^{n-j} f(x_i)f(x_j) \tag{22}$$

where

$$C = \frac{n!}{(i-1)!(j-i-1)!(n-j)!}.$$

We defined the first order statistics $X_{(1)} = \min(X_1, X_2, ..., X_n)$, the the last order statistics as $X_{(n)} = \max(X_1, X_2, \ldots, X_n)$ and median order X_{m+1} .

4.4 Distribution of Minimum, Maximum and Median

Let $X_{(1:n)} \leq X_{(2:n)} \leq \ldots \leq X_{(n:n)}$ be independently identically distributed order random variables from the transmuted generalized Lindley distribution having first , last and median order probability density function are given by the following

$$\begin{aligned} f_{1:n}(x) &= n\left[1 - F(x_{(1)})\right]^{n-1} f(x_{(1)}) \\ &= n\left[\frac{n\alpha\theta^2}{\theta+1}(1 + x_{(1)}^{\alpha})x_{(1)}^{\alpha-1} e^{-\theta x_{(1)}^{\alpha}}\right. \\ &\quad \left.\times \left[1 + \lambda - 2\lambda(1 - \zeta_{(1)})\right]\left[(1 - (1 - \zeta_{(1)}))(1 - \lambda\zeta_{(1)})\right]^{n-1}\right], \end{aligned} \tag{23}$$

where

$$\zeta_{(i)} = \left(1 + \frac{\theta x_{(i)}^{\alpha}}{\theta+1}\right) e^{-\theta x_{(i)}^{\alpha}}$$

$$\begin{aligned} f_{n:n}(x) &= n\left[F(x_{(n)}, \Phi)\right]^{n-1} f(x_{(n)}), \Phi) \\ &= n\left[\frac{n\alpha\theta^2}{\theta+1}(1 + x_{(n)}^{\alpha})x_{(n)}^{\alpha-1} e^{-\theta x_{(n)}^{\alpha}}\right. \\ &\quad \left.\times \left[1 + \lambda - 2\lambda(1 - \zeta_{(n)})\right]\left[(1 - (1 - \zeta_{(n)}))(1 - \lambda\zeta_{(n)})\right]^{n-1}\right] \end{aligned} \tag{24}$$

and

$$\begin{aligned} f_{m+1:n}(\widetilde{x}) &= \frac{(2m+1)!}{m!m!}(F(\widetilde{x}))^m(1 - F(\widetilde{x}))^m f(\widetilde{x}) \\ &= \frac{(2m+1)!}{m!m!}\left[(1 - \zeta_{(m+1)})(1 + \lambda\zeta_{(m+1)})\right]^m \\ &\quad \times \left[\frac{\alpha\theta^2}{\theta+1}(1 + x_{(m+1)}^{\alpha})x_{(m+1)}^{\alpha-1} e^{-\theta x_{(m+1)}^{\alpha}}\right. \\ &\quad \left.\times \left[1 + \lambda - 2\lambda(1 - \zeta_{(m+1)})\right]\left[(1 - (1 - \zeta_{(m+1)}))(1 - \lambda\zeta_{(m+1)})\right]^m\right]. \end{aligned} \tag{25}$$

We notice that the minimum, maximum and median order statistics of three parameters transmuted power Lindley distribution have different life time distributions when its parameters are changed.

5. Inference

In this section we consider the maximum likelihood estimators (MLE's) of transmuted power Lindley distribution. Let $\phi = (\alpha, \theta, \lambda)^T$,in order to estimate the parameters α, θ,and λ of transmuted power Lindley distribution, let $x_1, \ldots, x_n$ be a random sample of size n from TPL$(\alpha, \theta, \lambda)$, we obtain the likelihood function as follows

$$\begin{aligned} L(\alpha, \theta, \lambda) &= \left(\frac{\alpha\theta^2}{\theta+1}\right)^n \prod_{i=1}^{n}(1 + x_i^{\alpha})x_i^{\alpha-1} e^{-\theta x_i^{\alpha}} \\ &\quad \times \left[1 + \lambda - 2\lambda\left(1 - \left(1 + \frac{\theta x_i^{\alpha}}{\theta+1}\right) e^{-\theta x_i^{\alpha}}\right)\right], \end{aligned} \tag{26}$$

then the log likelihood function can be written as

$$\begin{aligned} \ln L(\alpha, \theta, \lambda) &= n \ln \alpha + 2n \ln \theta - n \ln(1 + \theta) + \sum_{i=1}^{n} \ln(1 + x_i^{\alpha}) \\ &\quad + (\alpha - 1)\sum_{i=1}^{n} \ln x_i - \theta \sum_{i=1}^{n} x_i^{\alpha} + \\ &\quad + \sum_{i=1}^{n} \ln\left[1 + \lambda - 2\lambda\left(1 - \omega_i e^{-\theta x_i^{\alpha}}\right)\right], \end{aligned} \tag{27}$$

where

$$\omega_i = \left(1 + \frac{\theta x_i^\alpha}{\theta + 1}\right).$$

Differentiating $\ln L(\alpha, \theta, \lambda)$ with respect to each parameter α, θ, and λ and setting the result equals to zero, we obtain maximum likelihood estimates. The partial derivatives of $\ln L(\alpha, \theta, \lambda)$ with respect to each parameter or the score function is given by

$$U_n(\phi) = (U_\alpha, U_\theta, U_\lambda)^T$$

where

$$\begin{aligned} U_\alpha &= \frac{\partial \ln L}{\partial \alpha} = \frac{n}{\alpha} + \sum_{i=1}^{n} \frac{x_i^\alpha \ln(x_i)}{(1 + x_i^\alpha)} + (1-\theta) \sum_{i=1}^{n} \ln x_i \\ &-2\lambda \sum_{i=1}^{n} \left\{ \frac{\theta x_i^\alpha \ln(x_i) e^{-\theta x_i^\alpha} \left[\omega_i - \frac{1}{1+\theta}\right]}{1 + \lambda - 2\lambda\left(1 - \omega_i e^{-\theta x_i^\alpha}\right)} \right\}, \end{aligned} \tag{28}$$

$$\begin{aligned} U_\theta &= \frac{\partial \ln L}{\partial \theta} = \frac{n(2+\theta)}{\theta(1+\theta)} - \sum_{i=1}^{n} \ln x_i^\alpha \\ &-2\lambda \sum_{i=1}^{n} x_i^\alpha e^{-\theta x_i^\alpha} \left\{ \frac{\omega_i - (1+\theta)^{-2}}{1 + \lambda - 2\lambda\left(1 - \omega_i e^{-\theta x_i^\alpha}\right)} \right\} \end{aligned} \tag{29}$$

and

$$U_\lambda = \frac{\partial \ln L}{\partial \lambda} = \sum_{i=1}^{n} \frac{2\omega_i e^{-\theta x_i^\alpha} - 1}{1 + \lambda - 2\lambda\left(1 - \omega_i e^{-\theta x_i^\alpha}\right)}. \tag{30}$$

The maximum likelihood estimation $\widehat{\phi} = (\widehat{\alpha}, \widehat{\theta}, \widehat{\lambda})^T$ of $\phi = (\alpha, \theta, \lambda)^T$ is obtained by solving the non linear equations $U_n(\phi) = 0$. These equations cannot be solved analytically but statistical software can be used to solve them numerically. For interval estimation and hypothesis tests on the model parameters, we require the information matrix. The 3×3 observed information matrix is given by

$$I_n(\varphi) = -\begin{bmatrix} I_{\alpha\alpha} & I_{\alpha\theta} & I_{\alpha\lambda} \\ I_{\theta\alpha} & I_{\theta\theta} & I_{\theta\lambda} \\ I_{\lambda\alpha} & I_{\lambda\theta} & I_{\lambda\lambda} \end{bmatrix},$$

where $I_n(\phi) = \frac{\partial^2 \ln L}{\partial \phi \partial \phi^T}$. Applying the usual large sample approximation, MLE of ϕ, i.e $\widehat{\phi}$ can be treated as being approximately $N_3(\phi, J_n(\phi)^{-1})$, where $J_n(\phi) = E\left[I_n(\phi)\right]$. Under conditions that are fulfilled for parameters in the interior of the parameter space but not on the boundary, the asymptotic distribution of $\sqrt{n}(\widehat{\phi} - \phi)$ is $N_3(0, J(\varphi)^{-1})$, where $J(\phi) = \lim_{n\to\infty} n^{-1} I_n(\phi)$ is the unit information matrix. This asymptotic behavior remains valid if $J(\phi)$ is replaced by the average sample information matrix evaluated at $\widehat{\phi}$, say $n^{-1} I_n(\widehat{\phi})$. The estimated asymptotic multivariate normal $N_3(\phi, I_n(\widehat{\phi})^{-1})$ distribution of $\widehat{\phi}$ can be used to construct approximate confidence intervals for the parameters and for the hazard rate and survival functions. An $100\%(1-\gamma)$ asymptotic confidence interval for each parameter ϕ_r is given by

$$ACI_r = \left(\widehat{\phi}_r - z_{\frac{\gamma}{2}} \sqrt{\widehat{I_{rr}}}, \widehat{\phi}_r + z_{\frac{\gamma}{2}} \sqrt{\widehat{I_{rr}}}\right),$$

where $\widehat{I_{rr}}$ is the (r, r) diagonal element of $I_n(\widehat{\varphi})^{-1}$ for $r = 1, 2, 3$, and $z_{\frac{\gamma}{2}}$ is the quantile $1 - \frac{\gamma}{2}$ of the standard normal distribution.

In order to compare the models seven different statistics of fit were used as selection criteria in Section 6.2: $-2\times$ log-likelihood (Neg2LogLike), Akaike's information criterion (AIC), corrected Akaike's information criterion (AICC), Kolmogorov-Smirnov statistic (KS), Anderson-Darling statistic (AD) and Cramér-von-Mises statistic (CvM).

The first ones, AIC and AICC, are widely used in reliability analysis as a selection criteria. The AIC can be obtained by using the following expression:

$$AIC = -2 \ln L_M(\zeta) + 2p,$$

with L_M the likelihood of the model M, ζ the vector of parameters to the model M and p the number of parameters to the model M. The AICC is given, respectively, by

$$AICC = -2\ln L_M(\zeta) + \frac{2p(p+1)}{n-p-1},$$

with n the number of observations.

Further, the Anderson-Darling and the Cramér-von Mises statistics are widely utilized to determine how closely a specific distribution whose associated cumulative distribution function fits the empirical distribution associated with a given data set. These statistics are

$$A^* = \left(\frac{9}{4n^2} + \frac{3}{4n} + 1\right)\left[n + \frac{1}{n}\sum_{j=1}^{n}(2j-1)\log\left[z_i\left(1 - z_{n-j+1}\right)\right]\right]$$

and

$$W^* = \left(\frac{1}{2n} + 1\right)\left[\sum_{j=1}^{n}\left(z_i - \frac{2j-1}{2n}\right)^2 + \frac{1}{12n}\right],$$

respectively, $z_i = F\left(y_j\right)$, where the y_j values being the *ordered observations*. The smaller these statistics are, the better the fit. Upper tail percentiles of the asymptotic distributions of these goodness–of–fit statistics were tabulated in Nichols and Padgett (2006).

6. Data Experiments

This section presents the results of a Monte Carlo experiment on the finite sample behavior of the MLEs as well as illustrate the applicability of the proposed distribution in various real data sets on rainfall.

6.1 Simulation Study

The Monte Carlo simulation results were obtained from 1,000 Monte Carlo replications. The sample sizes n range from 30 to 500, generated according to a transmuted power Lindley distribution for each combination of the parameter values α, λ and $\theta = 2$ fixed. Table 1 shows that the estimates and BIAS of the MLEs and Table 2 shows us the coverage probabilities of a 95% two sided confidence intervals for the model parameters and the mean square error which decrease with the increasing of the sample size.

Table 1. Parameters estimated and BIAS of the MLEs.

Sample Size	Generated		Estimate			BIAS		
	α	λ	θ	α	λ	θ	α	λ
30	3.0	−0.5	2.191	2.877	−0.278	0.191	0.123	0.222
50	3.0	−0.5	2.173	2.783	−0.292	0.173	0.217	0.208
100	3.0	−0.5	2.126	2.753	−0.258	0.126	0.247	0.242
300	3.0	−0.5	2.119	2.747	−0.255	0.119	0.253	0.245
500	3.0	−0.5	2.103	2.757	−0.237	0.103	0.243	0.263
30	8.0	−0.5	2.010	8.396	−0.421	0.010	0.396	0.079
50	8.0	−0.5	2.004	8.245	−0.442	0.004	0.245	0.058
100	8.0	−0.5	1.988	8.142	−0.445	0.012	0.142	0.055
300	8.0	−0.5	1.986	8.066	−0.458	0.014	0.066	0.042
500	8.0	−0.5	1.977	8.089	−0.448	0.023	0.088	0.052
30	5.0	0.5	2.12267	5.0985	0.47831	0.12267	0.09846	0.02169
50	5.0	0.5	2.11252	5.0225	0.45167	0.11252	0.0225	0.04833
100	5.0	0.5	2.08432	4.9765	0.44898	0.08432	0.02345	0.05102
300	5.0	0.5	2.05484	4.9635	0.46933	0.05484	0.03649	0.03067
500	5.0	0.5	2.05772	4.9686	0.46895	0.05772	0.03143	0.03105
30	3.0	−0.8	1.93496	3.2891	−0.75086	0.06504	0.28911	0.04914
50	3.0	−0.8	1.92625	3.2141	−0.76083	0.07375	0.21406	0.03917
100	3.0	−0.8	1.93109	3.1594	−0.77346	0.06891	0.15942	0.02654
300	3.0	−0.8	1.94909	3.0946	−0.81174	0.05091	0.0946	0.01174
500	3.0	−0.8	1.96397	3.0639	−0.83572	0.03603	0.06387	0.03572
30	8.0	0.2	2.20698	8.0716	0.40377	0.20698	0.07161	0.09623
50	8.0	0.2	2.21645	7.9576	0.36802	0.21645	0.04238	0.13198
100	8.0	0.2	2.24932	7.9094	0.3096	0.24932	0.09061	0.1904
300	8.0	0.2	2.30488	7.8905	0.25473	0.30488	0.10948	0.24527
500	8.0	0.2	2.31874	7.8921	0.2423	0.31874	0.10792	0.2577

Table 2. Mean square error (mse) of the MLEs and Coverage Probability.

Sample	Generated		MSE			Coverage Probability		
Size	α	λ	θ	α	λ	θ	α	λ
30	3.0	−0.5	0.715	0.921	0.834	0.998	0.881	0.859
50	3.0	−0.5	0.579	0.713	0.719	0.988	0.860	0.864
100	3.0	−0.5	0.439	0.524	0.591	0.972	0.900	0.920
300	3.0	−0.5	0.263	0.328	0.367	0.986	0.904	0.966
500	3.0	−0.5	0.201	0.251	0.301	0.987	0.920	0.992
30	8.0	−0.5	0.637	3.916	0.863	0.977	0.971	0.749
50	8.0	−0.5	0.527	3.186	0.722	0.985	0.939	0.797
100	8.0	−0.5	0.410	2.478	0.560	0.988	0.910	0.857
300	8.0	−0.5	0.272	1.501	0.374	0.982	0.919	0.891
500	8.0	−0.5	0.228	1.199	0.317	0.974	0.930	0.916
30	5.0	0.5	0.84692	1.4552	0.79661	0.942	0.978	0.752
50	5.0	0.5	0.74045	1.036	0.69788	0.913	0.969	0.82
100	5.0	0.5	0.60588	0.7043	0.58288	0.886	0.973	0.86
300	5.0	0.5	0.43729	0.3893	0.41779	0.89	0.95	0.886
500	5.0	0.5	0.3756	0.3079	0.35627	0.916	0.949	0.908
30	3.0	−0.8	0.45726	1.0588	0.82622	0.966	0.96	0.453
50	3.0	−0.8	0.35881	0.7789	0.63423	0.947	0.961	0.515
100	3.0	−0.8	0.27257	0.5424	0.42326	0.955	0.938	0.676
300	3.0	−0.8	0.15113	0.2901	0.18577	0.953	0.956	0.811
500	3.0	−0.8	0.10251	0.1957	0.11576	0.97	0.983	0.834
30	8.0	0.2	0.89199	2.8335	0.8573	0.924	0.981	0.723
50	8.0	0.2	0.78179	1.9179	0.76533	0.928	0.971	0.799
100	8.0	0.2	0.67178	1.3264	0.65662	0.920	0.961	0.859
300	8.0	0.2	0.52796	0.7569	0.48225	0.920	0.973	0.908
500	8.0	0.2	0.4532	0.5759	0.40603	0.956	0.97	0.924

6.2 Applications

In this section we fit, by using the maximum likelihood method, the transmuted power Lindley distribution (TPL) to rainfall data from six weather stations located in Santa Catarina state, Brazil. The data consist of monthly rainfall for the years from 1971 to 2014. and were obtained from the National Institute of Meteorology at website http://www.inmet.gov.br/portal/index.php?r=bdmep/bdmep. Table 3 gives the latitude, longitude, observed period and the number of valid observations in that period.

Also, for comparison proposes, we have considered four alternative distributions: the one parameter Lindley distribution (L) with $f(y \mid \theta) = \frac{\theta^2}{1+\theta}(1+y)e^{-\theta y}$, the weighted Lindley distribution (WL) with $f(y \mid \theta, \lambda) = \frac{\theta^{\lambda+1}}{(\theta+\lambda)\Gamma(\lambda)} y^{\lambda-1}(1+y)e^{-\theta y}$, the power Lindley distribution (PL) with $f(y \mid \theta, \lambda) = \frac{\lambda\theta^2}{1+\theta}\left(1+y^\lambda\right)y^{\lambda-1}e^{-\theta y^\lambda}$ and the transmuted Lindley distribution (TL) with $f(y \mid \theta, \lambda) = \frac{\theta^2}{1+\theta}(1+y)\,e^{-\theta y}\left[1-\lambda+2\lambda\left(1+\frac{\theta y}{1+\theta}\right)e^{-\theta y}\right]$. As a complement, Figure 3 presents a PPlot of the adjusted models.

In Table 5 we presented, for all models and data sets, the maximum likelihood and standard errors estimates for θ, α and λ. The maximum likelihood estimates were obtained by *SAS/SEVERITY* procedure, SAS (2011). The *SAS/SEVERITY* procedure can fit multiple distributions at the same time and choose the best distribution according to a specified selection criterion. Seven different statistics of fit were used as selection criteria: $-2\times$ log-likelihood (Neg2LogLike), AIC, AICC, KS, AD and CvM. The calculated values of theses statistics are reported in Table 4 which present the superscripts that indicates the rank obtained by the distribution according to the selection criteria (the smaller the better). The column labeled as "RT" shows the sum of the ranks. From the values of "RT" column we can see that the TPL distribution is judged as being the most appropriate for five data sets. The fitted transmuted power Lindley density is displayed in Figure 2.

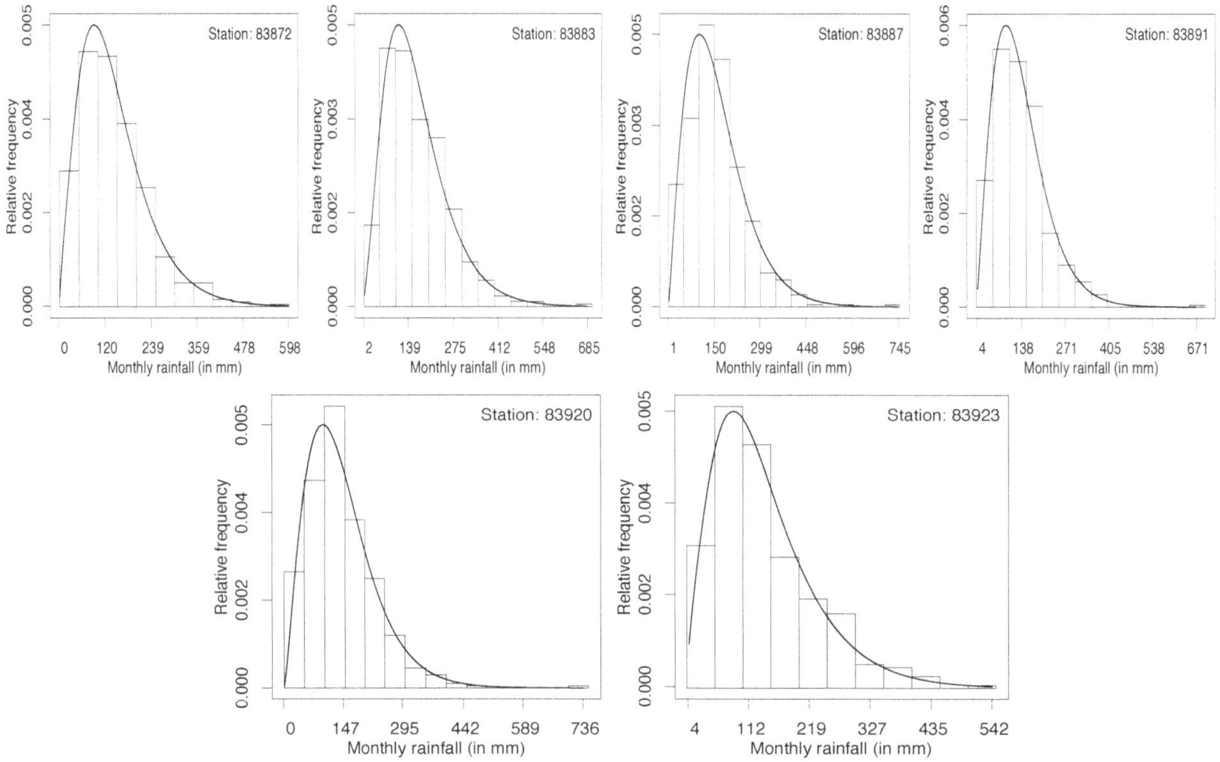

Figure 2. Adjusted TPL model.

Table 3. Description of meteorological stations.

Location	Latitude	Longitude	begin	end	n
Indaial (OMM: 83872)	-26.90	-49.21	31/12/1970	31/12/2014	443
Chapeco (OMM: 83883)	-27.11	-52.61	31/07/1973	31/12/2014	434
Campos Novos (OMM: 83887)	-27.38	-51.20	31/01/1970	31/12/2014	468
Lages (OMM: 83891)	-27.81	-50.33	31/01/1970	31/12/2014	467
São Joaquim (OMM: 83920)	-28.30	-49.93	31/01/1970	31/12/2014	447
Urussanga (OMM: 83923)	-28.51	-49.31	31/01/1970	31/07/2014	344

Table 4. Model selection.

Station	Model	Neg2LogLike	AIC	AICC	KS	AD	CvM	RT
	L	5142.88^5	5144.88^5	5144.89^5	1.36^5	2.92^5	0.50^5	34^5
	WL	5137.86^4	5141.86^4	5141.89^4	0.88^3	1.03^3	0.16^3	26^4
83872	PL	5134.41^3	5138.41^3	5138.44^3	0.73^2	0.61^2	0.08^2	17^2
	TL	5134.26^2	5138.26^2	5138.29^2	0.95^4	1.25^4	0.20^4	19^3
	TPL	5128.66^1	5134.66^1	5134.71^1	0.67^1	0.35^1	0.05^1	9^1
	L	5124.86^5	5126.86^5	5126.87^5	1.68^5	4.87^5	0.71^5	35^5
	WL	5101.58^3	5105.58^3	5105.61^3	0.68^2	0.37^3	0.04^1	17^3
83883	PL	5100.08^2	5104.08^2	5104.10^2	0.59^1	0.34^1	0.05^2	11^1
	TL	5111.68^4	5115.68^4	5115.70^4	1.28^4	2.62^4	0.35^4	28^4
	TPL	5097.86^1	5103.86^1	5103.92^1	0.75^3	0.35^2	0.05^3	14^2
	L	5529.37^5	5531.37^5	5531.38^5	2.31^5	6.37^5	1.21^5	35^5
	WL	5520.61^4	5524.61^4	5524.64^4	1.64^3	3.11^3	0.54^3	25^4
83887	PL	5514.79^2	5518.79^2	5518.82^2	1.36^2	2.13^2	0.33^2	13^2
	TL	5516.47^3	5520.47^3	5520.50^3	1.90^4	3.91^4	0.73^4	23^3
	TPL	5510.67^1	5516.67^1	5516.72^1	1.28^1	1.83^1	0.28^1	9^1
	L	5345.73^5	5347.73^5	5347.74^5	1.87^5	5.55^5	0.97^5	35^5
	WL	5328.22^3	5332.22^3	5332.25^3	1.11^3	1.39^3	0.23^3	20^3
83891	PL	5324.89^2	5328.89^2	5328.92^2	0.95^2	0.97^2	0.15^2	13^2
	TL	5332.44^4	5336.44^4	5336.46^4	1.45^4	3.14^4	0.55^4	28^4
	TPL	5322.09^1	5328.09^1	5328.14^1	0.92^1	0.83^1	0.13^1	9^1
	L	5170.00^5	5172.00^5	5172.01^5	2.01^5	5.97^5	1.05^5	35^5
	WL	5154.14^3	5158.14^3	5158.17^3	1.12^3	1.64^3	0.27^3	21^3
83920	PL	5148.31^2	5152.31^2	5152.34^2	0.87^2	1.01^2	0.15^2	13^2
	TL	5156.30^4	5160.30^4	5160.32^4	1.62^4	3.51^4	0.62^4	28^4
	TPL	5144.55^1	5150.55^1	5150.60^1	0.79^1	0.85^1	0.12^1	8^1
	L	3980.20^5	3982.20^5	3982.21^5	1.23^5	1.15^5	0.20^5	31^5
	WL	3977.86^4	3981.86^3	3981.89^3	0.86^4	0.54^3	0.09^3	24^4
83923	PL	3977.42^3	3981.42^2	3981.46^2	0.81^2	0.52^2	0.09^2	16^1
	TL	3977.10^2	3981.10^1	3981.13^1	0.84^3	0.56^4	0.10^4	17^2
	TPL	3976.00^1	3982.00^4	3982.07^4	0.68^1	0.42^1	0.07^1	17^2

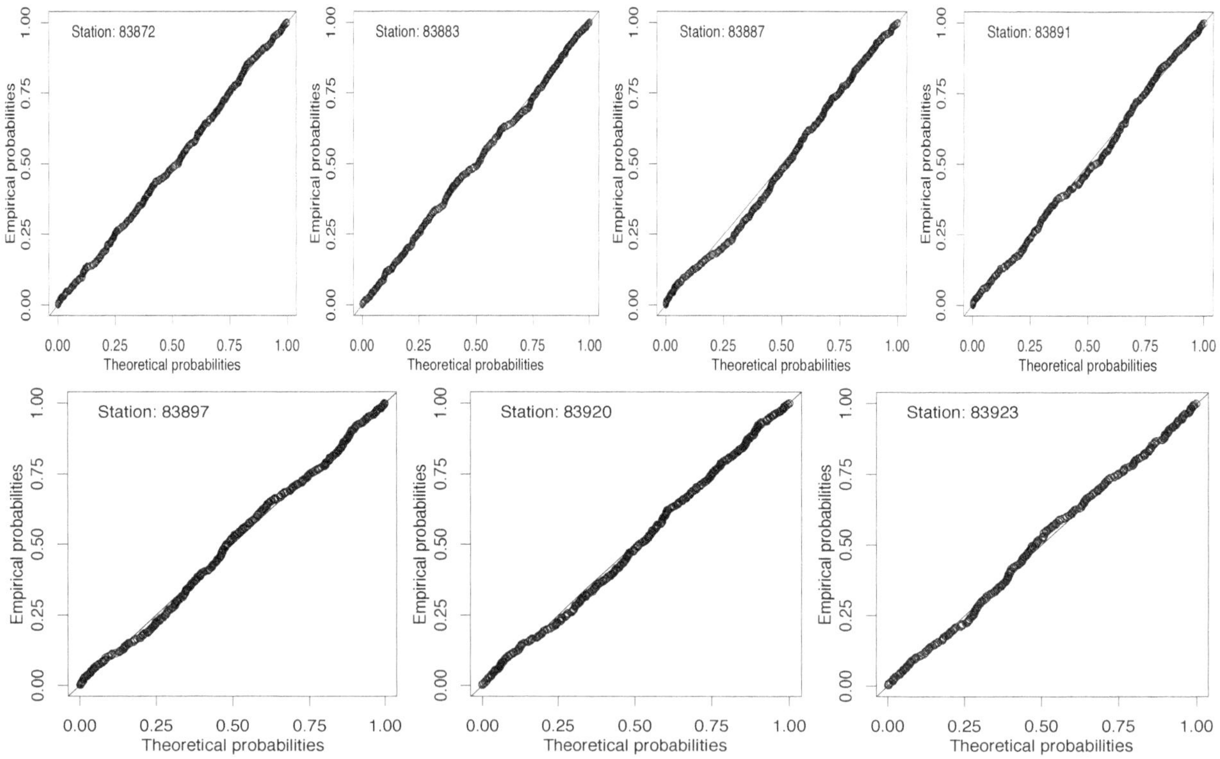

Figure 3. PPlot of the adjusted TPL model.

Table 5. Maximum likelihood and standard errors estimates for θ, α and λ.

Station	θ	α	λ
83872	0.0224	0.9502	-0.6632
	(0.0076)	(0.0588)	(0.1556)
83883	0.0183	0.9718	-0.7973
	(0.0065)	(0.0615)	(0.1399)
83887	0.0047	1.0979	0.9786
	(0.0010)	(0.0433)	(0.0447)
83891	0.0049	1.1314	0.9647
	(0.0010)	(0.0429)	(0.0495)
83920	0.0043	1.1436	0.9621
	(0.0009)	(0.0438)	(0.0474)
83923	0.0275	0.9157	-0.6510
	(0.0127)	(0.0789)	(0.2341)

7. Conclusion

In this paper we have introduced a new generalization of the power Lindley distribution, the transmuted power Lindley model. The proposed distribution was constructed by using a quadratic rank transmutation map and taking the power Lindley, which was formulated by using a power transformation $X = T^{1/\alpha}$, as the baseline distribution. Some mathematical properties along with order statistics and estimation issues are addressed.

A real data was considered in order to illustrate the usefulness and effectiveness of the new model. In addition to the transmuted model, four different models were fitted and seven different statistics of fit were used as selection criteria: $-2\times$ log-likelihood, AIC, AICC, BIC, KS, AD and CvM.

Furthermore, in this paper we showed that the TPL model, despite having a small number of parameters, still interpretable (the key parameters of the power Lindley distribution is kept) and flexible, i.e, the TPL model can be used in several areas of application.

Acknowledgments

The authors are grafeful the Editorial Boarding as well as to the referees for their comments and suggestions. The research is partially funded by the Brazilian organizations CNPq and FAPESP.

References

Aryal, G. R., & Tsokos, C. P. (2009). On the transmuted extreme value distribution with applications. *Nonlinear Analysis, 71*, 1401-1407.

Aryal, G. R., & Tsokos, C. P. (2011). Transmuted Weibull distribution: a generalization of the Weibull probability distribution. *European Journal of Pure and Applied Mathematics, 4*(2), 89-102.

Ghitany, M. E., Atieh, B., & Nadarajah, S. (2008). Lindley distribution and its applications. *Mathematics and Computers in Simulation, 78*(4), 493-506.

Ghitany, M. E., Al-Mutairi, D. K., Balakrishnan, N., & Al-Enezi, L. J. (2013). Power Lindley distribution and associated inference. *Computational Statistics and Data Analysis*, 20-33.

Granzotto, D. C. T., & Louzada, F. (2014). The Transmuted Log-Logistic Distribution: Modeling, Iference and an application to a Polled Tabapua Race Time up to First Calving Data. *Communications in Statistics - Theory and Methods* (Accepted).

Lindley, D. V. (1958). Fiducial distributions and Bayes theorem. *Journal of the Royal Statistical Society, 20*(1), 102-107.

Louzada, F., & Granzotto, D. C. T. (2015). The transmuted log-logistic regression model: a new model for time up to first calving of cows. Statistical Papers.

Merovci, F., & Elbatal, I. (2014). Transmuted Lindley-Geometric Distribution and its Applications. *Journal of Statistics Applications and Probability, 3*(1), 77-91.

Nadarajah, S., Bakouch, H., & Tahmasbi, R. (2011). A generalized Lindley distribution. *Sankhya B, 73*, 331-359.

SAS. (2011). SAS/ETSR® Userąŕs Guide, Version 9.33. Cary, NC: SAS Institute Inc.

Shaw, W. T., & Buckley, I. R. C. (2007). The alchemy of probability distributions: beyond Gram-Charlier expansions, and a skew-kurtotic-normal distribution from a rank transmutation map. UCL Discovery Repository, pages 1-16.

Warahena-Liyanage, G., & Pararai, M. (2014). A generalized power Lindley distribution with applications. *Asian Journal of Mathematics ans Applications,* 23.

Zakerzadeh, H., & Dolati, A. (2009). Generalized Lindley Distribution. *Journal of Mathematical Extension, 3*(2), 13-25.

15

Estimating the Area under the ROC Curve with Modified Profile Likelihoods

Giuliana Cortese[1]

[1] Department of Statistical Sciences, University of Padua, Italy

Correspondence: Giuliana Cortese, Department of Statistical Sciences, University of Padua, Via Cesare Battisti 241, 35121 Padua, Italy. E-mail: gcortese@stat.unipd.it

Abstract

Receiver operating characteristic (ROC) curves are a frequent tool to study the discriminating ability of a certain characteristic. The area under the ROC curve (AUC) is a widely used measure of statistical accuracy of continuous markers for diagnostic tests, and has the advantage of providing a single summary index of overall performance of the test. Recent studies have shown some critical issues related to traditional point and interval estimates for the AUC, especially for small samples, more complex models, unbalanced samples or values near the boundary of the parameter space, i.e., when the AUC approaches the values 0.5 or 1. Parametric models for the AUC have shown to be powerful when the underlying distributional assumptions are not misspecified. However, in the above circumstances parametric inference may be not accurate, sometimes yielding misleading conclusions. The objective of the paper is to propose an alternative inferential approach based on modified profile likelihoods, which provides more accurate statistical results in any parametric settings, including the above circumstances. The proposed method is illustrated for the binormal model, but can potentially be used in any other complex model and for any other parametric distribution. We report simulation studies to show the improved performance of the proposed approach, when compared to classical first-order likelihood theory. An application to real-life data in a small sample setting is also discussed, to provide practical guidelines.

Keywords: area under the ROC curve, binormal model, continuous diagnostic marker, modified profile likelihood, ROC curve, stress-strength model.

1. Introduction

Receiver operating characteristic (ROC) curves are frequently used to study the ability of a certain characteristic in discriminating and classifying units under study. One of the most popular summary measures based on the ROC curve is the area under the curve (AUC) (Krzanowski & Hand, 2009), which was originally developed in radar signal detection (Bamber, 1975), and later it has been used in a broad range of applied contexts such as radiology, psychiatry, reliability theory and industrial inspection systems, earthquake resistance.

The AUC is also widely applied in medicine as a measure of statistical accuracy of continuous markers for diagnostic tests (Faraggi & Reiser, 2002; Pepe, 2003; Zhou, McClish, & Obuchowski, 2009). A diagnostic test based on a continuous marker provides usually a response about the possible clinical status of subjects, identifying them as diseased (test positive) or non-diseased (test negative) patients. Such test requires that a certain cut-off point t is chosen. The probabilities that the test correctly classifies subjects as diseased and non-diseased, are called, respectively, the sensitivity and specificity of the test associated with t.

To formalize the problem more generally, denote with $\bar{D}$ and D, respectively, the true negative and true positive status of units in the population of interest (e.g., real condition of being non-diseased or diseased). Let us define two continuous random variables Y and X that describe a continuous characteristic of interest in the two distinct groups $\bar{D}$ and D, respectively. Let $F_Y(\cdot)$ and $F_X(\cdot)$ be the corresponding cumulative distribution functions, and $f_Y(t)$ and $f_X(t)$ the associated probability density functions. Consider a classification rule based on a certain cut-off point t (e.g., a diagnostic test that classifies subjects as 'non-diseased' if the observed value of the characteristic is below t, and as 'diseased' if the observed value is above t). The probability that a unit with true status $\bar{D}$ is correctly classified by the diagnostic test (test negative) is called 'specificity' and defined as $p(t) = F_Y(t)$, while the probability that a unit with true status D is correctly classified by the test (test positive) is called 'sensitivity' and defined as $q(t) = 1 - F_X(t)$. Sensitivity and specificity vary when different choices of t are made over the continuous scale of the characteristic. The ROC curve is then obtained by plotting $p(t)$ versus $1 - q(t)$ for all possible values of t.

The AUC has the advantage of providing a single index that summarizes the overall performance of the test (or rule) based on the continuous characteristic, rather than an entire curve, and it is particularly useful for comparisons under different populations or different tests. The aim is often to minimize the error $1 - q(t)$ committed by the test, and simultaneously increase the efficacy in discovering units from the D population. Therefore, values of the AUC close to 1 indicate very high accuracy of the test, while very low accuracy corresponds to values closer to 0.5. Bamber (1975) showed that the AUC based on continuous distributions is a probabilistic measure that is equal to

$$A = P(Y \leq X) = \int_{-\infty}^{\infty} F_Y(t)\, dF_X(t) = \int_0^1 q \circ \bar{p}^{-1}(z)\, dz, \tag{1}$$

where $\bar{p}(t) = 1 - p(t)$. The quantity A can also be interpreted as the probability that, in a randomly selected pair of $\bar{D}$ and D subjects, the test value is higher for a subject from the D population. In more general contexts, the AUC is used as a measure of difference between distributions (Wolfe & Hogg, 1971). It is often used in engineering and reliability theory with the name of stress-strength model (Johnson, 1988; Kotz, Lumelskii, & Pensky, 2003). When X represents the strength of a certain component and Y is the applied stress, then A measures the probability that a component would not fail if it is put under a systematic stress.

Inference for the AUC has been studied under different modeling assumptions, following mainly a nonparametric, a parametric or a Bayesian approach. In practical applications, it has been suggested that all these approaches are useful and the comparison of their results may provide additional information on the consistency among them. Moreover, the AUC has been also investigated under various relevant settings, such as presence of explanatory variables, measurement errors and clustered data (Pardo-Fernández, Rodríguez-Álvarez, Van Keilegom, et al., 2013; Reiser, 2000; Zou, Carlsson, & Yu, 2012).

Recently, a special attention has been devoted to interval estimation of A and some related critical issues have been widely discussed in the literature (Feng, Cortese, & Baumgartner, 2015). Some of these issues concern a bad performance of confidence intervals for the AUC especially for small samples, more complex models, unbalanced samples or values near the boundary of the parameter space (i.e, A approaching 0.5 or 1). In particular, classical parametric approaches have the general problem that the smaller the sample size and the higher the number of parameters, less accurate they are in the interval and point estimation. On the other hand, nonparametric methods tend also to perform poorly when the sample size is small. Moreover, in general the parametric methods seem to outperform the nonparametric ones when the underlying distributional assumptions are not misspecified, and in presence of samples that show a nearly perfect separation between subjects in the two groups $\bar{D}$ and D (Obuchowski & Lieber, 2002).

In the current papers we restrict our attention to the parametric framework for inference on the AUC. For the binormal model, where Y and X are assumed to follow normal distribution with different means and variances, Reiser and Guttman (1986) and Reiser and Faraggi (1997) proposed a method for the construction of confidence intervals based on a standard approximate t of Student solution. Although their procedure appears to work well also for unbalanced or small samples, it is not extendible to different parametric models for Y and X, such as Weibull, Gamma or any other more general parametric distributions not in the location-scale family, or to e.g. mixture model in presence of bimodal distributions. Moreover, it is not clear how to handle presence of explanatory variables or clustered data. Classical asymptotic methods based on parametric likelihood theory can easily be applied for constructing confidence intervals or test of hypothesis for the AUC for any type of assumed parametric model. However, it is well known from the general likelihood theory that the resulting Wald type statistic and likelihood ratio statistic do not show a good performance in all situations, especially in the coverage probability of 95% confidence intervals (Severini, 2000). A recent parametric approach was based on higher-order asymptotic likelihood theory (Cortese & Ventura, 2013). However, it has been shown that such method has some limitations: it may easily fail in presence of very small or unbalanced samples or when the samples produce a nearly perfect observed discrimination, it is computationally unstable near the maximum likelihood estimate of A. Some of these problems have been underlined in Feng et al. (2015).

To overcome these drawbacks, the current paper addresses the problem of inaccurate parametric inference in case of small or unbalanced sample sizes, with special attention to confidence intervals and test of hypothesis for the AUC. Also the problem of correct inference near the limit values 0.5 and 1, which represent the situations of, respectively, lowest and maximal accuracy of the continuous characteristic under study, is investigated. In regard of these objectives, we present inference for the AUC based on a modified version of the profile likelihood function, denoted in the literature as 'modified profile likelihood' (Cox & Reid, 1992). In this setting, the parameter identifying the AUC is treated as parameter of interest, whereas the remaining parameters related to the underling parametric distributions of Y and X are treated as nuisance parameter. The proposed approach is very general, applicable to any type of parametric distribution assumptions and to any data setting, such as clustered data or additional data on explanatory variables (Sartori, 2003).

It has been widely studied that standard likelihood inference for a parameter of interest could be misleading in presence of relatively many nuisance parameters, with respect to the sample size, or for small samples. The classical approach for making inference on a parameter of interest in presence of nuisance parameters is based on profile likelihoods. The profile likelihood function is the likelihood in which the nuisance parameters are maximized out, for every fixed value of the parameter of interest. This likelihood is not a proper likelihood and therefore, the derived score function is biased (Severini, 2000). Consequently, this bias may increase with the dimension of the nuisance parameter and produce inaccurate estimation. The modified profile likelihoods are an interesting alternative to the profile likelihoods, since they correct for the presence of nuisance parameters (Cox & Barndorff-Nielsen, 1994; Cox & Reid, 1992) showing an improved performance.

The scope of the paper is to investigate the performance of modified profile likelihoods for inference on the AUC based on a general parametric model. The inferential procedure is presented in the general setting. Then, the methodological aspects are illustrated for the binormal model. In order to show how to obtain point estimates, confidence intervals and test of hypothesis based on the modified profile likelihood, we consider an application to real data in a setting of small samples.

The paper is organized as follows. Section 2 provides the general notation and introduces the inferential problem in parametric models for the AUC. Here the classical approach and the proposed approach based on modified profile likelihoods are described. In Section 3, the theory is applied to the specific case of a binormal model and computations are illustrated. Section 4 reports simulation studies comparing the different methods and Section 5 shows the application to real-life data on imaging for detecting brain tumor. Finally, conclusions and future directions are given in Section 6.

2. Notation and the Inferential Problem

In this section we consider a generic parametric model for the AUC, where the Y and X components are assumed to follow the parametric distributions $F_Y(t;\theta_Y)$ and $F_X(t;\theta_X)$, identified by the finite-dimensional parameter vectors θ_Y and θ_X, respectively. Let us define $\theta = (\theta_Y, \theta_X)$ be the entire parameter vector of the model of dimension p, with $\theta \in \Theta \subseteq \mathbb{R}^p$. The AUC is then obtained as

$$A = \int_{-\infty}^{\infty} F_Y(t;\theta_Y)\, dF_X(t;\theta_X) \equiv g(F_Y(t;\theta_Y), F_X(t;\theta_X)), \tag{2}$$

where the functional relation between A and $(F_Y(\cdot), F_X(\cdot))$ is defined with $g(\cdot)$, for ease of notation.

With the scope of making inference on the AUC, let $y = (y_1, \ldots, y_{n_1})$ be a random sample of size n_1 of i.i.d. observations drawn from Y, and $x = (x_1, \ldots, x_{n_2})$ be a random sample of size n_2 of i.i.d. observations drawn from X. Assume also that Y and X are independent. Let $f_Y(y;\theta_Y)$ and $f_X(x;\theta_X)$ be the probability density functions associated to Y and X, respectively. The log-likelihood function for θ is defined as $\ell(\theta) = \ell(\theta; y, x) = \sum_{i=1}^{n_1} \log f_Y(y_i;\theta_Y) + \sum_{i=1}^{n_2} \log f_X(x_i;\theta_X)$, and under broad conditions, $\hat{\theta}$ is the maximum likelihood estimator (MLE) obtained as unique solution to the score equation $\ell_\theta(\theta) = \partial\ell(\theta)/\partial\theta = 0$. The MLE of the AUC can be directly obtained as $\hat{A} = g(\hat{\theta})$, due to the likelihood invariance property.

In the proposed approach, we intend to treat the parameter A as a scalar parameter of interest, while the remaining parameters that identify the parametric distributions of Y and X are considered as nuisance parameter. Then, the original model needs to be reparameterized so that $\psi = \psi(\theta) = A$ is the parameter of interest, as defined in (2), and $\lambda = \lambda(\theta)$ is a nuisance parameter vector of length $(p-1)$, obtained by a transformation of the original parameter θ. Therefore, we can write the likelihood function for the new parameters (ψ, λ) as

$$\ell(\psi, \lambda) = \sum_{i=1}^{n_1} \log f_Y(y_i;\psi,\lambda) + \sum_{i=1}^{n_2} \log f_X(x_i;\psi,\lambda)\,.$$

The MLEs $\hat{A} = \hat{\psi}$ and $\hat{\lambda}$ are the unique solutions to, respectively, the score equations $\ell_\psi(\psi,\lambda) = \partial\ell(\psi,\lambda)/\partial\psi = 0$ and $\ell_\lambda(\psi,\lambda) = \partial\ell(\psi,\lambda)/\partial\lambda = 0$.

2.1 Inference Based on the Profile Likelihood

From $\ell(\psi,\lambda)$, classical likelihood inference for the parameter of interest $\psi = A$ in presence of nuisance parameters, can be based on profile likelihood procedures, which require to eliminate the nuisance parameter λ by replacing it by the constrained MLE, $\hat{\lambda}_\psi$, obtained by maximizing $\ell(\psi,\lambda)$ with respect to λ for fixed ψ. This method is based on the profile log-likelihood $\ell_p(\psi) = \ell(\psi, \hat{\lambda}_\psi)$, which can then be easily maximized to get the estimated AUC, $\hat{\psi} = \hat{A}$. The related standard error can be computed as $(J_p(\hat{\psi}))^{-1/2}$, where $J_p(\psi) = -\partial^2\ell_p(\psi)/\partial\psi^2$ is the corresponding profile observed Fisher information.

Confidence intervals and test of hypothesis can rely on first-order approximations. Specifically, inference on A can be based on the Wald statistic

$$W_p(\psi) = J_p(\hat{\psi})^{1/2}(\hat{\psi} - \psi)\,, \tag{3}$$

or on the signed log-likelihood ratio statistic

$$R_p(\psi) = \mathrm{sign}(\hat{\psi} - \psi)\left(2(\ell_p(\hat{\psi}) - \ell_p(\psi))\right)^{1/2}, \tag{4}$$

which have asymptotic standard normal distributions.

A $100(1-\alpha)\%$ confidence interval for ψ based on the Wald statistic is given as $[\hat{\psi} - z_{1-\alpha/2}\, j_p(\hat{\psi})^{-1/2},\ \hat{\psi} + z_{1-\alpha/2}\, j_p(\hat{\psi})^{-1/2}]$, where $z_{1-\alpha}$ is the $(1-\alpha)$-quantile of the standard normal distribution. Alternatively, a $100(1-\alpha)\%$ confidence interval for ψ can be constructed from the $R_p(\psi)$ statistic, and can be written as $\{\psi : |R_p(\psi)| \leq z_{1-\alpha/2}\}$. The Wald-type confidence interval is often preferred because it is very simple and immediate to be computed, as compared to the likelihood ratio confidence interval, which typically requires a numerical solution. However, it is well-known that in general inferential procedures based on the Wald statistics have a general poor performance and are less accurate than the procedures based on the signed log-likelihood ratio statistic, especially at the boundaries of the parameter space (Severini, 2000).

2.2 Inference Based on the Modified Profile Likelihood

The profile likelihood is a standard method for inference in large-sample situations, and does not always perform well in small-sample problems. When the focus of the inferential interest is a parameter ψ, while the remaining parameters are not of central concern (nuisance), an interesting alternative approach is based on the modified profile likelihoods. With the scope to improve inferences, these likelihoods consist of an adjustment to the classical profile likelihoods by the inclusion of a penalization term for the possible presence of nuisance parameters. The amount of the penalization depends on the information available for λ, and increases when this information is large. Modified profile likelihoods have also the appealing property of being invariant to interest-preserving reparametrizations. This last property means that inferential results obtained for (ψ, λ) are also valid for $(\eta(\psi), \xi(\psi, \lambda))$, where η and ξ are one-to-one transformations.

The general expression for a modified profile log-likelihood (Severini, 2000) is

$$\ell_{mp}(\psi) = \ell_p(\psi) + M(\psi), \tag{5}$$

where $\ell_p(\psi)$ is the profile log-likelihood and $M(\psi)$ is the modification term. For this term, a high degree of accuracy is obtained when it has the expression

$$M(\psi) = \frac{\left|J_{\lambda\lambda}(\psi, \hat{\lambda}_\psi)\right|^{1/2}}{\left|\ell_{\lambda;\hat{\lambda}}(\psi, \hat{\lambda}_\psi; \hat{\psi}, \hat{\lambda})\right|}, \tag{6}$$

where

$$J_{\lambda\lambda}(\psi, \lambda) = -\ell_{\lambda\lambda}(\psi, \lambda) = -\partial^2 \ell(\psi, \lambda)/\partial\lambda\partial\lambda^T, \qquad \ell_{\lambda;\hat{\lambda}}(\psi, \lambda; \hat{\psi}, \hat{\lambda}) = -\partial^2 \ell(\psi, \lambda; \hat{\psi}, \hat{\lambda})/\partial\lambda\partial\hat{\lambda}^T.$$

In practice, the first term $J_{\lambda\lambda}(\psi, \hat{\lambda}_\psi)$ is easily computed numerically or analytically by differentiation of $\ell_{\lambda\lambda}(\psi, \lambda)$. When the log-likelihood can be written in terms of the MLE, $\hat{\psi}$ and $\hat{\lambda}$, and an ancillary statistic a, i.e., as $\ell(\psi, \lambda; y, x) = \ell(\psi, \lambda; \hat{\psi}, \hat{\lambda}, a)$, computation of the term $\ell_{\lambda;\hat{\lambda}}(\psi, \lambda; \hat{\psi}, \hat{\lambda})$ is also straightforward. When differentiating with respect to $\hat{\lambda}$, the quantities ψ, $\hat{\psi}$ and a need to be held fixed. However, we have here omitted the conditioning to the ancillary a because it is not needed explicitly for computations and, in our context of parametric models for the AUC, in most of the cases the modification term in (6) can be obtained without specifying a.

Inference for ψ can be easily performed by treating (5) as a standard log-likelihood for ψ, without the burden of dealing with nuisance parameters. The solution to the maximization of $\ell_{mp}(\psi)$ provides a maximum modified profile likelihood estimate (MMLE), defined as $\hat{\psi}_{mp}$. In particular, the standard error associated to $\hat{\psi}_{mp}$ is computed as $(J_{mp}(\hat{\psi}_{mp}))^{-1/2}$, where $J_{mp}(\psi) = -\partial^2 \ell_{mp}(\psi)/\partial\psi^2$. Therefore, using the normal approximation, it is possible to use a Wald-type confidence interval, e.g., $[\hat{\psi}_{mp} - z_{1-\alpha/2}\, J_{mp}(\hat{\psi}_{mp})^{-1/2},\ \hat{\psi}_{mp} + z_{1-\alpha/2}\, J_p(\hat{\psi}_{mp})^{-1/2}]$.

Moreover, the resulting signed modified log-likelihood ratio statistic, defined as

$$R_{mp}(\psi) = \mathrm{sign}(\hat{\psi}_{mp} - \psi)\left(2(\ell_{mp}(\hat{\psi}_{mp}) - \ell_{mp}(\psi))\right)^{1/2}, \tag{7}$$

has asymptotic standard normal distribution, and has properties that are superior to those of the usual signed likelihood ratio statistic (Sartori, 2003). The statistic $R_{mp}(\psi)$ is then preferred, with respect to the Wald-type statistic, for construction of confidence intervals and test of hypothesis. In practice, a $100(1-\alpha)\%$ confidence interval based on $R_{mp}(\psi)$ is given as $\{\psi : |R_{mp}(\psi)| \leq z_{1-\alpha/2}\}$. A one-sided statistical test with null hypothesis $H_0 : \psi = \psi_0$ can be performed using the test-statistic $R_{mp}(\psi_0)$.

3. An Important Example: the Binormal Model

The main example about possible applications of the theory described in Subsections 2.2, is given for the popular binormal model, where Y and X are normally distributed with different means and variances, e.g., $Y \sim N(\mu_Y, \sigma_Y^2)$ and $X \sim N(\mu_X, \sigma_X^2)$. Under this assumption, it is known (Kotz et al., 2003) that the AUC can be written as

$$A = \Phi(\delta) = \Phi\left(\frac{\mu_X - \mu_Y}{\sqrt{\sigma_X^2 + \sigma_Y^2}}\right), \tag{8}$$

where $\Phi(\cdot)$ is the cumulative probability function of the standard normal distribution. Denote with $\delta = (\mu_X - \mu_Y)/\sqrt{\sigma_X^2 + \sigma_Y^2}$ the quantile of the standard normal which provide an area equal to A. Here, there are two possible interesting choices for the parameter of interest ψ. We may have either $\psi = A$ or $\psi = \delta$. These two choices are equivalent in terms of inferential results because both the profile likelihood and the modified profile likelihood are invariant for interest-preserving reparameterizations, and thus for the transformation $A = \Psi(\delta)$. In the current paper, for practical reasons, we illustrate the procedures for the second choice $\psi = \delta$, since this case is relatively simpler to implement. Moreover, in this case, convergence in the corresponding parameter space $\Psi = \mathbb{R}$ is always obtained, whereas the choice $\psi = A$ with parameter space $\Psi = [0, 1]$ may yield computational problems on the boundaries.

We study the parameter of interest $\psi = \delta$, while the nuisance parameter can be chosen to be, e.g., $\lambda = (\lambda_1, \lambda_2, \lambda_3)$, with $\lambda_1 = \mu_Y$, $\lambda_2 = \sqrt{\sigma_Y^2}$, and $\lambda_3 = \sqrt{\sigma_Y^2 + \sigma_X^2}$. Other choices are also possible, where the parameter space is $\Psi \times \Lambda$, and thus the range of λ is independent of the range of ψ.

Given the MLE $\hat{\theta}$ computed from the original likelihood $\ell(\theta)$, by the invariance property, the MLE for the AUC is

$$\hat{A} = \Phi\left(\hat{\delta}\right) = \Phi\left(\frac{\hat{\mu}_X - \hat{\mu}_Y}{\sqrt{\hat{\sigma}_Y^2 + \hat{\sigma}_X^2}}\right),$$

where $\hat{\mu}_Y = \sum_i y_i/n_1$, $\hat{\mu}_X = \sum_i x_i/n_2$ and $\hat{\sigma}_Y^2 = \sum_i (y_i - \hat{\mu}_Y)^2/n_1$, $\hat{\sigma}_X^2 = \sum_i (x_i - \hat{\mu}_X)^2/n_2$.

Consider now the likelihood function for the new parameters (ψ, λ),

$$\ell(\psi, \lambda) = -\frac{1}{2}\left[n_1 \log \lambda_2^2 + n_2 \log(\lambda_3^2 - \lambda_2^2)\right] - \frac{n_1\left[\hat{\lambda}_2^2 + (\hat{\lambda}_1 - \lambda_1)^2\right]}{2\lambda_2^2} - \frac{n_2\left[\hat{\lambda}_3^2 - \hat{\lambda}_2^2 + (\hat{\lambda}_1 + \hat{\psi}\hat{\lambda}_3 - \lambda_1 - \psi\lambda_3)^2\right]}{2(\lambda_3^2 - \lambda_2^2)}, \tag{9}$$

and observe that it is a function only of the unknown parameters and the minimal sufficient statistic $(\hat{\psi}, \hat{\lambda})$, where $\hat{\psi} = \hat{\delta}$, and $\hat{\lambda}_1 = \hat{\mu}_Y$, $\hat{\lambda}_2 = \sqrt{\hat{\sigma}_Y^2}$ and $\hat{\lambda}_3 = \sqrt{\hat{\sigma}_Y^2 + \hat{\sigma}_X^2}$, and thus, depends on the data only through the MLEs.

The constrained MLE $\hat{\lambda}_\psi = (\hat{\lambda}_{1\psi}, \hat{\lambda}_{2\psi}, \hat{\lambda}_{3\psi})$ for fixed ψ is found by numerical procedures as solution to the system of score equations $\ell_{\lambda_i}(\psi, \lambda) = \partial\ell(\psi, \lambda)/\partial\lambda_i = 0$, for $i = 1, 2, 3$. Their analytic expressions is given in the Appendix. The profile log-likelihood $\ell_p(\psi, \hat{\lambda}_\psi)$ is then obtained by replacing λ with $\hat{\lambda}_\psi$ in (9).

For the binormal model, computation of the signed log-likelihood ratio statistic $R_p(\psi)$ given in (4) is then straightforward. The Wald statistic $W_p(\psi)$ in (3) requires to find the observed information $J_p(\hat{\psi})$, which can be computed analytically or by a numerical procedure, for example by using the function `hessian` of package `numDeriv` in the `R` software.

The key parameter of interest is the AUC, therefore we can easily obtain inferential conclusions on A from those obtained from ψ. For example, the Delta method can be applied to find the standard error of $\hat{A} = \Phi(\hat{\psi})$, which is then equal to $\hat{s}_A = \Phi'(\hat{\psi})(J_p(\hat{\psi}))^{-1/2} = f_Z((\hat{\psi})(J_p(\hat{\psi}))^{-1/2}$, with $f_Z(\cdot)$ being the p.d.f. of the standard normal. Therefore, a Wald-type confidence interval for A is given as $[\hat{A} - z_{1-\alpha/2}\,\hat{s}_A,\ \hat{A} + z_{1-\alpha/2}\,\hat{s}_A]$, and a hypothesis testing concerning A can be based on the test-statistic $(\hat{A} - A_0)/\hat{s}_A$.

In addition, to specify the modified profile log-likelihood in (5) and (6), we need to compute the modification term $M(\psi)$. In doing so, the block of the observed information matrix, $J_{\lambda\lambda}(\psi, \hat{\lambda}_\psi)$ is equal to minus the Hessian matrix, which can be easily obtained by numerical procedures in the `R` software, as above. The analytic expressions of the sample space derivatives $\ell_{\lambda;\hat{\lambda}}(\psi, \lambda; \hat{\psi}, \hat{\lambda})$ for the binormal model are provided in the Appendix. The signed modified log-likelihood ratio statistic $R_{mp}(\psi)$ given in (7) can then be constructed to solve test of hypothesis concerning key values of the AUC, such as e.g. $A = 0.5$ or $A = 1$. For example, the one-sided test with hypotheses $H_0 : \psi = \psi_0 = 0$ versus $H_0 : \psi > 0$ is equivalent to testing whether the AUC is significantly higher than 0.5, and can be performed using the test-statistic

$R_{mp}(\psi_0) = \text{sign}(\hat{\psi}^0_{mp} - \psi_0)\left(2(\ell_{mp}(\hat{\psi}^0_{mp}) - \ell_{mp}(\psi_0))\right)^{1/2}$, where $\hat{\psi}^0_{mp}$ denotes the maximum modified likelihood estimate of ψ in the parameter space $\Psi_0 = \{\psi \in \Psi : \psi > \psi_0\}$.

4. Simulation Studies

The performance of the proposed method for constructing confidence intervals and point estimates for A is illustrated through a simulation study, based on 5000 Monte Carlo trials. We considered different values of ψ ($\psi = 0.6, 0.8, 0.95, 0.99$) and many different combinations of sample sizes $(n_1, n_2) = (5,5), (10,10), (20,20), (30,30), (15,5), (5,15), (30,5), (5,30), (80,10), (10,80)$. Note that the case of very unbalanced samples is also taken into account.

First, the simulation studies investigated the coverage probabilities of 95% confidence intervals based on the signed profile log-likelihood ratio statistic $R_p(\psi)$, the signed modified profile log-likelihood ratio statistic $R_{mp}(\psi)$, and the Wald statistic $W_p(\psi)$. All these statistics are asymptotically distributed as standard normal, and the approximation is often more accurate for $R_{mp}(\psi)$. Results in Table 1 show that $R_{mp}(\psi)$ is more accurate than $R_p(\psi)$ and $W_p(\psi)$, in terms of both central coverage probability and symmetry of the error rates, for all the considered AUC values and sample sizes. Of course, for all methods, we observe a less accurate coverage when the sample sizes are very small ($(n_1, n_2) = (5,5), (10,10)$), which then increases for higher sample sizes. However, the $R_{mp}(\psi)$ coverage is observed to reach nearly the 95% nominal level, being slightly affected by low values of sample sizes (see, e.g., for $A = 0.95, 0.99$), in contrast to the $W_p(\psi)$ and $R_p(\psi)$ that provide seriously poor performance for small samples. Very interestingly, this poor performance becomes even worse for higher values of the AUC, such as $A = 0.95, 0.99$. On the contrary, the good performance of the $R_{mp}(\psi)$ seems to be very stable for all values of the AUC.

An important result is observed for unbalanced samples: $W_p(\psi)$ and $R_p(\psi)$ seem to be negatively affected by the sample unbalance, since their coverage decreases even more with respect to the nominal level, whereas, the $R_{mp}(\psi)$ coverage keeps stable and enough accurate in all the unbalanced settings. In particular, we note that the coverages are lower for samples with high n_1 and low n_2 (e.g., $(n_1, n_2) = (30,5), (80,10)$) as compared to the inverse case of low n_1 and high n_2. This fact may depend on the reparameterization chosen for the nuisance parameters λ, since we have that the MLEs $\hat{\lambda}_2$ and $\hat{\lambda}_3$ are both affected by the small sample size n_2 and then would be poorly estimated.

The very poor performance shown by the Wald statistic for high values of the AUC is expected. It is well known that when the profile log-likelihood is not quadratic around the MLE, as it happens in our AUC study (see Figure 1 of data example in Section 5), the Wald statistic may lead to very asymmetric confidence intervals. In fact, in Table 1 we observe a nearly null empirical lower error and a higher empirical upper error than expected. Asymmetric errors are also seen for the $R_p(\psi)$ statistic, although the discrepancy from the expected errors is negligible.

Simulation studies were also used to evaluate the properties of the $R_{mp}(\psi)$-based estimator of A, in comparison with the MLE $\hat{\psi}$. The two estimators are compared in terms of median bias and results are shown in Table 2, where estimated standard errors and simulations-based (empirical) median absolute deviation (MAD) are also reported. The choice of a median-bias criteria is due to the median unbiasedness property of the $\hat{\psi}_{mp}$-based estimator, and it is more robust under model misspecification. It can be noted that the estimator based on modified likelihood, $\hat{\psi}_{mp}$, is preferable to the MLE in terms of the considered criteria, since it is less median-biased than the MLE, in particular for small sample sizes and unbalanced samples. Estimates seem to be more biased for unbalanced samples with high n_1 and low n_2. However, this problem is attenuated when the AUC value increases, and the bias of $\hat{\psi}_{mp}$ reduces to about the half of the bias of the MLEs.

5. A Worked Data Example

In this section, an application of the inferential approaches discussed in the current paper to real-life data is presented. We consider data from imaging studies used for brain tumor grading. The data have been originally collected in Tsuchida, Takeuchi, Okazawa, Tsujikawa, and Fujibayashi (2008), and were also discussed in the paper by Feng et al. (2015). This data are also available in the R package auRoc (Feng, 2015). The objective of the study was to evaluate the clinical significance of 1-11C-acetate (ACE) positron emission tomography (PET) in 10 patients with brain glioma, in comparison with 18F-fluorodeoxyglucose (FDG) PET. FDG and ACE are two different imaging techniques for detecting brain glioma. The aim of this section is to examine again the diagnostic accuracy of both techniques in discriminating between patients with low grade (grades I or II) and patients with high grade (grades III and IV). Patients grading was previously determined by magnetic resonance imaging, a gold-standard method used to classify patients with brain glioma in low and high grade classes. Five patients were characterized as low grade and the other five patients as high grade. All patients underwent FDG and ACE diagnostic measurements and the standard uptake value (SUV) was calculated for the same regions of interest in the brain. These SUV values were compared between low grade and high grade patients. The diagnostic accuracy of FDG and ACE was investigated by estimating the area under the ROC curve. Point estimates, confidence intervals and test of hypothesis were performed following the three approached presented in the paper.

Table 1. Two-sided empirical coverage of confidence intervals with 95% nominal levels for A based on the Wald statistic $W_p(\psi)$, the profile log-likelihood ratio statistic $R_p(\psi)$ and the modified profile log-likelihood ratio statistic $R_{mp}(\psi)$, under the binormal model. The central coverage probabilities and the non-coverage probabilities on the left and right tails, which represent, respectively, the lower and upper errors, are reported.

		$W_p(\psi)$			$R_p(\psi)$			$R_{mp}(\psi)$		
A	(n_1, n_2)	coverage	lower	upper	coverage	lower	upper	coverage	lower	upper
	(5, 5)	0.841	0.049	0.110	0.920	0.032	0.049	0.940	0.027	0.033
	(10, 10)	0.894	0.030	0.075	0.933	0.026	0.041	0.942	0.024	0.034
	(20, 20)	0.920	0.032	0.048	0.938	0.031	0.031	0.942	0.031	0.027
	(30, 30)	0.936	0.023	0.040	0.947	0.024	0.029	0.950	0.023	0.027
0.6	(15, 5)	0.830	0.057	0.114	0.908	0.038	0.054	0.938	0.029	0.033
	(5, 15)	0.890	0.037	0.073	0.926	0.031	0.042	0.946	0.025	0.029
	(30, 5)	0.826	0.057	0.116	0.908	0.040	0.052	0.938	0.031	0.030
	(5, 30)	0.882	0.047	0.071	0.919	0.040	0.041	0.946	0.028	0.026
	(80, 10)	0.891	0.040	0.069	0.928	0.034	0.038	0.943	0.029	0.028
	(10, 80)	0.920	0.031	0.050	0.939	0.028	0.033	0.953	0.022	0.025
	(5, 5)	0.776	0.018	0.206	0.912	0.026	0.062	0.935	0.026	0.039
	(10, 10)	0.863	0.011	0.127	0.934	0.021	0.045	0.944	0.022	0.034
	(20, 20)	0.905	0.011	0.084	0.945	0.020	0.036	0.948	0.022	0.030
	(30, 30)	0.912	0.011	0.077	0.944	0.020	0.036	0.947	0.022	0.031
0.8	(15, 5)	0.765	0.013	0.221	0.910	0.023	0.068	0.938	0.023	0.039
	(5, 15)	0.881	0.013	0.106	0.934	0.023	0.042	0.948	0.023	0.028
	(30, 5)	0.771	0.022	0.207	0.896	0.031	0.072	0.931	0.028	0.041
	(5, 30)	0.907	0.014	0.079	0.939	0.024	0.037	0.949	0.023	0.027
	(80, 10)	0.863	0.014	0.123	0.928	0.027	0.046	0.938	0.027	0.035
	(10, 80)	0.935	0.014	0.051	0.951	0.021	0.029	0.956	0.020	0.023
	(5, 5)	0.648	0.001	0.351	0.905	0.013	0.082	0.941	0.017	0.042
	(10, 10)	0.764	0.002	0.235	0.924	0.017	0.059	0.942	0.021	0.037
	(20, 20)	0.840	0.001	0.159	0.939	0.019	0.042	0.946	0.023	0.032
	(30, 30)	0.874	0.002	0.124	0.945	0.021	0.034	0.950	0.024	0.026
0.95	(15, 5)	0.656	0.001	0.343	0.896	0.018	0.086	0.936	0.021	0.043
	(5, 15)	0.809	0.001	0.190	0.939	0.015	0.046	0.954	0.019	0.027
	(30, 5)	0.653	0.002	0.345	0.896	0.016	0.088	0.936	0.018	0.046
	(5, 30)	0.857	0.002	0.141	0.939	0.020	0.040	0.958	0.018	0.024
	(80, 10)	0.770	0.001	0.229	0.926	0.015	0.059	0.942	0.018	0.040
	(10, 80)	0.908	0.006	0.086	0.944	0.023	0.033	0.949	0.025	0.026
	(5, 5)	0.562	0.000	0.438	0.898	0.011	0.091	0.944	0.017	0.038
	(10, 10)	0.687	0.000	0.313	0.922	0.014	0.064	0.940	0.019	0.041
	(20, 20)	0.774	0.000	0.226	0.939	0.017	0.044	0.947	0.021	0.032
	(30, 30)	0.824	0.000	0.176	0.945	0.020	0.035	0.953	0.023	0.024
0.99	(15, 5)	0.749	0.000	0.251	0.937	0.015	0.048	0.948	0.018	0.034
	(5, 15)	0.562	0.000	0.438	0.898	0.011	0.091	0.947	0.015	0.039
	(30, 5)	0.817	0.000	0.183	0.942	0.016	0.042	0.946	0.021	0.032
	(5, 30)	0.563	0.000	0.437	0.898	0.010	0.092	0.948	0.014	0.038
	(80, 10)	0.890	0.001	0.109	0.942	0.022	0.037	0.944	0.024	0.032
	(10, 80)	0.674	0.000	0.326	0.922	0.014	0.064	0.941	0.019	0.040

Table 2. Empirical median biases (Bias), median absolute deviations (MAD) and estimated standard errors (SE) of the estimators for the AUC obtained from the profile likelihood ($\hat{A} = \Phi(\hat{\psi})$) and from the modified profile likelihood ($\hat{A}_{mp} = \Phi(\hat{\psi}_{mp})$), in the binormal model.

		$\hat{A} = \Phi(\hat{\psi})$			$\hat{A}_{mp} = \Phi(\hat{\psi}_{mp})$		
A	(n_1, n_2)	Bias	MAD	SE	Bias	MAD	SE
	(5, 5)	0.0170	0.1938	0.1561	0.0099	0.1811	0.1574
	(10, 10)	0.0121	0.1323	0.1173	0.0083	0.1281	0.1178
	(20, 20)	0.0029	0.0903	0.0853	0.0010	0.0890	0.0854
	(30, 30)	0.0014	0.0720	0.0702	0.0002	0.0713	0.0703
0.6	(15, 5)	0.0131	0.1757	0.1400	0.0082	0.1672	0.1437
	(5, 15)	0.0115	0.1314	0.1144	0.0075	0.1267	0.1165
	(30, 5)	0.0131	0.1763	0.1343	0.0094	0.1693	0.1404
	(5, 30)	0.0060	0.1131	0.0969	0.0039	0.1103	0.1018
	(80, 10)	0.0078	0.1140	0.1023	0.0058	0.1122	0.1046
	(10, 80)	0.0019	0.0728	0.0684	0.0007	0.0722	0.0709
	(5, 5)	0.0378	0.1450	0.1177	0.0203	0.1432	0.1234
	(10, 10)	0.0193	0.1017	0.0917	0.0112	0.1001	0.0941
	(20, 20)	0.0090	0.0725	0.0682	0.0050	0.0721	0.0690
	(30, 30)	0.0075	0.0572	0.0565	0.0048	0.0570	0.0569
0.8	(15, 5)	0.0374	0.1444	0.0812	0.0222	0.1433	0.1201
	(5, 15)	0.0132	0.0873	0.1130	0.0040	0.0861	0.0829
	(30, 5)	0.0390	0.1401	0.1125	0.0249	0.1391	0.1191
	(5, 30)	0.0071	0.0682	0.0631	-0.0034	0.0687	0.0643
	(80, 10)	0.0140	0.1004	0.0895	0.0072	0.0992	0.0922
	(10, 80)	0.0025	0.0416	0.0414	-0.0002	0.0417	0.0420
	(5, 5)	0.0263	0.0340	0.0511	0.0152	0.0469	0.0610
	(10, 10)	0.0119	0.0399	0.0416	0.0055	0.0436	0.0458
	(20, 20)	0.0063	0.0321	0.0315	0.0030	0.0333	0.0331
	(30, 30)	0.0049	0.0257	0.0263	0.0027	0.0262	0.0272
0.95	(15, 5)	0.0242	0.0367	0.0514	0.0138	0.0487	0.0610
	(5, 15)	0.0085	0.0362	0.0368	0.0019	0.0407	0.0398
	(30, 5)	0.0239	0.0373	0.0515	0.0136	0.0494	0.0609
	(5, 30)	0.0052	0.0283	0.0282	0.0008	0.0295	0.0299
	(80, 10)	0.0116	0.0396	0.0410	0.0059	0.0431	0.0449
	(10, 80)	0.0019	0.0192	0.0186	0.0004	0.0195	0.0192
	(5, 5)	0.0075	0.0037	0.0197	0.0045	0.0082	0.0272
	(10, 10)	0.0043	0.0078	0.0150	0.0022	0.0104	0.0181
	(20, 20)	0.0025	0.0081	0.0108	0.0013	0.0091	0.0120
	(30, 30)	0.0016	0.0074	0.0089	0.0008	0.0079	0.0095
0.99	(15, 5)	0.0030	0.0083	0.0124	0.0014	0.0098	0.0142
	(5, 15)	0.0076	0.0036	0.0198	0.0047	0.0078	0.0277
	(30, 5)	0.0014	0.0077	0.0089	0.0006	0.0082	0.0096
	(5, 30)	0.0074	0.0038	0.0200	0.0045	0.0082	0.0282
	(80, 10)	0.0006	0.0050	0.0054	0.0003	0.0051	0.0056
	(10, 80)	0.0047	0.0074	0.0148	0.0027	0.0099	0.0179

Table 3. Point estimates (estimated A) and 95% confidence intervals (95% CI) based on the Wald, profile log-likelihood and modified profile log-likelihood statistics, for the FDG and ACE imaging techniques.

	FDG			ACE		
Method	Estimated A	SE	95% CI	Estimated A	SE	95% CI
Wald	0.726	0.160	(0.413, 1)	0.897	0.096	(0.709,1)
Profile lik.	0.726	0.160	(0.368, 0,939)	0.897	0.096	(0.585,0.990)
Modified profile lik.	0.714	0.163	(0.355, 0.934)	0.879	0.107	(0.548,0.986)

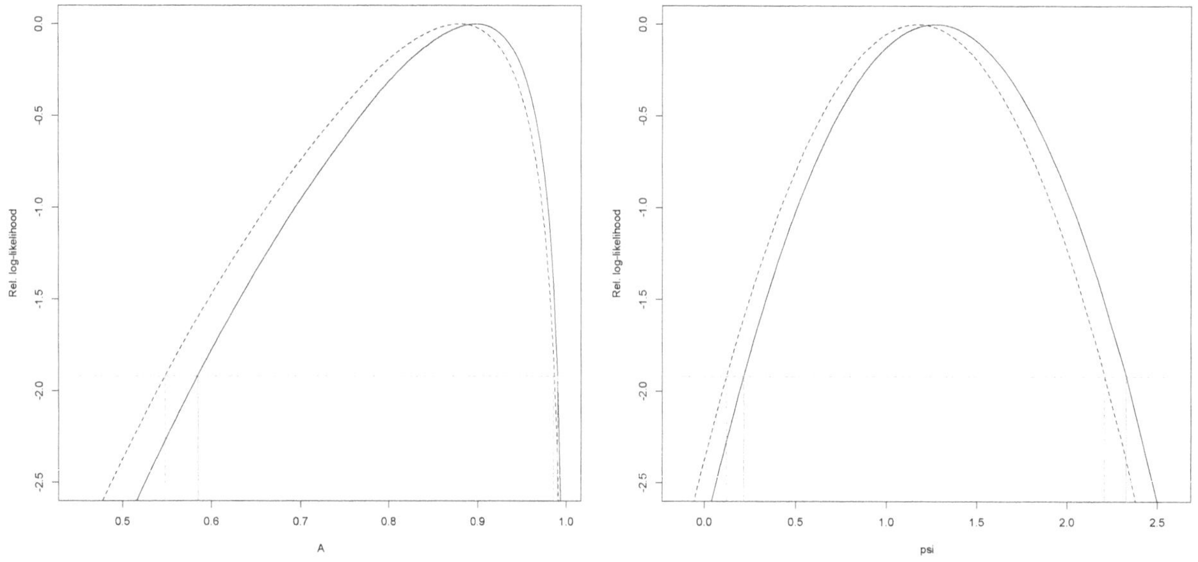

Figure 1. Plot of r_p (thick solid line) and r_p^* (thick dashed line) for a range of values of the parameter R. Vertical lines are drawn to identify confidence intervals for R based on r_p (thin solid line) and r_p^* (thin dashed line)

Assumption of normality in the distributions of SUVs from FDG and ACE in the low-grade and high-grade patients has been shown not to be violated (Feng et al., 2015). For FDG, the average SUV values in the low and high groups were, respectively, 4.714 and 7.124, while for ACE, the average SUV values in the low and high groups were, respectively, 1.850 and 2.626. Lower SUV values are associated to the low grade patients. Therefore, here the random variable Y represents the FDG SUV values in the low grade population, while the random variable X represents the FDG SUV values in the high grade population.

Table 3 summaries the main inferential results for the AUC computed, separately, for the FDG SUV values and the ACE SUV values. For the FDG, the different statistical methods gave very similar estimates of the area under the ROC curve, equal to ~0.7, showing that the FDG has poor discrimination accuracy between the low and high grade populations. The standard errors are also very similar, whereas the Wald confidence interval equal to $(0.413, 1)$ is right-shifted as compared to the confidence intervals based on the $R_p(\psi)$ and $R_{mp}(\psi)$ statistics, which are virtually identical. In addition, we performed a test for the null hypothesis $H_0 : A = 0.5$ versus the alternative $H_1 : A > 0.5$, and found that the $R_p(\psi)$ and $R_{mp}(\psi)$ statistics produce similar non significant p-values ($p = 0.105$ and $p = 0.119$, respectively), then suggesting that there is no evidence of any discriminatory power in the FDG technique.

From Table 3, we observe that also for the ACE, the different statistical methods gave similar estimates of the area under the ROC curve, equal to ~0.9. Thus, it was found that the ACE is much more accurate at discriminating. The Wald confidence interval equal to $(0.709, 1)$ is extremely and erroneously right-shifted, and thus it deviates from the other confidence intervals based on the $R_p(\psi)$ and $R_{mp}(\psi)$ statistics. These latter two differ in particular at the lower limit, as also illustrated in Figure 1. This fact is due to the skewed shape of the profile log-likelihood. In the case of ACE, the test of hypothesis with null $H_0 : A = 0.5$ gave significant results, which differ between the two $R_p(\psi)$- and $R_{mp}(\psi)$-based approaches ($p = 0.009$ and $p = 0.015$, respectively). This result indicates that the ACE technique has the ability to discriminate. When testing the null $H_0 : A = 0.6$, the resulting p-values ($p = 0.030$ and $p = 0.043$, respectively) provide evidence of a discrimination accuracy above 0.6, but the more correct $R_{mp}(\psi)$-based approach gives less evidence for this conclusion. Note that here the power of the tests is low due to very small sample sizes.

Figure 1 reports the relative log-likelihoods, defined as $\ell(\theta) - \ell(\hat{\theta})$, for both the parameters A and ψ. It is noted that relative modified profile log-likelihoods are shifted to the left with respect to the relative profile log-likelihoods, due the adjustment term $M(\psi)$. Moreover, we observe that the $\Phi(\cdot)$ reparameterization on the parameter of interest has the natural effect to make the quadratic functions for $\psi = \delta$ become skewed to the left.

6. Discussion

The paper has presented the performance of a new inferential approach in parametric models for the AUC, which was shown to be useful and easy to implement. The proposed method was applied to make inference for the binormal model, and can immediately be adapted to any other parametric distribution. Alternatively, when the normality assumption is violated, a Box-Cox type power transformation to the original data can also be applied (Box & Cox, 1964; Faraggi & Reiser, 2002). The additional unknown parameters concerning the Box-Cox transformation may be either treated within the entire model as nuisance parameters, or one may, first, apply the appropriate transformation to the original data, and then use inference for the normal theory presented in this paper. We note that the presence of additional nuisance parameters to the model is not expected to affect the accuracy of the inferential results when a modified profile likelihood approach is adopted.

Profile likelihoods have a biased score function of order $O(1)$, which does not typically disappear asymptotically. Modified profile likelihoods have properties very similar to those of usual full likelihoods, and their adjustment term reduces the bias to order $O(n^{-1})$. Consequently, the signed likelihood ratio statistic based on the modified profile likelihood has properties that are superior to those of the usual signed likelihood ratio statistic (Cox & Barndorff-Nielsen, 1994).

The results from simulation studies show that inference based on the modified profile log-likelihood approach has superior performance compared to the classical profile log-likelihood approach, in terms of central coverage probability, symmetry of error rates, and median bias. Wald statistics can lead to seriously misleading inferential conclusions for small or unbalanced samples, especially at the boundaries of the parameter space (Molenberghs & Verbeke, 2007). Moreover, Wald-type tests of hypothesis may lead to erroneous significant results. For example, in the real data application for the ACE technique, it was found that $A_0 = 0.7$ falls outside the Wald confidence interval, and a test of hypothesis of the null $H_0 : A = 0.7$ yields a significant p-value of 0.02, in contrast to the profile likelihood approach that shows no evidence for a discrimination ability above 0.7.

The proposed approach has the potential to be applicable to any general parametric setting, for example, in settings where Y and X follow two different parametric distributions, or in models with mixture distributions, which are often used when the empirical distribution shows a bimodal behaviour. Moreover, future developments concerning the modified likelihood approach could be very relevant in the context of AUC estimation, especially when the data are stratified, or the interest of the inquiry is on modeling different stratum-specific AUC, for example by including stratum-specific fixed effects as nuisance parameters. Another important case of application of the proposed approach may be when the two random variables X and Y depend on covariates, or in general when the AUC models rely on many nuisance parameters (Sartori, 2003).

Acknowledgements

The author was supported by 'Progetto di Ateneo 2015' (CPDA153257), University of Padua.

References

Bamber, D. (1975). The area above the ordinal dominance graph and the area below the receiver operating characteristic graph. *Journal of Mathematical Psychology*, *12*(4), 387–415. https:/doi.org/10.1016/0022-2496(75)90001-2

Box, G. EP., & Cox, D.R. (1964). An analysis of transformations. *Journal of the Royal Statistical Society. Series B (Methodological)*, *26*, 211–252.

Cortese, G., & Ventura, L. (2013). Accurate higher-order likelihood inference on $P(Y < X)$. *Computational statistics*, *28*(3), 1035–1059. https:/doi.org/10.1007/s00180-012-0343-z

Cox, D., & Barndorff-Nielsen, O. (1994). *Inference and asymptotics* (Vol. 52). CRC Press.

Cox, D., & Reid, N. (1992). A note on the difference between profile and modified profile likelihood. *Biometrika*, *79*(2), 408–411. https:/doi.org/10.1093/biomet/79.2.408

Faraggi, D., & Reiser, B. (2002). Estimation of the area under the roc curve. *Statistics in Medicine*, *21*(20), 3093–3106. https:/doi.org/10.1002/sim.1228

Feng, D. (2015). auRoc: Various methods to estimate the AUC [Computer software manual]. Retrieved from `https://CRAN.R-project.org/package=auRoc` (R package version 0.1-0)

Feng, D., Cortese, G., & Baumgartner, R. (2015). A comparison of confidence/credible interval methods for the area under the ROC curve for continuous diagnostic tests with small sample size. *Statistical Methods in Medical Research*, Published online before print. https:/doi.org/10.1177/0962280215602040

Johnson, R. A. (1988). Stress-strength models for reliability. *Handbook of Statistics*, *7*, 27–54. https:/doi.org/10.1016/S0169-7161(88)07005-1

Kotz, S., Lumelskii, Y., & Pensky, M. (2003). The stress-strength model and its generalizations. *Theory and Applications. Singapore: World Scientific*, *43*, 44.

Krzanowski, W. J., & Hand, D. J. (2009). *ROC curves for continuous data*. CRC Press.

Molenberghs, G., & Verbeke, G. (2007). Likelihood ratio, score, and wald tests in a constrained parameter space. *The American Statistician*, *61*(1), 22–27. https:/doi.org/10.1198/000313007X171322

Obuchowski, N. A., & Lieber, M. L. (2002). Confidence bounds when the estimated ROC area is 1.0. *Academic Radiology*, *9*(5), 526–530. https:/doi.org/10.1016/s1076-6332(03)80329-x

Pardo-Fernández, J. C., Rodríguez-Álvarez, M. X., Van Keilegom, I., et al. (2013). *A review on ROC curves in the presence of covariates* (Tech. Rep.). UCL.

Pepe, M. S. (2003). *The statistical evaluation of medical tests for classification and prediction*. Oxford University Press, USA.

Reiser, B. (2000). Measuring the effectiveness of diagnostic markers in the presence of measurement error through the use of ROC curves. *Statistics in Medicine*, *19*(16), 2115–2129.

Reiser, B., & Faraggi, D. (1997). Confidence intervals for the generalized ROC criterion. *Biometrics*, 644–652. https:/doi.org/10.2307/2533964

Reiser, B., & Guttman, I. (1986). Statistical inference for $Pr(Y < X)$: the normal case. *Technometrics*, *28*(3), 253–257.

Sartori, N. (2003). Modified profile likelihoods in models with stratum nuisance parameters. *Biometrika*, *90*(3), 533–549. https:/doi.org/10.1093/biomet/90.3.533

Severini, T. A. (2000). *Likelihood methods in statistics*. Oxford University Press.

Tsuchida, T., Takeuchi, H., Okazawa, H., Tsujikawa, T., & Fujibayashi, Y. (2008). Grading of brain glioma with 1-11 C-acetate PET: comparison with 18 F-FDG PET. *Nuclear Medicine and Biology*, *35*(2), 171–176. https:/doi.org/10.1016/j.nucmedbio.2007.11.004

Wolfe, D. A., & Hogg, R. V. (1971). On constructing statistics and reporting data. *The American Statistician*, *25*(4), 27–30.

Zhou, X.-H., McClish, D. K., & Obuchowski, N. A. (2009). *Statistical methods in diagnostic medicine* (Vol. 569). John Wiley & Sons.

Zou, K. H., Carlsson, M. O., & Yu, C.-R. (2012). Comparison of adjustment methods for stratified two-sample tests in the context of ROC analysis. *Biometrical Journal*, *54*(2), 249–263. https:/doi.org/10.1002/bimj.201000251

Appendix

The analytic expressions of the partial derivatives of $\ell(\psi, \lambda)$ with respect to the nuisance parameters for the binormal model are given hereafter:

$$\ell_{\lambda_1}(\psi, \lambda) = \frac{n_1(\hat{\lambda}_1 - \lambda_1)}{\lambda_2^2} + \frac{n_2(\hat{\lambda}_1 + \hat{\psi}\hat{\lambda}_3 - \lambda_1 - \psi\lambda_3)}{\lambda_3^2 - \lambda_2^2}$$

$$\ell_{\lambda_2}(\psi, \lambda) = -\frac{n_1}{\lambda_2} + \frac{n_2\lambda_2}{\lambda_3^2 - \lambda_2^2} - \frac{n_2\lambda_2\left[\hat{\lambda}_3^2 - \hat{\lambda}_2^2 + (\hat{\lambda}_1 + \hat{\psi}\hat{\lambda}_3 - \lambda_1 - \psi\lambda_3)^2\right]}{(\lambda_3^2 - \lambda_2^2)^2} + \frac{n_1\left[\hat{\lambda}_2^2 + (\hat{\lambda}_1 - \lambda_1)^2\right]}{\lambda_2^3}$$

$$\ell_{\lambda_3}(\psi, \lambda) = -\frac{n_2\lambda_3}{\lambda_3^2 - \lambda_2^2} + \frac{n_2\lambda_3\left[\hat{\lambda}_3^2 - \hat{\lambda}_2^2 + (\hat{\lambda}_1 + \hat{\psi}\hat{\lambda}_3 - \lambda_1 - \psi\lambda_3)^2\right]}{(\lambda_3^2 - \lambda_2^2)^2} + \frac{n_2\psi(\hat{\lambda}_1 + \hat{\psi}\hat{\lambda}_3 - \lambda_1 - \psi\lambda_3)}{\lambda_3^2 - \lambda_2^2}$$

The analytic expressions of the sample space derivatives $\ell_{\lambda;\hat{\lambda}}(\psi, \lambda)$ for the binormal model follow hereafter:

$$
\begin{aligned}
\ell_{\lambda_1;\hat{\lambda}_1}(\psi, \lambda) &= \frac{n_1}{\lambda_2^2} + \frac{n_2}{\lambda_3^2 - \lambda_2^2} \\
\ell_{\lambda_1;\hat{\lambda}_2}(\psi, \lambda) &= 0 \\
\ell_{\lambda_1;\hat{\lambda}_3}(\psi, \lambda) &= \frac{n_2\hat{\psi}}{\lambda_3^2 - \lambda_2^2} \\
\ell_{\lambda_2;\hat{\lambda}_1}(\psi, \lambda) &= -\frac{2n_2\lambda_2(\hat{\lambda}_1 + \hat{\psi}\hat{\lambda}_3 - \lambda_1 - \psi\lambda_3)}{(\lambda_3^2 - \lambda_2^2)^2} + \frac{2n_1(\hat{\lambda}_1 - \lambda_1)}{\lambda_2^3} \\
\ell_{\lambda_2;\hat{\lambda}_2}(\psi, \lambda) &= \frac{2n_2\lambda_2\hat{\lambda}_2}{(\lambda_3^2 - \lambda_2^2)^2} + \frac{2n_1\hat{\lambda}_2}{\lambda_2^3} \\
\ell_{\lambda_2;\hat{\lambda}_3}(\psi, \lambda) &= -\frac{2n_2\lambda_2\left[\hat{\lambda}_3 + \hat{\psi}(\hat{\lambda}_1 + \hat{\psi}\hat{\lambda}_3 - \lambda_1 - \psi\lambda_3)\right]}{(\lambda_3^2 - \lambda_2^2)^2} \\
\ell_{\lambda_3;\hat{\lambda}_1}(\psi, \lambda) &= \frac{2n_2\lambda_3(\hat{\lambda}_1 + \hat{\psi}\hat{\lambda}_3 - \lambda_1 - \psi\lambda_3)}{(\lambda_3^2 - \lambda_2^2)^2} + \frac{n_2\psi}{\lambda_3^2 - \lambda_2^2} \\
\ell_{\lambda_3;\hat{\lambda}_2}(\psi, \lambda) &= -\frac{2n_2\lambda_3\hat{\lambda}_2}{(\lambda_3^2 - \lambda_2^2)^2}
\end{aligned}
$$

16

Characterizations of Extreme Value Extended Marshall-Olkin Models with Exponential Marginals

Nikolai Kolev[1] & Jayme Pinto[1]

[1] Department of Statistics, University of São Paulo, Brazil

Correspondence: Nikolai Kolev, Department of Statistics, University of São Paulo. Rua do Matão, 1010 - ZIP CODE 05508-090 - São Paulo (SP), Brazil. E-mail: kolev.ime@gmail.com

Abstract

We construct and characterize bivariate extreme value distributions with exponential marginals generated by the stochastic representation $(X_1, X_2) = (\min(T_1, T_3), \min(T_2, T_3))$ where the random variable T_3 is independent of random variables T_1 and T_2 which are assumed to be dependent. A building procedure is suggested when the joint distribution of (T_1, T_2) is absolutely continuous and T_i's are not necessarily exponentially distributed, $i = 1, 2, 3$. The Pickands representation of the vector (X_1, X_2) is computed. We illustrate the general relations by examples.

Keywords: bivariate extreme value distribution, extended Marshall-Olkin model, Pickands measure and dependence function.

1. Introduction

Let us consider the fatal shock model defined by the stochastic representation

$$(X_1, X_2) = (\min(T_1, T_3), \min(T_2, T_3)), \tag{1}$$

where non-negative continuous random variables T_1 and T_2 identify the occurrence of independent "individual shocks" affecting two devices and T_3 is their "common shock". The random vector (X_1, X_2) presents the joint distribution of both lifetimes.

Denote by $S_{X_1,X_2}(x_1, x_2) = P(X_1 > x_1, X_2 > x_2)$ the joint survival function of the vector (X_1, X_2) for all $x_1, x_2 \geq 0$. If the shocks are governed by independent homogeneous Poisson processes, then T_i's in (1) are exponentially distributed with parameters $\lambda_i > 0$, $i = 1, 2, 3$, and we obtain the classical Marshall-Olkin's (MO) bivariate exponential distribution

$$S_{X_1,X_2}(x_1, x_2) = \exp\{-\lambda_1 x_1 - \lambda_2 x_2 - \lambda_3 \max(x_1, x_2)\}, \quad x_1, x_2 \geq 0, \tag{2}$$

see Marshall and Olkin (1967).

To give a probability interpretation of model (1), consider a system composed by two items, to be denoted by 1 and 2. We associate with each item j, $j = 1, 2$, a Bernoulli random variable Z_j, indicating whether the item is operational ($Z_j = 1$) or failed ($Z_j = 0$). The bivariate Bernoulli random vector (Z_1, Z_2) represents the state of the system. It is specified in terms of MO construction (1). The vector (X_1, X_2) exhibits the latent state of the system, since the MO model (2) is defined in terms of vector (T_1, T_2, T_3) of latent variables that identify independent exponential shock times. Each shock takes down a given subset of items ({1}, {2} or {both 1 and 2}) and occurs at an exponential time with constant rates λ_1, λ_2 and λ_3, respectively.

The stochastic relation (1) is widely used in literature. For example, Li and Pellerey (2011) launched the Generalized MO (GMO) model considering non exponential independent random variables T_i in (1), $i = 1, 2, 3$. The corresponding joint distributions do not possess bivariate lack of memory property, i.e., are "aging". As a further step, Pinto and Kolev (2015) introduced the Extended MO (EMO) model assuming dependence between latent variables T_1 and T_2, but keeping T_3 independent of them. The motivation is that the individual shocks might be dependent if the items share a common environment. In this case however, the EMO distributions can be "aging" or "non-aging" depending on the parameters of joint distribution of (T_1, T_2) and the distribution of T_3.

We will assume further that T_1 and T_2 are no more independent, but defined by their joint survival function $S_{T_1,T_2}(x_1, x_2) = P(T_1 > x_1, T_2 > x_2)$ and the random variable T_3 is independent of T_1 and T_2. Let $S_{T_i}(x) = P(T_i > x)$ be the survival functions of T_i, $i = 1, 2, 3$ for $x \geq 0$. Thus, the joint survival function of EMO model generated by (1) can be written as

$$S_{X_1,X_2}(x_1, x_2) = S_{T_1,T_2}(x_1, x_2) S_{T_3}(\max\{x_1, x_2\}). \tag{3}$$

In Section 2 we establish the extreme value representation and characterizations of a subclass of EMO distributions defined by (3) whose marginals X_1 and X_2 are exponentially distributed, see Theorem 1 and the most general Theorem 3. We suggest a procedure to construct EMO distributions with exponential marginals even if T_i's in (1) are not exponentially distributed under the corresponding additional restriction, e.g., Theorem 2. We obtain in Section 3 the Pickands dependence function and Pickands measure corresponding to (3) if the joint distribution of (T_1, T_2) is absolutely continuous. We illustrate the general relationships with two examples. As a particular case one can find the corresponding representations associated to the MO's bivariate exponential distribution (2) obtained by Mai and Scherer (2010). We finish the article with a short discussion.

2. Extreme Value EMO Distributions

There is a number of mathematical results characterizing multivariate extreme value distributions and extreme value copulas, see Chapter 6 in Joe (1997) for example. The marginals of any multivariate extreme value distribution must be members of the class of univariate generalized extreme value distributions, i.e., location-scale families of distributions based on Weibull, Fréchet and Gumbel distributions, see Theorem 2.4.1 in Galambos (1978).

Without loss of generality, we will assume hereafter marginal exponential distributions in (3), i.e., $X_i \sim Exp(\lambda_{X_i}),\ i = 1,2$. Such a choice does not have influence on the corresponding survival copula to be obtained, being invariant on strict monotone transformations. It is also well known that these transformations can be used to move from one member to the other in the class of univariate generalized extreme value distributions, see Galambos (1978) and Beirlant et al. (2005) for a related discussion.

The joint survival function of the EMO distributions specified by (3) can be equivalently represented by

$$S_{X_1,X_2}(x_1, x_2) = S_{T_1,T_2}(x_1, x_2)\min\{S_{T_3}(x_1), S_{T_3}(x_2)\}. \tag{4}$$

Note that the right hand side in (4) is a product of two bivariate distributions. The first one is defined by $S_{T_1,T_2}(x_1, x_2)$, and the second one refers to a bivariate random vector with comonotonic components sharing the same marginal distribution as T_3. Such a product construction technique in terms of copula has been discussed by Genest et al. (1998), see their Proposition 2. Consult Liebscher (2008) for a general power function based approach as well.

Starting from the latent random vector (T_1, T_2) we "modify" it via (4) to get (X_1, X_2), which can be interpreted as follows: the "modified" joint distribution (X_1, X_2) can be used to model a complementary amount of bivariate asymmetry induced by (T_1, T_2). Note that, in general, this additional asymmetry does not necessarily imply an increase of upper tail dependence (if exists) governed by (T_1, T_2), see supporting comments in Joe (2015), page 184.

The simplest way to ensure exponentially distributed marginals X_1 and X_2 in (4) is to advocate that $T_i \sim Exp(\lambda_i),\ i = 1, 2, 3$. This distributional choice has nice properties and will be justified in Theorem 1.

Other distributional possibilities for $T_i,\ i = 1, 2, 3$, do exist in order to construct EMO model with exponentially distributed marginals X_j with parameter $\lambda_{X_j},\ j = 1, 2$. To believe, denote by $r_{T_i}(x) = \frac{d}{dx}[-\ln S_{T_i}(x)]$ the failure rates of $T_i,\ i = 1, 2, 3$. Note that the marginal survival functions in (4) are given by $S_{X_i}(x_i) = S_{T_i}(x_i)S_{T_3}(x_i)$ and the only condition in terms of failure rate functions to get exponential marginals in (4) is $r_{T_i}(x_i) + r_{T_3}(x_i) = \lambda_{X_i}$ for all $x_i \geq 0,\ i = 1, 2$. For example, consider $r_{T_1}(x) = r_{T_2}(x) = 2 + \sin(x)$ and $r_{T_3}(x) = 1 - \sin(x)$ for $x \geq 0$, to obtain $\lambda_{X_i} = 3,\ i = 1, 2$.

Denote by BEV_E the set of bivariate extreme value distributions with exponential marginals and let $BEV_E - EMO$ be a subclass of BEV_E satisfying (4), where $(T_1, T_2) \subset BEV_E$ and $T_3 \sim Exp(\lambda_3)$. This means that the following functional equation is fulfilled

$$S_{T_1,T_2}(tx_1, tx_2) = [S_{T_1,T_2}(x_1, x_2)]^t \quad \text{for all} \quad t > 0. \tag{5}$$

Joe (1997) observed that when multivariate copulas take on univariate generalized extreme value distributions one obtains multivariate extreme value distributions for maxima, but when the copulas take on generalized extreme value survival margins for minima, multivariate extreme value survival functions for minima results, see page 176. A general related discussion in terms of functional equations involving joint survival functions can be found in Marshall and Olkin (1991) as well.

Relation (5) implies that the survival copula $\overline{C}_{T_1,T_2}(u, v)$ associated to $S_{T_1,T_2}(x_1, x_2)$ is also an extreme value copula for all $t > 0$, i.e.,

$$\overline{C}_{T_1,T_2}(u^t, v^t) = [\overline{C}_{T_1,T_2}(u, v)]^t, \quad u, v \in [0, 1). \tag{6}$$

It follows our first characterization statement.

Theorem 1 *Let $T_3 \sim Exp(\lambda_3)$ in (4). Then $(X_1, X_2) \subset BEV_E - EMO$ if and only if $(T_1, T_2) \subset BEV_E$.*

Proof. Let $(X_1, X_2) \subset BEV_E - EMO$. Taking into account that X_i and T_3 are exponentially distributed and relations $S_{X_i}(x_i) = S_{T_i}(x_i)S_{T_3}(x_i)$, we conclude that T_i are exponentially distributed, $i = 1, 2$. Since $S_{X_1,X_2}(x_1, x_2)$ is an extreme value survival function with exponential marginals, then $S_{X_1,X_2}(tx_1, tx_2) = [S_{X_1,X_2}(x_1, x_2)]^t$ and applying representation (3) we get

$$\begin{aligned} S_{X_1,X_2}(tx_1, tx_2) &= S_{T_1,T_2}(tx_1, tx_2)S_{T_3}(\max\{tx_1, tx_2\}) \\ &= [S_{T_1,T_2}(x_1, x_2)S_{T_3}(\max\{x_1, x_2\})]^t. \end{aligned}$$

But T_3 is exponentially distributed, so that $[S_{T_3}(\max\{x_1, x_2\})]^t = S_{T_3}(\max\{tx_1, tx_2\})$ and we arrive to (5).

Now assume that $(T_1, T_2) \subset BEV_E$ and $T_3 \sim Exp(\lambda_3)$ in (4). Since the comonotonic copula $M(u, v) = \min(u, v)$ is an extreme value copula, applying Sklar's theorem in (4) and taking into account (6) we obtain

$$\begin{aligned} S_{X_1,X_2}(tx_1, tx_2) &= \overline{C}_{T_1,T_2}(\exp(-\lambda_1 tx_1), \exp(-\lambda_2 tx_2)) \min\{\exp(-\lambda_3 tx_1), \exp(-\lambda_3 tx_2)\} \\ &= \{\overline{C}_{T_1,T_2}[\exp(-\lambda_1 x_1), \exp(-\lambda_2 x_2)]\}^t[\min\{\exp(-\lambda_3 x_1), \exp(-\lambda_3 x_2)\}]^t \\ &= [S_{X_1,X_2}(x_1, x_2)]^t. \end{aligned}$$

Thus, the functional equation (5) for the vector (X_1, X_2) is satisfied and (X_1, X_2) belongs to the class $BEV_E - EMO$.

Obviously, if both $(T_1, T_2) \subset BEV_E$ and $(X_1, X_2) \subset BEV_E - EMO$ in (4) then $T_3 \sim Exp(\lambda_3)$.

To proceed, we will need to recall that any bivariate survival function $S_{T_1,T_2}(x_1, x_2)$ can be represented as

$$S_{T_1,T_2}(x_1, x_2) = \exp\{-H_{T_1}(x_1) - H_{T_2}(x_2) + D_{T_1,T_2}(x_1, x_2)\}, \tag{7}$$

where $H_{T_i}(x_i) = -\ln[S_{T_i}(x_i)]$ are the cumulative hazard functions of random variables T_i, $i = 1, 2$, and $D_{T_1,T_2}(x_1, x_2) = \ln\left[\frac{S_{T_1,T_2}(x_1,x_2)}{S_{T_1}(x_1)S_{T_2}(x_2)}\right]$ is the dependence function introduced by Sibuya (1960), to be referred as *Sibuya's dependence function*. It exhibits interesting relationships with dependence phenomena, see Kolev (2016) or Pinto and Kolev (2015a) for a deep discussion.

With an additional assumption of absolute continuity of the joint distribution of (T_1, T_2), we justify in the next Lemma 1 the choice of exponential T_i's in order to get a member of $BEV_E - EMO$ via (4), $i = 1, 2, 3$. Let us denote by $BEV_E - EMO^{AC}$ this subclass.

Lemma 1 *If* $(X_1, X_2) \subset BEV_E - EMO^{AC}$ *in (4) and* $S_{T_1,T_2}(x_1, x_2)$ *is absolutely continuous, then* $(T_1, T_2) \subset BEV_E$ *and* $T_3 \sim Exp(\lambda_3)$.

Proof. Suppose $x_1 > x_2 \geq 0$ and $t > 0$. Since $S_{X_1,X_2}(x_1, x_2)$ is an extreme value survival function we have $S_{X_1,X_2}(tx_1, tx_2) = [S_{X_1,X_2}(x_1, x_2)]^t$. Applying the exponential representation for bivariate survival functions (7) in (3) we obtain

$$\begin{aligned} &\exp\{-H_{T_1}(tx_1) - H_{T_3}(tx_1) - H_{T_2}(tx_2) + D_{T_1,T_2}(tx_1, tx_2)\} \\ &= \exp\{-tH_{T_1}(x_1) - tH_{T_3}(x_1) - tH_{T_2}(x_2) + tD_{T_1,T_2}(x_1, x_2)\}. \end{aligned}$$

Taking logarithms in both sides of former equation and calculating the mixed partial derivatives, we get

$$t\frac{\partial^2}{\partial x_1 \partial x_2} D_{T_1,T_2}(tx_1, tx_2) = \frac{\partial^2}{\partial x_1 \partial x_2} D_{T_1,T_2}(x_1, x_2).$$

Integrating we have

$$\int_0^{x_1}\int_0^{x_2} t\frac{\partial^2}{\partial u \partial v} D_{T_1,T_2}(tu, tv) dv du = \int_0^{x_1}\int_0^{x_2} \frac{\partial^2}{\partial u \partial v} D_{T_1,T_2}(u, v) dv du.$$

The boundary conditions $D_{T_1,T_2}(x_1, 0) = D_{T_1,T_2}(0, x_2) = 0$ imply equalities

$$\frac{\partial}{\partial x_1} D_{T_1,T_2}(x_1, 0) = \frac{\partial}{\partial x_2} D_{T_1,T_2}(0, x_2) = 0.$$

Thus, we arrive to the functional equation

$$D_{T_1,T_2}(tx_1, tx_2) = tD_{T_1,T_2}(x_1, x_2) \quad \text{for all} \quad t > 0. \tag{8}$$

Since $S_{X_1,X_2}(x_1,x_2)$ is an extreme value survival function with exponential marginals, we deduce that

$$\left[\frac{S_{X_1,X_2}(x_1,x_2)}{S_{X_1}(x_1)S_{X_2}(x_2)}\right]^t = \frac{S_{X_1,X_2}(tx_1,tx_2)}{S_{X_1}(tx_1)S_{X_2}(tx_2)}. \tag{9}$$

Applying the exponential representation (7) in (9) and taking into account (8), we obtain the homogeneous of order 1 functional equation $H_{T_3}(tx_2) = tH_{T_3}(x_2)$, for all $x_2, t > 0$, where $H_{T_3}(x_2)$ is the cumulative hazard of T_3. Substitute $\lambda_3 = H_{T_3}(1)$ to get the general solution $H_{T_3}(x_2) = \lambda_3 x_2$ with $\lambda_3 > 0$, since $H_{T_3}(x_2)$ is nonnegative for all $x_2 \geq 0$, see Aczel (1966) as well.

The conclusion is the same when $x_2 \geq x_1 \geq 0$. Thus, $T_3 \sim Exp(\lambda_3)$. Applying the "only if" branch of Theorem 1 we finish the proof.

Linking Theorem 1 and Lemma 1 we obtain the following characterization.

Theorem 2 *Let the joint distribution of* (T_1, T_2) *be absolutely continuous. Then* $(X_1, X_2) \subset BEV_E - EMO^{AC}$ *if and only if* $(T_1, T_2) \subset BEV_E$ *and* $T_3 \sim Exp(\lambda_3)$.

The absolute continuity assumption for (T_1, T_2) in Theorem 2 is an important condition. In the next we suggest a building procedure showing that it is possible to construct a member of $BEV_E - EMO$ class where the random variables T_i in (3) are not necessarily exponentially distributed, $i = 1, 2, 3$, but as a compensation we have to relax the assumption of absolutely continuity of the survival function $S_{T_1,T_2}(x_1, x_2)$.

Example 1 (Building procedure). Let us consider three nonnegative absolutely continuous and independent random variables defined by $S_{Y_1}(x) = S_{Y_2}(x) = \exp\{-x\}$ and $S_{Y_3}(x) = \exp\{-bx + f(x) - a\}$, where $a, b \geq 0$ and the continuous function $f(x)$ is such that $a - bx < f(x) < a + bx$ for all $x > 0$ with $f(0) = a$.

The following two-step procedure generates a vector $(X_1, X_2) \subset BEV_E - EMO$ with non-exponentially distributed T_i's:

1. Construct a GMO distribution $(T_1, T_2) = (\min(Y_1, Y_3), \min(Y_2, Y_3))$. Its joint survival function is $S_{T_1,T_2}(x_1, x_2) = \exp\{-x_1 - x_2 - H_{Y_3}(\max(x_1, x_2))\}$;
2. Select a random variable T_3, independent of (T_1, T_2), with a survival function $S_{T_3}(x) = \exp\{-bx - f(x) + a\}$. Apply (3) to get the corresponding EMO survival function $S_{X_1,X_2}(x_1, x_2) = \exp\{-x_1 - x_2 - 2b\max(x_1, x_2)\}$ with exponentially distributed marginals.

Notice that $S_{T_1,T_2}(x_1, x_2)$ obtained in the first procedure step is neither an extreme value survival function nor absolutely continuous, since has a singular component with support on the set $\{(x_1, x_2) \in [0, \infty)^2 \mid x_1 = x_2 = x\}$. In addition, T_3 defined in the second step is not exponentially distributed. Finally, inequalities $a - bx < f(x) < a + bx$ and $f(0) = a$ guarantee that $S_{Y_3}(x)$ and $S_{T_3}(x)$ are proper survival functions.

Now observe that the functional equation (8) involving Sibuya's dependence function $D_{T_1,T_2}(x_1, x_2)$ is homogeneous of order 1 for all $x_1, x_2 \geq 0$ and $t > 0$. As a consequence of Theorem 6.2 in Joe (1997), equation (8) can serve as a characterization of bivariate extreme value survival functions $S_{T_1,T_2}(x_1, x_2)$ with exponential marginals even when absolute continuity for (T_1, T_2) does not hold true. Hence, we deduce the following characterization.

Lemma 2 *A bivariate distribution is* BEV_E *if an only if its Sibuya's dependence function is homogeneous of order 1.*

To characterize distributions belonging to the subclass $BEV_E - EMO$ when the joint distribution of (T_1, T_2) is not absolutely continuous, we will need an additional assumption since $BEV_E - EMO \subset BEV_E$. It is given below.

Theorem 3 $(X_1, X_2) \subset BEV_E - EMO$ *if and only if the functional equation*

$$D_{T_1,T_2}(tx_1, tx_2) + H_{T_3}(t\min(x_1, x_2)) = tD_{T_1,T_2}(x_1, x_2) + tH_{T_3}(\min(x_1, x_2))$$

is satisfied for all $x_1, x_2 \geq 0$ *and* $t > 0$.

Proof. Under conditions in the theorem, the Sibuya's dependence function for the EMO model (3) is given by

$$D_{X_1,X_2}(x_1, x_2) = D_{T_1,T_2}(x_1, x_2) + H_{T_3}(\min(x_1, x_2)).$$

To finalize the proof, just apply Lemma 2.

Example 2 ($BEV_E - EMO$ distributions when (T_1, T_2) is not absolutely continuous). The Marshall-Olkin survival function

$$S_{X_1,X_2}(x_1, x_2) = \exp\{-x_1 - x_2 - 2\max(x_1, x_2)\}$$

is of EMO-type, being an example of extreme value survival function as well. We use relation (4) and the two step procedure from Example 1: first selecting random variables (T_1, T_2) with survival function $S_{T_1,T_2}(x_1, x_2) = \exp\{-x_1 - x_2 - H_{Y_3}(\max(x_1, x_2))\}$, where $H_{Y_3}(x) = x - \cos(x) + 1$, and second, choosing T_3 independent of (T_1, T_2) with cumulative failure rate $H_{T_3}(x) = x + \cos(x) - 1$. Notice that $S_{T_1,T_2}(x_1, x_2)$ is neither an extreme value survival function nor absolutely continuous, as well as T_3 is not exponentially distributed. In this case we have $D_{T_1,T_2}(x_1, x_2) = H_{Y_3}(\min(x_1, x_2))$, so that the functional equation given in Theorem 3 is satisfied.

3. Pickands Representation and Examples

Given the characterization established in Theorem 2, in the sequel we assume absolutely continuous distributions for (T_1, T_2) to ensure uniqueness in our construction of extreme value EMO distributions. In the next theorem we obtain the general form of the corresponding survival function and related copula. We will need to remind basic facts related to Pickands measure involved first.

Pickands (1981) proves that each multivariate extreme value distribution is uniquely characterized by a finite measure satisfying boundary conditions and related dependence function. In the bivariate case, the bivariate extreme value (survival) copulas can be completely characterized by the relation

$$\overline{C}(u, v) = \exp\left\{(\ln uv)\mathcal{A}\left(\frac{\ln u}{\ln uv}\right)\right\}, \tag{10}$$

where $\mathcal{A}(w) = \int_0^1 \max((1-x)w, x(1-w))d\mathbb{H}(x)$, for a positive finite measure $\mathbb{H}$ on $[0, 1]$, denominated Pickands measure, see Joe (1997).

The so-called Pickands dependence function $\mathcal{A}(w) : [0, 1] \to [\frac{1}{2}, 1]$ must be convex and should satisfy $\max(w, 1-w) \leq \mathcal{A}(w) \leq 1$. The lower bound of $\mathcal{A}(w)$ corresponds to the comonotonic copula and related Pickands measure puts mass 2 at $w = \frac{1}{2}$; the upper bound of $\mathcal{A}(w)$ is associated to the independence copula with Pickands measure assigning mass 1 to both $w = 0$ and $w = 1$.

The Pickands dependence function can be recovered from the copula $\overline{C}$ by setting

$$\mathcal{A}(w) = -\ln \overline{C}(\exp(-w), \exp(-(1-w))), \quad w \in [0, 1] \tag{11}$$

and is uniquely related to the measure $\mathbb{H}$ via equation

$$\mathbb{H}([0, w]) = \begin{cases} 1 + \frac{d}{dw}\mathcal{A}(w), & \text{if } w \in [0, 1), \\ 2, & \text{if } w = 1, \end{cases} \tag{12}$$

where $\frac{d}{dw}\mathcal{A}(w)$ is the right-hand derivative of $\mathcal{A}(w)$, see Theorem 3.1 in Pickands (1981) and section 8.5.3 in Beirlant et al. (2005). Moreover, the point masses of $\mathbb{H}$ at 0 and 1 are given by

$$\mathbb{H}(\{0\}) = 1 + \frac{d}{dw}\mathcal{A}(0) \quad \text{and} \quad \mathbb{H}(\{1\}) = 1 - \frac{d}{dw}\mathcal{A}(1),$$

where $\frac{d}{dw}\mathcal{A}(1) = \sup_{0 \leq w < 1} \frac{d}{dw}\mathcal{A}(w)$.

Theorem 4 *Suppose $(X_1, X_2) \in BEV_E - EMO^{AC}$. Then, the survival function of (X_1, X_2) has the form*

$$S_{X_1,X_2}(x_1, x_2) = \exp\left\{-(\lambda_1 x_1 + \lambda_2 x_2)\mathcal{A}_{T_1,T_2}\left(\frac{\lambda_1 x_1}{\lambda_1 x_1 + \lambda_2 x_2}\right) - \lambda_3 \max(x_1, x_2)\right\}, \tag{13}$$

where $\mathcal{A}_{T_1,T_2}(w)$ is the Pickands dependence function corresponding to $\overline{C}_{T_1,T_2}(u, v)$ and $\lambda_i > 0$, $i = 1, 2, 3$.

The associated survival copula writes as

$$\begin{aligned} &\overline{C}_{X_1,X_2}(u, v) = \\ &= \exp\left\{(\ln uv)\left[(\Lambda_1(w) + \Lambda_2(w))\mathcal{A}_{T_1,T_2}\left(\tfrac{\Lambda_1(w)}{\Lambda_1(w)+\Lambda_2(w)}\right) + \max(\Lambda_1(w), \Lambda_2(w))\right]\right\}, \end{aligned} \tag{14}$$

where $u, v \in [0, 1)$, $w = \frac{\ln u}{\ln uv}$, $\Lambda_1(w) = \frac{\lambda_1 w}{\lambda_1+\lambda_3}$ and $\Lambda_2(w) = \frac{\lambda_2(1-w)}{\lambda_2+\lambda_3}$.

The Pickands measure $\mathbb{H}$ is given by

$$\mathbb{H}([0, w]) = \begin{cases} 1 + \left(\frac{\lambda_1}{\lambda_1+\lambda_3} - \frac{\lambda_2}{\lambda_2+\lambda_3}\right)\mathcal{A}_{T_1,T_2}\left(\frac{\Lambda_1(w)}{\Lambda_1(w)+\Lambda_2(w)}\right) \\ +\left(\frac{\lambda_1 w}{\lambda_1+\lambda_3} + \frac{\lambda_2(1-w)}{\lambda_2+\lambda_3}\right)\frac{d}{dw}\mathcal{A}_{T_1,T_2}\left(\frac{\Lambda_1(w)}{\Lambda_1(w)+\Lambda_2(w)}\right) - \frac{\lambda_3}{\lambda_2+\lambda_3}, & \text{if } 0 \leq w < \frac{\lambda_1+\lambda_3}{\lambda_1+\lambda_2+2\lambda_3}, \\ 1 + \left(\frac{\lambda_1}{\lambda_1+\lambda_3} - \frac{\lambda_2}{\lambda_2+\lambda_3}\right)\mathcal{A}_{T_1,T_2}\left(\frac{\Lambda_1(w)}{\Lambda_1(w)+\Lambda_2(w)}\right) \\ +\left(\frac{\lambda_1 w}{\lambda_1+\lambda_3} + \frac{\lambda_2(1-w)}{\lambda_2+\lambda_3}\right)\frac{d}{dw}\mathcal{A}_{T_1,T_2}\left(\frac{\Lambda_1(w)}{\Lambda_1(w)+\Lambda_2(w)}\right) + \frac{\lambda_3}{\lambda_1+\lambda_3}, & \text{if } \frac{\lambda_1+\lambda_3}{\lambda_1+\lambda_2+2\lambda_3} \leq w < 1, \\ 2 & \text{if } w = 1, \end{cases} \tag{15}$$

where $\frac{d}{dw}\mathcal{A}_{T_1,T_2}(.)$ *refers to the right-hand derivative.*

Proof. Since $(X_1, X_2) \subset BEV_E - EMO^{AC}$, according to Lemma 1 we have $(T_1, T_2) \subset BEV_E$ and $T_3 \sim Exp(\lambda_3)$. From Sklar's theorem and the expression for bivariate copulas written in terms of Pickands dependence function (11) we have

$$S_{T_1,T_2}(x_1, x_2) = \overline{C}_{T_1,T_2}(S_{T_1}(x_1), S_{T_2}(x_2)) = \overline{C}_{T_1,T_2}(\exp\{-\lambda_1 x_1\}, \exp\{-\lambda_2 x_2\})$$
$$= \exp\left\{-(\lambda_1 x_1 + \lambda_2 x_2)\mathcal{A}_{T_1,T_2}\left(\frac{\lambda_1 x_1}{\lambda_1 x_1 + \lambda_2 x_2}\right)\right\}.$$

Considering the expression of EMO survival functions (3) we obtain relation (13), i.e., the survival function of (X_1, X_2).

Let $u = S_{X_1}(x_1) = \exp\{-(\lambda_1 + \lambda_3)x_1\}$ and $v = S_{X_2}(x_2) = \exp\{-(\lambda_2 + \lambda_3)x_2\}$. From Sklar's theorem, $\overline{C}_{X_1,X_2}(u, v) = S_{X_1,X_2}(\frac{-\ln u}{\lambda_1+\lambda_3}, \frac{-\ln v}{\lambda_2+\lambda_3})$. Applying (11) we have $\mathcal{A}_{X_1,X_2}(w) = -\ln S_{X_1,X_2}\left(\frac{w}{\lambda_1+\lambda_3}, \frac{1-w}{\lambda_2+\lambda_3}\right)$. Substituting $S_{X_1,X_2}(x_1, x_2)$ by its expression given by (13) we obtain

$$\mathcal{A}_{X_1,X_2}(w) = (\Lambda_1(w) + \Lambda_2(w))\,\mathcal{A}_{T_1,T_2}\left(\frac{\Lambda_1(w)}{\Lambda_1(w) + \Lambda_2(w)}\right) + \max\left(\Lambda_1(w), \Lambda_2(w)\right). \quad (16)$$

Considering again expression (11) and relation (16) we get (14), i.e., the survival copula of the bivariate extreme value EMO distribution. By its turn, from the expression of Pickands measure in terms of the derivative of Pickands dependence function (12) and from (16) we obtain (15), the Pickands measure $\mathbb{H}$.

Whenever the derivative of $\mathcal{A}_{T_1,T_2}(w)$ is continuous from (15) we obtain

$$\mathbb{H}\left(\left\{\frac{\lambda_1 + \lambda_3}{\lambda_1 + \lambda_2 + 2\lambda_3}\right\}\right) = \frac{\lambda_3}{\lambda_1 + \lambda_3} + \frac{\lambda_3}{\lambda_2 + \lambda_3}. \quad (17)$$

We will apply relations established in Theorem 4 in the next two examples.

Example 3 (MO bivariate exponential distribution). Consider the MO distribution (2). In this case T_1 and T_2 are independent and hence $\overline{C}_{T_1,T_2}(u, v) = uv$. Therefore, $\mathcal{A}_{T_1,T_2}(w) = 1$ when $w \in [0, 1]$. Substituting $w = \frac{\ln u}{\ln uv}$ for $u, v \in [0, 1)$ in (16), the Pickands dependence function $\mathcal{A}_{X_1,X_2}(w)$ can be presented as

$$\mathcal{A}_{X_1,X_2}(u, v) = \begin{cases} 1 - \frac{\lambda_3}{\lambda_1+\lambda_3}\frac{\ln u}{\ln uv}, & \text{if} \quad 1 > u^{\frac{\lambda_3}{\lambda_1+\lambda_3}} > v^{\frac{\lambda_3}{\lambda_2+\lambda_3}} > 0, \\ \frac{\lambda_2}{\lambda_2+\lambda_3} + \frac{\lambda_3}{\lambda_2+\lambda_3}\frac{\ln u}{\ln uv}, & \text{if} \quad 0 < u^{\frac{\lambda_3}{\lambda_1+\lambda_3}} \leq v^{\frac{\lambda_3}{\lambda_2+\lambda_3}} < 1. \end{cases}$$

Applying (15) one can get the corresponding Pickands measure, confirming the representation obtained by Mai and Scherer (2010).

Example 4 (EMO extreme value distribution where the dependence structure of (T_1, T_2) is represented by Gumbel-Hougaard survival copula). Let T_i be exponentially distributed with parameter λ_i, $i = 1, 2$ and consider the Gumbel's Type III bivariate exponential survival function for (T_1, T_2) given by

$$S_{T_1,T_2}(x_1, x_2) = \exp\{-[(\lambda_1 x_1)^m + (\lambda_2 x_2)^m]^{\frac{1}{m}}\}, \quad m > 0,$$

see Gumbel (1960). The corresponding survival copula is

$$\overline{C}(u, v) = \exp\{-[(-\ln u)^m + (-\ln v)^m]^{\frac{1}{m}}\},$$

being an example of bivariate extreme value copula. Select T_3 exponentially distributed with parameter λ_3, independent of (T_1, T_2) and consider EMO survival function (3). We obtain the following bivariate extreme value EMO distribution

$$S_{X_1,X_2}(x_1, x_2) = \exp\{-[(\lambda_1 x_1)^m + (\lambda_2 x_2)^m]^{\frac{1}{m}}\}\exp\{-\lambda_3 \max(x_1, x_2)\}.$$

From (11), the Pickands dependence function for (T_1, T_2) is

$$\mathcal{A}_{T_1,T_2}(w) = [w^m + (1 - w)^m]^{\frac{1}{m}},$$

and from (16) we get the Pickands dependence function

$$\mathcal{A}_{X_1,X_2}(w) = \left[\left(\frac{\lambda_1 w}{\lambda_1 + \lambda_3}\right)^m + \left(\frac{\lambda_2(1 - w)}{\lambda_2 + \lambda_3}\right)^m\right]^{\frac{1}{m}} + \max\left(\frac{\lambda_3 w}{\lambda_1 + \lambda_3}, \frac{\lambda_3(1 - w)}{\lambda_2 + \lambda_3}\right).$$

Applying relation (15), we obtain the Pickands measure

$$\mathbb{H}([0,w]) = \begin{cases} 1 - \frac{\lambda_3}{\lambda_2+\lambda_3} + \left[\left(\frac{\lambda_1 w}{\lambda_1+\lambda_3}\right)^m + \left(\frac{\lambda_2(1-w)}{\lambda_2+\lambda_3}\right)^m\right]^{\frac{1}{m}-1} \\ \times\left[\left(\frac{\lambda_1}{\lambda_1+\lambda_3}\right)^m w^{m-1} - \left(\frac{\lambda_2}{\lambda_2+\lambda_3}\right)^m (1-w)^{m-1}\right], & \text{if } 0 \le w < \frac{\lambda_1+\lambda_3}{\lambda_1+\lambda_2+2\lambda_3}, \\ 1 + \frac{\lambda_3}{\lambda_1+\lambda_3} + \left[\left(\frac{\lambda_1 w}{\lambda_1+\lambda_3}\right)^m + \left(\frac{\lambda_2(1-w)}{\lambda_2+\lambda_3}\right)^m\right]^{\frac{1}{m}-1} \\ \times\left[\left(\frac{\lambda_1}{\lambda_1+\lambda_3}\right)^m w^{m-1} - \left(\frac{\lambda_2}{\lambda_2+\lambda_3}\right)^m (1-w)^{m-1}\right], & \text{if } \frac{\lambda_1+\lambda_3}{\lambda_1+\lambda_2+2\lambda_3} \le w < 1, \\ 2, & \text{if } w = 1. \end{cases}$$

Finally, using (17) we get

$$\mathbb{H}\left(\left\{\frac{\lambda_1+\lambda_3}{\lambda_1+\lambda_2+2\lambda_3}\right\}\right) = \frac{\lambda_3}{\lambda_1+\lambda_3} + \frac{\lambda_3}{\lambda_2+\lambda_3}.$$

Notice that when $m = 1$, the Gumbel-Hougaard survival copula becomes the independence copula and we repeat the results of Example 3.

For $1 < m < \infty$, the remaining mass $2 - \left(\frac{\lambda_3}{\lambda_1+\lambda_3} + \frac{\lambda_3}{\lambda_2+\lambda_3}\right)$ is spread over the interval $[0, \frac{\lambda_1+\lambda_3}{\lambda_1+\lambda_2+2\lambda_3}) \bigcup (\frac{\lambda_1+\lambda_3}{\lambda_1+\lambda_2+2\lambda_3}, 1]$ and, in particular, $\mathbb{H}(\{0\}) = \mathbb{H}(\{1\}) = 0$.

4. Conclusions

In this note, we found representation of extreme value EMO distributions with exponential marginals. We characterize it when the vector (T_1, T_2) in (3) is absolutely continuous or not, see Theorem 2 and Theorem 3, respectively. The corresponding Pickands representation is obtained in Theorem 4, having as a particular case earlier conclusions of Mai and Scherer (2010) regarding the MO's bivariate exponential distribution (2).

We would like to call attention that Theorem 2 shows the way to generate extreme value EMO distributions with exponential marginals when the vector (T_1, T_2) in (3) is absolutely continuous, see the Example 2. If (T_1, T_2) in (3) has a singular component, one may follow a procedure given after Theorem 2 by using GMO models as a first step. The only restriction in the choice of the failure rate of T_3 in the second step is that $r_{T_i}(x_i)+r_{T_3}(x_i) = \lambda_{X_i}$, $i = 1, 2$, being the constant failure rates of the marginals of extreme value EMO distribution of (X_1, X_2). Another general option is to select dependence function $H_{T_1,T_2}(x_1, x_2)$ and cumulative hazard $H_{T_3}(x)$, satisfying the functional equation in Theorem 3, since it is valid even when the joint distribution of (T_1, T_2) is not absolutely continuous and T_i' in (3) do not need to be exponentially distributed.

Acknowledgements

The first author is partially supported by FAPESP Grant No. 2013/07375-0.

References

Aczel, J. (1966). *Lectures on Functional Equations and Their Applications.* New York, NY: Academic Press.

Beirlant, J., Goegebeur, Y., Segers, J., & Teugels, J. (2005). *Statistics of Extremes: Theory and Application.* Chichester, UK: John Wiley & Sons.

Galambos, J. (1978). *The Asymptotic Theory of Extreme Order Statistics.* New York, NY: John Wiley & Sons.

Genest, C., Ghoudi, K., & Rivest, L.-P. (1998). Discussion on the paper "Understanding relationships using copulas" by E. Frees and Valdez, E. *North American Actuarial Journal, 2*, 143-149. http://dx.doi.org/10.1080/10920277.1998.10595749

Gumbel, E. (1960). Bivariate exponential distributions. *Journal of American Statistical Association, 55*, 698-707. http://dx.doi.org/10.1080/01621459.1960.10483368

Joe, H. (1997). *Multivariate Models and Dependence Concepts.* London, UK: Chapman & Hall.

Joe, H. (2015). *Dependence Modeling with Copulas.* Boca Raton: CRC Press.

Kolev, N. (2016). Characterizations of the class of bivariate Gompertz distributions. *Journal of Multivariate Analysis, 148*, 173-179. http://dx.doi.org/10.1016/j.jmva.2016.03.004

Li, X., & Pellerey, F. (2011). Multivariate generalized Marshall-Olkin distributions and related bivariate properties. *Journal of Multivariate Analysis, 102*, 1399-1409. http://dx.doi.org/10.1016/j.jmva.2011.05.006

Liebscher, E. (2008). Construction of asymmetric multivariate copulas. *Journal of Multivariate Analysis, 99*, 2234-2250. http://dx.doi.org/10.1016/j.jmva.2008.02.025

Mai, J.-F., & Scherer, M. (2010). The Pickands representation of survival Marshall-Olkin copulas. *Statistics and Probability Letters, 80*, 357-360. http://dx.doi.org/10.1016/j.spl.2009.11.010

Marshall, A., & Olkin, I. (1967). A multivariate exponential distribution. *Journal of American Statistical Association, 62*, 30-41. http://dx.doi.org/10.1080/01621459.1967.10482885

Marshall, A., & Olkin, I. (1991). Functional equations for multivariate exponential distributions. *Journal of Multivariate Analysis, 39*, 209-215. http://dx.doi.org/10.1016/0047-259X(91)90014-S

Pickands, J. (1981). Multivariate extreme value distributions. *Bulletin of the International Statistical Institute, Proceedings of the 43rd Session*, 859-878.

Pinto, J., & Kolev, N. (2015). Extended Marshall-Olkin model and its dual version. In U. Cherubini, F. Durante & S. Mulinacci (Eds.), *Springer Series in Mathematics & Statistics, 141*, (pp.87-113). http://dx.doi.org/10.1007/978-3-319-19039-6

Pinto, J., & Kolev, N. (2015a). Sibuya-type bivariate lack of memory property *Journal of Multivariate Analysis, 134*, 119-128. http://dx.doi.org/10.1016/j.jmva.2014.11.001

Sibuya, M. (1960). Bivariate extreme statistics I. *Annals of the Institute of Statistical Mathematics, 11*, 195-210. http://dx.doi.org/10.1007/BF01682329

Permissions

All chapters in this book were first published in IJSP, by Canadian Center of Science and Education; hereby published with permission under the Creative Commons Attribution License or equivalent. Every chapter published in this book has been scrutinized by our experts. Their significance has been extensively debated. The topics covered herein carry significant findings which will fuel the growth of the discipline. They may even be implemented as practical applications or may be referred to as a beginning point for another development.

The contributors of this book come from diverse backgrounds, making this book a truly international effort. This book will bring forth new frontiers with its revolutionizing research information and detailed analysis of the nascent developments around the world.

We would like to thank all the contributing authors for lending their expertise to make the book truly unique. They have played a crucial role in the development of this book. Without their invaluable contributions this book wouldn't have been possible. They have made vital efforts to compile up to date information on the varied aspects of this subject to make this book a valuable addition to the collection of many professionals and students.

This book was conceptualized with the vision of imparting up-to-date information and advanced data in this field. To ensure the same, a matchless editorial board was set up. Every individual on the board went through rigorous rounds of assessment to prove their worth. After which they invested a large part of their time researching and compiling the most relevant data for our readers.

The editorial board has been involved in producing this book since its inception. They have spent rigorous hours researching and exploring the diverse topics which have resulted in the successful publishing of this book. They have passed on their knowledge of decades through this book. To expedite this challenging task, the publisher supported the team at every step. A small team of assistant editors was also appointed to further simplify the editing procedure and attain best results for the readers.

Apart from the editorial board, the designing team has also invested a significant amount of their time in understanding the subject and creating the most relevant covers. They scrutinized every image to scout for the most suitable representation of the subject and create an appropriate cover for the book.

The publishing team has been an ardent support to the editorial, designing and production team. Their endless efforts to recruit the best for this project, has resulted in the accomplishment of this book. They are a veteran in the field of academics and their pool of knowledge is as vast as their experience in printing. Their expertise and guidance has proved useful at every step. Their uncompromising quality standards have made this book an exceptional effort. Their encouragement from time to time has been an inspiration for everyone.

The publisher and the editorial board hope that this book will prove to be a valuable piece of knowledge for researchers, students, practitioners and scholars across the globe.

List of Contributors

Zohdy M. Nofal, Ahmed Z. Afify and Haitham M. Yousof
Department of Statistics, Mathematics and Insurance, Benha University, Egypt

Daniele C. T. Granzotto
Department of Statistics, State University of Maringá, Maringá, Brazil

Francisco Louzada
Institute of Mathematical and Computing Sciences, University of São Paulo, São Carlos, Brazil

Nikolai Kolev and Jayme Pinto
Department of Statistics, University of São Paulo, Brazil

Imene Allab and Francois Watier
Department of mathematics, Université du Québec à Montréal, Montreal, Canada

Andriy Yurachkivsky
Taras Shevchenko National University, Kyiv, Ukraine

Oykum Esra Askin and Ali Hakan Buyuklu
Department of Statistics, Yildiz Technical University, Istanbul, Turkey

Deniz Inan
Department of Statistics, Marmara University, Istanbul, Turkey

Chunlin Ji
Kuang-Chi Institute of Advanced Technology, Shenzhen, China

Michael B Brimacombe
Department of Biostatistics, KUMC, Kansas City, Kansas, USA

Li Guan
College of Applied Sciences, Beijing University of Technology, Beijing, China

Gokarna R. Aryal
Department of Mathematics, Statistics and CS, Purdue University Northwest, USA

Edwin M. Ortega
Departamento de Ciências Exatas, Universidade de São Paulo, Piracicaba, SP, Brazil

G. G. Hamedani
Department of Mathematics, Statistics and Computer Science, Marquette University, USA

Haitham M. Yousof
Department of Statistics, Mathematics and Insurance, Benha University, Egypt

E. A. Appiah and G. S. Ladde
Department of Mathematics and Statistics, University of South Florida, Tampa, FL, USA

Elizabeth M. Hashimoto
Departamento Acadêmico de Matemática, UTFPR, Londrina, Brazil

Gauss M. Cordeiro
Departamento de Estatística, UFPE, Recife, Brazil

Edwin M.M. Ortega
Departamento Ciências Exatas, ESALQ-USP, Piracicaba, Brazil

G.G. Hamedani
Departament of Mathematics, Statistics and Computer Science, Milwaukee, USA

Tang Jiahui
Mathematical Modeling Innovative Practice Base, JiNan University, ZhuHai GuangDong,China

Giuliana Cortese
Department of Statistical Sciences, University of Padua, Italy

Daniele C. T. Granzotto and Josmar Mazucheli
Universidade Estadual de Maringá, DEs, PR, Brazil

Francisco Louzada
Universidade de São Paulo, ICMC, SP, Brazil

KanéLadji and DialloMoumouni
Faculty of Economics and Management (F.S.E.G) Bamako-Mali

DiawaraDaouda
Zhongnan University of Economics and Law Wuhan China

Tristan Guillaume
Université de Cergy-Pontoise, Laboratoire Thema, Cergy, France

Zhang Yuanbiao

Peng Churu and Huang Xinxin

Index

www.ingramcontent.com/pod-product-compliance
Lightning Source LLC
Chambersburg PA
CBHW080623200326
41458CB00013B/4482
9781632385802